Noise and Acoustic Fatigue in Aeronautics

Edited by

E. J. RICHARDS
Vice-Chancellor of The University of Technology, Loughborough formerly Professor of Applied Acoustics and Director, Institute of Sound and Vibration Research, University of Southampton

and

D. J. MEAD
Reader, Department of Aeronautics and Astronautics, University of Southampton

JOHN WILEY & SONS LTD
London · New York · Sydney

Library of Congress catalog card number 68–55813

SBN 471 71944 7

Printed in Great Britain by
William Clowes and Sons, Limited, London and Beccles

Noise and Acoustic Fatigue in Aeronautics

Contributing Authors

BULL, M. K.

Senior Lecturer, Department of Mechanical Engineering, University of Adelaide, Australia, formerly Shell-B.P. Research Fellow, Institute of Sound and Vibration Research, University of Southampton

CLARKSON, B. L.

Professor of Vibration Studies, and Director, Institute of Sound and Vibration Research, University of Southampton

CURLE, S. N.

Professor of Applied Mathematics, University of St. Andrews, formerly Hawker Siddeley Reader in Noise Research, Department of Aeronautics and Astronautics, University of Southampton

DOAK, P. E.

Hawker Siddeley Reader in Noise Research, Institute of Sound and Vibration Research, University of Southampton

FAHY, F. J.

Central Electricity Generating Board Lecturer, Institute of Sound and Vibration Research, University of Southampton

JAMES, D. O. N.

Principal Engineer, Dynamics Department, British Aircraft Corporation (Operating) Ltd., Weybridge, formerly Research Assistant, Institute of Sound and Vibration Research, University of Southampton

LEVERTON, J. W.

Research Fellow, Institute of Sound and Vibration Research, University of Southampton

MEAD, D. J.

Reader, Department of Aeronautics and Astronautics, University of Southampton

MIDDLETON, D. W.   *Lecturer, Department of Mathematics, University of Ibadan, formerly Research Assistant, Department of Aeronautics and Astronautics, University of Southampton*

MORFEY, C. L.   *Hawker Siddeley Lecturer, Institute of Sound and Vibration Research, University of Southampton*

SHARLAND, I. J.   *Acoustics Manager, Woods of Colchester Ltd., Colchester, Essex, formerly Lecturer, Institute of Sound and Vibration Research, University of Southampton*

RICE, C. G.   *Lecturer, Institute of Sound and Vibration Research, University of Southampton*

RICHARDS, E. J.   *Vice-Chancellor, Loughborough University of Technology, formerly Professor of Applied Acoustics and Director, Institute of Sound and Vibration Research, University of Southampton*

WARREN, C. H. E.   *Head, Flutter, Vibration and Noise Division, Structures Department, Royal Aircraft Establishment, Farnborough*

WILBY, J. F.   *Research Engineer, Acoustics Unit, Commercial Airplane Division, Boeing Airplane Company, Seattle, U.S.A., formerly Shell-Mex and B.P. Research Fellow, Institute of Sound and Vibration Research, University of Southampton*

WILLIAMS, T. R. G.   *Senior Lecturer, Department of Mechanical Engineering, University of Southampton*

YEOW, K. W.   *Research Assistant, Institute of Sound and Vibration Research, University of Southampton*

Introduction

The subject of noise in aeronautics needs little introduction; it has introduced itself to the world at large over the past two decades. The enormous increase in power of aeroplane engines over that period has led to aeronautical noise being a major social and scientific problem. Acoustic fatigue has followed in its wake, being the fatigue of the aeroplane structure due to the merciless hammering of the fluctuating sound pressures. Fortunately, the fatigue failures have not been catastrophic, but rather have constituted an intense nuisance. Nevertheless, acoustic fatigue is a serious enough problem, as well as noise itself, to warrant careful attention in the early design stages of aeroplanes or engines.

Although the science of noise and acoustic fatigue is a very small part of the whole science of aeroplane design, it draws upon several widely-different scientific disciplines. At the source of the problem is aerodynamic turbulence, which is a stochastic process in a viscous flow. The generation of noise at this source and its subsequent propagation involves fundamental acoustic theory. The vibration response of a structure to the noise pressure is a problem in applied mechanics and dynamics, while the subsequent fatigue failures in the structure involve metal physics and metallurgy. Experimental studies of acoustic fatigue and noise bring in the disciplines of instrument technology, while the testing techniques require a sound understanding of the principles of engineering design.

Since specialists in the problem of noise and acoustic fatigue must be acquainted with such a wide field, several short courses on the subject have been given at the University of Southampton over the past twelve years. The success of these courses (and also of an identical course given in Dayton, Ohio, U.S.A.) prompted us to present our lecture notes in the more permanent form of this book. In this, an attempt has been made to introduce the reader to the fundamentals of the various disciplines involved, and within a relatively short space to show how these disciplines have been developed and linked to study the whole problem. Much that is contained is necessarily but an outline of developments along these lines, but with several lists of references provided, the full details may be acquired by further reading.

The first three chapters of this book deal respectively with the elements of the theory of sound generation, propagation and transmission. The fourth chapter introduces the statistical theory required to handle random processes, after which the theory of aerodynamic noise is presented in Chapter 5. This outlines Lighthill's theory of sound due to turbulence, and subsequent developments.

Before specific noise sources are considered, the subjective assessment of noise is dealt with in Chapter 6. The different units of noise measurement are introduced and the annoyance and interference levels are discussed. Chapter 7 then reviews the practical results of noise measurements from jets and rockets.

The noise inside a high-subsonic or supersonic aeroplane derives primarily from the turbulent flow of air over the outer surface of the aeroplane. The associated pressure fluctuations on the surface can have magnitudes of the same order as those produced by the jet engine. The magnitudes and statistical properties of these boundary layer pressure fluctuations are discussed in Chapter 8.

Propellers of turbo-prop engines and helicopter rotors are prolific sources of noise. Accordingly, these sources are dealt with in Chapter 9.

From the newer generation of jet engines with high by-pass ratios the noise generated by the larger diameter compressor under landing conditions has become very important. Chapter 10 deals with the mechanisms of compressor and fan noise, the propagation of noise along the duct and radiation from the intake.

Chapter 11 deals in two main sections with the reduction of jet noise, firstly in relation to aeroplanes in flight (including jet-helicopters, V/STOL aircraft, etc.) and then in relation to aircraft on the ground. The use and limitations of ground mufflers, screens, test-pens, etc., are considered.

The final chapter on noise sources deals with the problem of sonic booms. The propagation and intensity of the boom are discussed, together with the effects of focussing, aircraft speed and weight.

The next eleven chapters deal with the two-fold effect of the incident noise on the aeroplane structure, viz. the vibration response and fatigue of the structure, and the noise transmitted by the vibration into the enclosed cabin. A theoretical approach to the prediction of stress levels is presented, utilizing the 'normal mode' method of vibration theory. From the theory, the principal structural and excitation parameters are identified, which govern the vibration stress levels. Thus it has been possible to outline some of the measures which can be taken to extend the fatigue life of the structure and to attenuate the transmitted sound pressure level. Aspects of the practical testing of structures for acoustic fatigue are also discussed.

Chapter 13 is in the first place, a 'revision chapter' of harmonic vibration theory, but then proceeds to introduce the subject of structural response to harmonic sound fields. Chapter 14 extends the theory to include random excitation and response. In both chapters, it is assumed that the normal modes of the structure are known, and are well-defined. Chapter 15 discusses the types of vibration mode that may be excited in an aeroplane structure consisting of thin stiffened plates.

There are very many modes which can participate in the structural motion to some degree or other. It is often found, however, that one or two modes dominate the response spectrum, and that a simple calculation based on just one appropriate mode of vibration can predict a stress level to an acceptable degree of accuracy. This simple approach is discussed in the first part of Chapter 16. On the other hand, some acoustically excited structures respond more-or-less equally in very many modes of vibration. This feature has led to the development of the so-called 'Statistical Energy Method' of analysing structural response which is becoming increasingly well-known. The second part of Chapter 16 introduces the underlying ideas of this method.

Chapter 17 considers the design of stiffened plate structures for maximum acoustic fatigue resistance. One method of achieving this resistance is to increase the structural damping, a subject to which the whole of Chapter 18 is given. In Chapter 19, the design and operation of acoustic test facilities are discussed.

The fundamental mechanisms of fatigue damage in structural materials are described briefly in Chapter 20. The cumulative damaging effect of random loading is dealt with in Chapter 21.

The concluding Chapters (22 and 23) deal respectively with the problems of calculating and reducing the noise transmitted into an aeroplane cabin. A simple, approximate theory is first presented and its limitations are emphasized. A review is then given of the more sophisticated methods of calculating the internal sound levels arising from the boundary layer pressure fluctuations on the cabin exterior, particular emphasis being laid upon their limitations. The final chapter then treats of the theoretical and practical aspects of sound-proofing aeroplane cabins.

Although the main thrust of this book is towards aeronautics the principles may be applied to many other technologies. For instance, they may be applied directly to the problem of acoustic fatigue in nuclear power plants, where fan noise can excite intense random vibration of the walls of heat-exchanger ducts. The chapters on the generation of aerodynamic noise, sound propagation and compressor noise are applicable to ventilation noise problems. The chapters on random processes and random vibration can be applied to the study of the vibration of tall chimneys excited by atmospheric gusts, or of dams shaken by earthquakes.

The chapters on sound transmission and sound-proofing are applicable to noise problems in passenger vehicles of all types; road, rail, sea and air. To this list other problems can be added, and the list is continually growing as new problems of noise and random vibration emerge.

This book has come into existence by a process of evolution extending over several years. New chapters were written and old ones were modified as each of the short courses were given. In more recent years, one half of the contributors left Southampton for the four corners of the earth, and some of the research studies they led and wrote about are no longer actively pursued at Southampton. However, their interest in the subject has continued, and by maintaining contact with us they have added to the strong international flavour of our work on noise and acoustic fatigue. This began with collaborative programmes of research work with overseas groups such as University of Minnesota, Sud-Aviation at Toulouse, and the U.S.A.F. Materials Laboratory at Wright Field, Ohio, and continues to this day.

The authors are indebted to many who have assisted or encouraged them in the preparation of this book. The work of many other investigators has provided much information and data which is included. The scale of the research work and its wide coverage at Southampton, out of which grew the courses in noise and acoustic fatigue, was only possible through the generous sponsorship of the Ministry of Aviation (now Technology) and of the European Office of the United States Air Force. Special thanks are due to Mr. Walter J. Trapp (of Wright Patterson Air Force Base, Dayton, Ohio) for his interest in the acoustic fatigue programme and for sponsoring the short course we gave in Dayton in 1961. Out of this course came the first requests for this book. Mr. Trapp has waited long and patiently for it!

We are indebted also to our long-suffering publisher, Messrs. John Wiley & Sons, and their printer, who have made such a fine production despite the insufferable delays and procrastinations we have caused.

Finally, those of us still at Southampton must record our indebtedness to our co-editor, Dr. E. J. Richards, who built up and led the noise and vibration team with inspiring zeal over seventeen years. His elevation to the Vice-Chancellorship at Loughborough University was more than well-deserved and we feel privileged to have worked under his direction. Sixteen of the contributors to this book would wish, if it were possible, to dedicate this book to him!

Contents

CHAPTER 1

An Introduction to Sound Radiation and its Sources

1.1 The nature of sound

If a taut string be disturbed laterally, a transverse wave will run along the string at a definite speed. Longitudinal pressure waves travel similarly down a rod, or through the air in a pipe. Such air-borne longitudinal pressure waves arriving at the ear cause vibrations of the eardrum, which are in turn interpreted as sound by the auditory nerves and brain.

The speed of these small pressure waves in a particular solid or fluid is called the 'speed of sound' in that material. To establish what properties of the material determine the speed of sound and the nature of sound fields associated with various types of sound sources, it is necessary to investigate the dynamics of sound waves in some detail.

1.1.1 *Dynamics of plane sound waves and the speed of sound*

Suppose that the material disturbed by the wave moves only in the direction of propagation, x. Consider a mass of material occupying a cylindrical volume of unit cross-sectional area, the ends of the cylinder being initially the planes x and $x + \delta x$. Let ξ be the displacement, measured from x, of the layer of material initially at x. Then at time t, when the displacement of this layer is ξ, the displacement of the material initially at $x + \delta x$ will be $\xi + (\partial \xi / \partial x) \delta x$. The increase in the volume occupied by the mass of material is $(\partial \xi / \partial x) \delta x$.

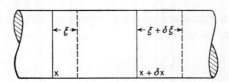

Fig. 1.1 Deformation of a material element by a plane acoustic disturbance.

1

The mass of the cylinder of material under consideration is constant so that there must be a density change, the density ρ being related to the initial density ρ_0 by

$$\rho_0 \delta x = \rho\left(1 + \frac{\partial \xi}{\partial x}\right)\delta x.$$

It is convenient to express the density in terms of the initial density ρ_0 and the 'condensation', or fractional change in density, $s = (\rho - \rho_0)/\rho_0$. Then

$$\rho_0 \delta x = \rho_0(1 + s)\left(1 + \frac{\partial \xi}{\partial x}\right)\delta x,$$

or, to first order for small values of the condensation,

$$s = -\frac{\partial \xi}{\partial x}. \tag{1.1.1}$$

For a wide range of conditions met in practice, it may be assumed that the expansions and compressions of the layer, caused by the wave passing through it, are so rapid that a negligible amount of heat is lost from the material element through thermal conduction (which is a slow process) and also that there are no other appreciable heat losses. In this case the pressure p and the density ρ are related by the adiabatic law

$$\frac{p}{\rho^\gamma} = \text{constant} = \frac{p_0}{\rho_0^\gamma},$$

where p_0 is the initial pressure, and γ is the ratio of the specific heat at constant pressure to that at constant volume. Thus the small pressure increment due to the wave can be found from

$$p = p_0 + (p - p_0) = \left(\frac{p_0}{\rho_0^\gamma}\right)\rho_0^\gamma(1 + s)^\gamma$$

$$= p_0\{1 + \gamma s + \gamma(\gamma - 1)s^2/2 + \cdots\},$$

or

$$p - p_0 = \gamma p_0 s\{1 + (\gamma - 1)s/2 + \cdots\}. \tag{1.1.2}$$

To first order in the condensation, equations 1.1.1 and 1.1.2 can be combined to give the equation

$$\frac{(p - p_0)}{\gamma p_0} = s = -\frac{\partial \xi}{\partial x}. \tag{1.1.3}$$

It is important to realize that p and ρ are the pressure and density of the layer of material *initially* at x. At time t this material has moved to the position $x + \xi$. Thus the expression $p(x, t)$ strictly means the pressure at time t of the material initially at x, and hence is the pressure at time t at the point $x + \xi$ in space, not at the point x.

As viscous forces and external forces are assumed to be absent or negligible, the motion of the mass of material is caused solely by the pressure* forces acting on it. Let $\eta = x + \xi$ denote the absolute position at time t of the layer initially at x. Then the forces on the element initially between x and $x + \delta x$ are $p(\eta)$ at η, and $p\{\eta + (\partial\eta/\partial x)\delta x\}$ at $\eta + (\partial\eta/\partial x)$ δx. The resultant force is thus $-(\partial p/\partial\eta)(\partial\eta/\partial x)\delta x$, or $-(\partial p/\partial x)\delta x$. The mass of the element is $\rho_0\delta x$ and its acceleration is $\partial^2\xi/\partial t^2$.

The equation of motion of the element is therefore

$$\rho_0\delta x \frac{\partial^2\xi}{\partial t^2} = -\frac{\partial p}{\partial x} \delta x$$

or

$$\rho_0 \frac{\partial^2\xi}{\partial t^2} = -\frac{\partial p}{\partial x}. \tag{1.1.4}$$

If the material was in a uniform condition before the arrival of the small disturbance, so that p_0 and ρ_0 are constants independent of x, then $\partial p/\partial x = \partial(p - p_0)/\partial x$, and equation 1.1.3 can be used to eliminate p from equation 1.1.4, giving

$$\frac{\partial^2\xi}{\partial t^2} = (\gamma p_0/\rho_0)\frac{\partial^2\xi}{\partial x^2}, \tag{1.1.5}$$

to first order. This is the wave equation in one space dimension.

The general solution of this equation is

$$\xi = \xi_+(x - a_0t) + \xi_-(x + a_0t), \tag{1.1.6}$$

where ξ_+ and ξ_- are arbitrary functions of the variables $x - a_0t$ and $x + a_0t$, respectively. The quantity $a_0 = \sqrt{\gamma p_0/\rho_0}$ is evidently the speed of sound, because $\xi = \xi_+(x - a_0t)$ represents a wave form travelling in the direction of increasing x at speed a_0 without change of shape, and $\xi = \xi_-(x + a_0t)$ similarly represents an unchanging wave form travelling at speed a_0 in the direction of decreasing x. In this way the analysis shows that small disturbances travel without change of form, and at a constant speed, a_0.

The velocity of the material particles in the layer, due to the disturbance caused by the passage of the wave, is $\partial\xi/\partial t + \dot{\xi}$. The particles themselves do not move at the speed of sound; only the wave form does. In

---

* 'Pressure' is used here in the sense of a general, internal normal stress.

acoustics the speed of sound in a material—fluid or solid—is often expressed in terms of the coefficient of volume elasticity of the material, κ (commonly called the 'bulk modulus'). This coefficient is defined as the ratio of the normal stress increment on an element of the material, δp, to the resulting change in volume of the element per unit volume, $\delta \rho / \rho$: that is,

$$\kappa = \frac{\delta p}{\delta \rho / \rho} = \frac{p - p_0}{s}, \tag{1.1.7}$$

for small pressure differences, $p - p_0$. This relationship is a form of Hooke's Law which asserts that stress is proportional to strain. In this case the strain is the volume strain, or condensation.

Thus the adiabatic law as given in equation 1.1.3, which can be written as $(p - p_0)/s = \gamma p_0$, is just a form of Hooke's Law, expressing the fact that the material is elastic and that its adiabatic bulk modulus is $\kappa_0 = \gamma p_0 = \rho_0 a_0{}^2$. The speed of sound in any elastic material conforming to Hooke's Law is evidently $a = \sqrt{\kappa/\rho}$. Thus the nature and propagation of small-amplitude, longitudinal, compressive waves are essentially the same in all elastic materials—gas, liquid or solid.

1.1.2 *Plane waves of finite amplitude*

If the condensation is not small then the wave does change form and does not travel at a constant speed. In such cases higher order terms in the adiabatic law, equation 1.1.2, cannot be neglected. To a second approximation in s this is

$$p - p_0 = \gamma p_0 s \left\{ 1 + \frac{(\gamma - 1)s}{2} \right\} = \rho_0 a_0^2 s \left\{ 1 + \frac{(\gamma - 1)s}{2} \right\}. \tag{1.1.8}$$

The expression for conservation of mass, $\rho_0 \delta x = \delta_0 (1 + s)(1 + \partial \xi / \partial x)\delta x$, gives exactly $s = -(1 + s)\partial \xi / \partial x$. Thus

$$p - p_0 = -\rho_0 a_0^2 \frac{\partial \xi}{\partial x} (1 + s) \left\{ 1 + \frac{(\gamma - 1)s}{2} \right\}$$

$$= -\delta_0 a_0^2 \frac{\partial \xi}{\partial x} \left\{ 1 + \frac{(\gamma + 1)s}{2} \right\},$$

to second order. The equation of motion now becomes

$$\rho_0 \frac{\partial^2 \xi}{\partial t^2} = -\frac{\partial}{\partial x}(p - p_0) = \rho_0 a_0^2 \frac{\partial^2 \xi}{\partial x^2} \left(1 + \frac{\gamma + 1}{2} s \right) + \rho_0 a_0^2 \frac{\gamma + 1}{2} \frac{\partial \xi}{\partial x} \frac{\partial s}{\partial x}.$$

Again, to second order $(\partial\xi/\partial x)(\partial s/\partial x) = (-s)(-\partial^2\xi/\partial x^2)$, so that finally, dividing both sides by ρ_0 and collecting terms gives

$$\frac{\partial^2\xi}{\partial t^2} = a_0^2\{1 + (\gamma + 1)s\}\frac{\partial^2\xi}{\partial x^2}. \qquad (1.1.9)$$

As now $a_0^2\{1 + (\gamma + 1)s\}$ appears in place of a_0^2 as in (1.1.5) this equation shows that to this order of approximation the speed of propagation of the wave from particle to particle is $a_0\{1 + (\gamma + 1)s/2\}$ and is hence greater the greater the compression. The peaks of a pressure wave of initially sinusoidal form will move at a speed faster than a_0 and the troughs will move at a speed slower than a_0, so that the wave form tends to a saw tooth shape as shown in Fig. 1.2. The eventual steepness is determined by the

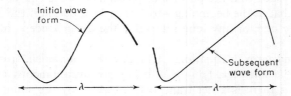

Fig. 1.2   Steepening of wave forms of finite amplitude.

rate of irreversible internal dissipation of energy characteristic of the material, which has not been considered here; for example, strong disturbances in gases may steepen up into shock waves[1].

1.1.3 *Intensity and energy in a plane travelling wave*

Travelling sound waves carry energy. Consider a wave of small amplitude travelling at speed a_0 in the direction of increasing x. The particle displacement is $\xi = \xi_+(x - a_0t)$. The particle velocity (which is also equal to the volume velocity across a surface of unit cross sectional area) is $\dot{\xi} = \partial\xi/\partial t = -a_0\xi'_+(x - a_0t)$, where the dash denotes differentiation with respect to the argument $x - a_0t$. The force per unit area on the material to the right of $x + \xi$ (exerted by the material to the left of $x + \xi$) is $p(x, t) = p_0 + (p - p_0)$. Using relationships obtained previously gives

$$p - p_0 = \rho_0 a_0^2 s = -\rho_0 a_0^2 \frac{\partial\xi}{\partial x} = -\rho_0 a_0^2 \xi'_+ = \rho_0 a_0 \dot{\xi}. \qquad (1.1.10)$$

This equation expresses the important result that the acoustic pressure in a plane, small-amplitude, travelling wave is proportional to the particle velocity in the direction of propagation. The constant of proportionality, $\rho_0 a_0$, is called the *characteristic specific acoustic resistance* of the material.

The rate at which material to the left of $x + \xi$ does work on the material to the right of $x + \xi$ is, per unit area, $p\dot{\xi} = p_0\dot{\xi} + \rho_0 a_0\dot{\xi}^2$. This is the *total intensity*, the total energy carried across unit area in unit time by the wave. The mean intensity over an interval $t_1 < t < t_2$ can be written

$$\frac{1}{t_2 - t_1} \int_{t_1}^{t_2} (p_0\dot{\xi} + \rho_0 a_0\dot{\xi}^2)dt = p_0 \frac{\xi(t_2) - \xi(t_1)}{t_2 - t_1} + \bar{J}, \qquad (1.1.11)$$

where

$$\bar{J} = \frac{1}{t_2 - t_1} \int_{t_1}^{t_2} \rho_0 a_0\dot{\xi}^2 \, dt = \rho_0 a_0\overline{\dot{\xi}^2} = \frac{\overline{(p - p_0)^2}}{\rho_0 a_0}, \qquad (1.1.12)$$

the bar denoting a mean value. The first term in equation 1.1.11 is zero for any interval such that the particle displacement at the end of the interval is equal to that at the beginning—for example, a period if the disturbance is a periodic one, or any long interval if the disturbance is of a random nature.

The second term in equation 1.1.11, $\bar{J}$, does not vanish. It is called the *mean acoustic intensity*, $(p - p_0)\dot{\xi}$ being the instantaneous acoustic intensity. Thus there is a net transport of energy by the sound wave, the mean energy flux being equal to the mean acoustic intensity,

$$\overline{(p - p_0)\dot{\xi}}.$$

The mean acoustic intensity is the time average of the product of the particle velocity by that part of the pressure fluctuation that is in phase with it. In the case of the plane progressive wave considered above, this flow of energy $\overline{(p - p_0)\dot{\xi}}$ is also proportional to the mean kinetic energy density, and in turn to the mean square pressure fluctuation, as can be seen from equation 1.1.12. This occurs because the pressure in such a wave is proportional to the particle velocity, the constant of proportionality being $\rho_0 a_0$, the characteristic specific acoustic resistance of the material.

[Note that the expression $(p - p_0) = (\rho_0 a_0)\dot{\xi}$ is of the same form as Ohm's Law for electrical circuits, namely $E = RI$, where E is the voltage, R the resistance and I the current. Further, the intensity relationships $\bar{J} = \overline{(p - p_0)\dot{\xi}} = \overline{\dot{\xi}^2}(\rho_0 a_0) = \overline{(p - p_0)^2}/\rho_0 a_0$ are also exactly analogous to the power relationships for an electric circuit, namely $\overline{W} = \overline{EI} = \overline{I^2}R = \overline{E^2}/R$. This is an illustration of the fact that the well-known analogies between electrical circuits and vibrating mechanical systems also apply in some measure to continuous systems where the vibrations are caused by propagating waves.]

Note that the mean acoustic intensity in a plane progressive wave is proportional to the mean square pressure fluctuation, and to the mean square particle velocity. For example, in a sinusoidal wave of peak pressure amplitude P_0

$$p - p_0 = \rho_0 a_0 \dot{\xi} = P_0 \sin 2\pi\left(\frac{ft - x}{\lambda}\right),$$

where f is the frequency and λ is the wavelength $\lambda = a_0/f$. It is common in acoustical work to use the radian frequency $\omega = 2\pi f = 2\pi/T$ (where T is the period) in place of the frequency, and also the wavenumber $k = \omega/a_0 = 2\pi/\lambda$ in place of the wavelength.

In terms of these quantities

$$p - p_0 = \rho_0 a_0 \dot{\xi} = P_0 \sin(\omega t - kx).$$

The mean intensity is then

$$\bar{J} = \frac{1}{2\pi/\omega} \int_{t_1}^{t_1 + 2\pi/\omega} (p - p_0)\dot{\xi}\, dt,$$

$$= \frac{\omega}{2\pi} \int_{t_1}^{t_1 + 2\pi/\omega} \frac{P_0^2}{\rho_0 a_0} \sin^2(\omega t - kx) = \frac{1}{2}\frac{P_0^2}{\rho_0 a_0} = (\rho_0 a_0)\frac{1}{2}\frac{P_0^2}{(\rho_0 a_0)^2};$$

that is, the mean intensity is equal to one-half the square of the peak pressure amplitude, divided by the characteristic specific acoustic resistance, which is the same thing as the mean square pressure amplitude divided by this resistance. Alternatively, the mean intensity in this case can be written as the product of the characteristic specific acoustic resistance by the mean square amplitude of the particle velocity (or one-half the square of the peak particle velocity amplitude).

1.2 Dynamics of three-dimensional sound waves

Acoustic waves are by no means necessarily plane waves. In the general case, an acoustic disturbance involves displacements of the elements of the material in all three directions. The dynamics of such general, three-dimensional, acoustic disturbances can be described in a manner closely analogous to that used above for plane disturbances.

1.2.1 *The three-dimensional wave equation*

Consider first the distortion, due to the acoustic disturbance, of a small material element. This may be taken, for convenience, as the material initially contained in a rectangular parallelepiped, as shown in Fig. 1.3.

When the element is distorted by the disturbance the material initially in its equilibrium position at, say, the corner (x, y, z) of the parallelepiped will be displaced to a point $(x + \xi, y + \eta, z + \zeta)$, where (ξ, η, ζ) are the components of the acoustic particle displacement at (x, y, z). The material initially at the corner $(x + \delta x, y, z)$ will be displaced to the point

$$\left(x + \xi + \delta x \frac{\partial \xi}{\partial x}, \quad y + \eta + x \frac{\partial \eta}{\partial x}, \quad z + \zeta + x \frac{\partial \xi}{\partial x} \right),$$

to first order in the strains $\partial \xi / \partial x, \ldots$, and similarly for the material initially in the neighbourhoods of the other corners of the parallelepiped.

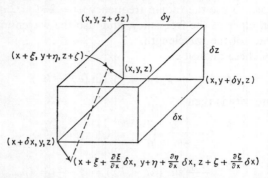

Fig. 1.3   Sketch indicating deformation of a small material element.

In general a small deformation of a material element consists of a rotation and shearing of the original parallelepiped together with an expansion of the volume occupied by the element. In a small-amplitude deformation by a longitudinal acoustic wave only the volume deformation is of interest. The expansion is such as to increase the volume of the material element from $\delta x \delta y \delta z$ to $\delta x \delta y \delta z \, (1 + \partial \xi / \partial x + \partial \eta / \partial y + \partial \xi / \partial z)$, again to first order in the strains. Thus, in place of equation 1.1.3 for the plane wave case, the relative pressure increment and condensation, for three-dimensional acoustic motion, are given by

$$\frac{(p - p_0)}{(\rho_0 a_0^2)} = s = -\left(\frac{\partial \xi}{\partial x} + \frac{\partial \eta}{\partial y} + \frac{\partial \xi}{\partial z} \right). \tag{1.2.1}$$

The left-hand part of this equality expresses the adiabatic nature of the process, just as before, and the right-hand side is the appropriate statement of the principle of conservation of mass.

The three components of the vector force balance equation, analogous to equation 1.1.4 for the plane wave case, are obtained, respectively, by exactly the same arguments as for equation 1.1.4. They are

$$\rho_0 \frac{\partial^2 \xi}{\partial t^2} = -\frac{\partial p}{\partial x}, \quad \rho_0 \frac{\partial^2 \eta}{\partial t^2} = -\frac{\partial p}{\partial y}, \quad \rho_0 \frac{\partial^2 \xi}{\partial t^2} = -\frac{\partial p}{\partial z}, \tag{1.2.2}$$

where again ρ_0 denotes the (constant) ambient mass density.

Equations 1.2.2 show that (apart from a linearly time-dependent part which is of no interest here) the displacement (ξ, η, ζ) is the gradient of a scalar quantity, namely $(1/\rho_0)\int(\int p\,dt)\,dt$. It is customary to use, as a basic dependent variable, the particle velocity potential (instead of a particle displacement potential),

$$\phi(x, y, z, t),$$

such that the particle velocity components (u, v, w) [equal to $\partial\xi/\partial t, \ldots$] are given by

$$u = -\frac{\partial\phi}{\partial x}, \quad v = -\frac{\partial\phi}{\partial y}, \quad w = -\frac{\partial\phi}{\partial z}. \tag{1.2.3}$$

The force balance equations 1.2.2 then require that

$$p - p_0 = \rho_0 \frac{\partial\phi}{\partial t}. \tag{1.2.4}$$

When differentiated with respect to time, equation 1.2.1 can be written

$$\frac{\partial(p - p_0)}{\partial t} = -\rho_0 a_0^2 \left(\frac{\partial u}{\partial x} + \frac{\partial v}{\partial y} + \frac{\partial w}{\partial z}\right),$$

and in turn, when equations 1.2.3 and 1.2.4 have been used to express u, v, w and $p - p_0$ in terms of ϕ, this becomes

$$a_0 \nabla^2 \phi - \frac{\partial^2\phi}{\partial t^2} = 0, \tag{1.2.5}$$

where ∇^2 is the Laplacian operator, $\partial^2/\partial x^2 + \partial^2/\partial y^2 + \partial^2/\partial z^2$, and the speed of sound, a_0, is again given by $a_0 = \sqrt{\kappa_0/\rho_0}$, in terms of the ambient mass density and adiabatic bulk modulus of elasticity of the material.

Equation 1.2.5 is called the homogeneous, scalar wave equation. Its solutions include all the functions describing mathematically the behaviour of all small-amplitude, longitudinal acoustic disturbances in lossless, elastic materials, fluid or solid. Since the pressure, condensation, and the rectangular Cartesian components of the particle velocity (and any partial derivatives of these, either with respect to time or to a cartesian co-ordinate) are all linearly related to the velocity potential, and the wave equation itself is linear, it follows that these quantities also satisfy the wave equation.

The usual procedure in calculating acoustic disturbances arising from movements of some portion of the boundary of the region of material under consideration is to first find a velocity potential which satisfies both the wave equation 1.2.5 and the appropriate boundary conditions. From this, the pressure (and hence the condensation) can be determined from equation 1.2.4, and the particle velocities (and hence particle displacements) from equation 1.2.3. Not surprisingly, the boundary conditions are of great importance in determining the solution, as they impart to it not only the information about any specified movements of portions of the boundary, but also that about the nature of the reflection and transmission by each portion of the boundary of any waves arriving there from whatever direction.

1.2.2 *Acoustic energy density, energy flow and mean square pressure*

Most microphones measure sound pressure, in most conditions, and again in most conditions it is the fluctuating pressure at an observer's ear that is most closely related to the sensation of loudness he experiences. Nevertheless, it cannot be taken for granted that this will be the case in all conditions met in practice. A practical microphone, or an ear, is usually a diaphragm with an associated electromechanical system, all having a certain mass, resistance and stiffness, and encased in a reflecting object (the head, or the microphone housing) the size of which may not always be negligible by comparison with the wavelength of the acoustic disturbance. Any reflecting object or movable diaphragm, inserted into an acoustic field, will to some extent reflect or re-radiate acoustic disturbances back into the field and its boundaries and back on to the source itself, thus altering the pressure and particle velocity distribution everywhere in the field and on its boundaries.

Sound in these respects is quite unlike light; the sound wavelengths are generally quite comparable with the dimensions of reflecting objects, measuring instruments and sound sources. The sound sources are often coherent over appreciable areas of the source surfaces, instead of incoherent, like most light sources. Furthermore, the internal mechanical impedances of both sources and receivers are not always effectively infinitely large (or small) by comparison with the acoustical impedances in which they are terminated when they are placed in an acoustic field.

All this results in the sound field being a complicated mixture of wave interference patterns, comprising both travelling waves and standing waves, sensitive to changes in boundary conditions, including on occasion the insertion of the measuring instruments themselves. It is for these reasons, of course, that good quality commercial microphones are always supplied with calibration information giving corrections to apply to the microphone readings when the instrument is used in certain commonly encountered types of acoustic fields. Even such corrections, of course, are not sufficient when the microphone is placed in an acoustic field where the consequent additional mechanical impedance arising from the coupling of the microphone to the acoustic field becomes comparable with the internal mechanical impedance. In such extreme cases, and they are unfortunately common enough—in highly resonant enclosures, for example —the source, acoustic field and receiver may have to be considered all together as a coupled electro-acoustical-mechanical system. A close and familiar, but much simpler, analogy is the essay of attempting to measure a voltage from a high-impedance source with a low-impedance voltmeter.

Because of these difficulties and experimental uncertainties it is necessary in interpreting the results of measurements in acoustics to have as clear as possible an understanding of the essential nature of acoustic fields, in particular of the relationships between particle velocity and pressure, between mean square pressure and acoustic intensity, and between acoustic intensity and energy density.

Some of these relationships have already been described and discussed in some detail for plane travelling waves in section 1.1.3. There it was shown that in a plane travelling wave the ratio of pressure to particle velocity in the direction of propagation was a constant—namely the characteristic specific acoustic resistance of the material—and that the ratio of the mean square pressure to the mean acoustic intensity was also equal to this characteristic resistance.

It will be shown later in this section that these relationships are also valid for another important class of acoustic disturbances, the outwardly propagating acoustic waves comprising the *radiation field* of a distribution of acoustic sources. The radiation field occurs in the region of space far from the source region (which is taken to include the regions occupied by any sound scattering objects), both in terms of acoustic wavelengths and in terms of a representative source region circumference. Ideally, a source distribution can have a radiation field only when it is embedded in an infinite expanse of surrounding material, so that at large distances all waves must appear to an observer to be travelling radially outwards from some effective origin in the source region. In practice there is often such a radiation field region around a source distribution, into which no appreciable scattered sound comes from reflecting objects lying even further away from the source region.

Nevertheless, the relationships are *not* generally valid. Close to a source, or to a reflecting object, or in a resonant enclosure, the particle velocity may be completely out of phase with the pressure and have a magnitude vastly different from $(p - p_0)/(\rho_0 a_0)$. When, as in a pure standing wave field, the particle velocity is completely out of phase with the pressure the field is said to be *reactive*. In regions close to a source distribution the acoustic field may be largely reactive. This part of the field of a source distribution is called the *near field*.

The mean acoustic intensity in a purely reactive field is identically zero. Thus there is no mean flow of energy into, or out of, the field and it can therefore be maintained without any net supply of energy*. Mean square pressures in a reactive field are by no means necessarily zero, however. In rooms and other relatively resonant enclosures, for example, the reactive part of the field is usually much larger than the active (or radiative) part, and the ear (or a good pressure microphone) usually acts like an ideal pressure transducer of nearly infinite internal impedance, hence absorbing very little energy from the field. Thus the loudness experienced by an observer in the room (or the microphone reading) is determined almost entirely by the relatively large reactive component of the pressure, being quite unrelated to the relatively small active component (and hence similarly unrelated to the mean acoustic intensity).

In electrical analogue terms the pressure may be regarded as the voltage across a circuit element and the mean acoustic intensity as the mean electrical power consumption of the element. Then the ear, or microphone, is acting like an ideal voltmeter, not like a wattmeter.

There are several expressions relating energy quantities in acoustic fields which are valuable both for the insight they provide into questions of energy densities and energy radiation generally and as formulas for use in practical calculations.

Consider an element of surface δS in the neighbourhood of a point (x, y, z) in the material. The rate at which the material on one side of the

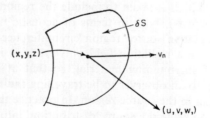

Fig. 1.4   Energy flux across a surface element.

surface does work on the material on the other side is $pv_n\delta S$, where v_n is the component of the particle velocity vector (u, v, w) normal to the surface. The energy flow per unit area (the energy flux) is the vector (pu, pv,

---

* In practice, only enough energy need be supplied to overcome small, irreversible loss of energy through internal friction or heat conduction.

pw). The average value of this vector over an interval $t_1 < t < t_2$ is, for the x-component,

$$\overline{pu} = \frac{1}{t_2 - t_1} \int_{t_1}^{t_2} pu \, dt = \frac{1}{t_2 - t_1} \left\{ \int_{t_1}^{t_2} p_0 u \, dt + \int_{t_1}^{t_2} (p - p_0) u \, dt \right\},$$

(1.2.6)

p_0 being the constant reference pressure. If either the velocity fluctuations are periodic and $t_2 - t_1$ is a period, or if the fluctuations are random and $t_2 - t_1$ is a very long time, then the first integral in the brackets vanishes and

$$\overline{pu} = \frac{1}{t_2 - t_1} \int_{t1}^{t2} (p - p_0) u \, dt = \bar{J}_x(x, y, z),$$

(1.2.7)

which defines the x component of the *mean acoustic intensity vector* $\bar{J}_x(x, y, z)$, $\bar{J}_y(x, y, z)$, $\bar{J}_z(x, y, z)$. The other two components are defined by

$$\bar{J}_y = \overline{pv}, \qquad \bar{J}_z = \overline{pw}.$$

The mean acoustic intensity vector is the mean rate of flow of energy per unit area due to the acoustic disturbances. The acoustic waves, on average, carry energy from one part of the material to another.

The material has been assumed to be non-dissipative, so that no acoustic energy can be permanently lost to the medium itself. Therefore, if S is any closed surface fixed in space and enclosing no acoustic sources then the mean flow of energy out of S must be zero. This physical argument can be put in a precise mathematical form. By Gauss' divergence theorem the vanishing of the integral of the normal component of a vector over any closed surface implies that the divergence of the vector must be zero at all points enclosed by the surface. Therefore

$$\frac{\partial \bar{J}_x}{\partial x} + \frac{\partial \bar{J}_y}{\partial y} + \frac{\partial \bar{J}_z}{\partial z} = 0.$$

(1.2.8)

Verification of this result is straightforward. Using equations 1.2.7 and 1.2.3 gives for the term $\partial (\overline{pu})/\partial x$,

$$\frac{\partial (\overline{pu})}{\partial x} = \frac{\overline{\partial (p - p_0) u}}{\partial x} = -\frac{\partial}{\partial x} \overline{\left\{ p_0 \frac{\partial \phi}{\partial t} \frac{\partial \phi}{\partial x} \right\}}$$

$$= -p_0 \overline{\left\{ \frac{\partial^2 \phi}{\partial x \partial t} \frac{\partial \phi}{\partial x} + \frac{\partial \phi}{\partial t} \frac{\partial^2 \phi}{\partial x^2} \right\}}.$$

The first term in the final bracket above is

$$p_0 \frac{\partial}{\partial t} \left\{ \frac{1}{2} \left(\frac{\partial \phi}{\partial x} \right)^2 \right\} = \frac{\partial}{\partial t} \left(\frac{1}{2} p_0 u^2 \right).$$

Using this result and similar expressions for the other terms of equation 1.2.8 gives

$$\frac{\partial(\overline{pu})}{\partial x} + \frac{\partial(\overline{pv})}{\partial y} + \frac{\partial(\overline{pw})}{\partial z} = -\overline{\left\{\frac{\partial}{\partial t}\left(\frac{1}{2}\rho_0 U^2\right) + \rho_0\frac{\partial\phi}{\partial t}\nabla^2\phi\right\}}, \qquad (1.2.9)$$

where $U^2 = u^2 + v^2 + w^2$. Now ϕ satisfies the homogeneous wave equation 1.2.5 and so the second term in the brackets of (1.2.9) above is equal to

$$\frac{\partial}{\partial t}\overline{\left\{\frac{1}{2}\frac{\rho_0}{a_0{}^2}\left(\frac{\partial\phi}{\partial t}\right)^2\right\}} = \frac{\overline{(p-p_0)^2}}{2\rho_0 a_0^2}.$$

Thus the divergence of the mean intensity vector is

$$-\frac{\partial}{\partial t}\overline{\left\{\frac{1}{2}\rho_0 U^2 + \frac{(p-p_0)^2}{2\rho_0 a_0^2}\right\}} = 0. \qquad (1.2.10)$$

The divergence vanishes as the time average of the time derivative of a fluctuating quantity must be zero.

This verifies that the mean acoustic intensity vector is divergenceless. The result can be stated in the form

$$\int_S \bar{J}_n(x, y, z)\, dS = 0. \qquad (1.2.11)$$

Physically equation 1.2.11 shows that the integral of the normal component of the mean acoustic intensity over any closed surface S enclosing no sources is zero.

The two terms in the bracket in equation 1.2.10 are physically significant. The first, $(\frac{1}{2})\rho_0 U^2$, is the kinetic energy density (kinetic energy per unit volume) associated with the acoustic disturbances. The second term, therefore, might be expected to be the potential energy density. It can be verified that this is true.

In an acoustic compression the work per unit mass done on the material at a point by the surrounding material is $-\int (p - p_0)dV$, where V is the specific volume, $V = 1/\rho$. Hence the work done per unit volume is $-\rho_0\int (p - p_0)d(1/\rho) = (1/\rho_0)\int (p - p_0)d(\rho - \rho_0)$. The process is reversible so the work must be stored by the material as potential energy, the mechanism of storage being the elasticity of the material. The acoustic potential energy density is, therefore,

$$\frac{1}{\rho_0}\int (p - p_0)d(\rho - \rho_0) = \frac{1}{\rho_0 a_0^2}\int (p - p_0)d(p - p_0) = \frac{(p - p_0)^2}{2\rho_0 a_0^2}.$$

The total energy density associated with an acoustic disturbance is thus

$$\frac{1}{2}(\rho_0 U^2) + \frac{(p - p_0)^2}{2\rho_0 a_0^2}. \qquad (1.2.12)$$

The total energy density is not necessarily constant at a point (even though the material is non-dissipative) because acoustic waves arriving at the point may bring with them an excess of energy density, or a defect thereof. As is shown by equation 1.2.10, however, the time average of this total energy density must be zero, because the acoustic motion is a reversible fluctuation about equilibrium conditions.

The fact that the mean acoustic intensity is divergenceless in regions containing no sources means that the mean energy flow into such regions must be exactly balanced by the radiation of energy out. Let R be a region of finite size containing sources, in an unbounded material. Let S_R be the bounding surface of R. Also let S be a large surface containing S_R

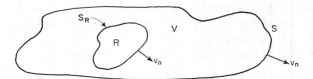

Fig. 1.5   Total mean acoustic energy in the volume V is conserved.

and let V be the region contained between S and S_R. If v_n is the normal component of the acoustic particle velocity, v_n on S_R being directed into V and v_n on S being directed out of V, then the energy flow balance is explicitly

$$\int_{S_R} (p - p_0)v_n \, dS = \int_S (p - p_0)v_n \, dS. \qquad (1.2.13)$$

This relationship shows that the acoustic power output of the sources (the surface integral of the mean intensity) can be found by integrating the normal mean intensity over either the boundary of the source region itself or any other larger surface enclosing it. In particular the larger surface may be one lying wholly in the radiation field.

It follows that the acoustic power output can be calculated from a knowledge of the radiation field alone. This means that the near field must be a reactive load on the source. Energy flowing into the near region during one interval must be returned from the field to the source in some subsequent interval. The energy in the radiation field, on the other hand, travels on out to regions infinitely distant from the source and is thus 'dissipated', as it were, in the form of acoustic radiation, although no energy is dissipated internally in the material itself in the absence of irreversible processes like viscosity and heat conduction.

The source distribution radiates energy only insofar as it produces on its boundary, S_R, fluctuations in which the pressure and normal velocity are in phase. If the pressure and normal velocity are everywhere out of phase on this boundary a reactive near field may be produced in regions near the source but there will be no radiation field and no energy radiated away.

It is shown in section 1.4 that in the radiation field the acoustic pressure from any source distribution is of the form

$$\frac{(p - p_0)}{(\rho_0 a_0^2)} = s(r, \theta, \varphi, t) \sim \frac{1}{4\pi r} f\left(\theta, \varphi, t - \frac{r}{a_0}\right), \qquad (1.2.14)$$

where (r, θ, φ) are spherical polar co-ordinates and $f(\ )$ is a function of the indicated variables. For convenience, and without loss of generality, one may write $f = \partial g/\partial t$ (that is, express f as the time derivative of some appropriate function g), and then equation 1.2.14 can be written as $p - p_0 = \rho_0 a_0^2 s \sim \rho_0 a_0^2 (1/4\pi r)\partial g(\theta, \varphi, t - r/a_0)/\partial t \sim \rho_0 \partial \phi/\partial t$, so that in the radiation field the velocity potential ϕ must be of the form

$$\phi(r, \theta, \varphi, t) \sim a_0^2\left(\frac{1}{4\pi r}\right)g\left(\theta, \varphi, t - \frac{r}{a_0}\right). \qquad (1.2.15)$$

The radial velocity component is $v_r = -\partial\phi/\partial r$.

In the radiation field $-\partial\phi/\partial r$ is approximately equal to $(1/a_0)\partial\phi/\partial t$ because differentiation of the factor $1/r$ in equation 1.2.15 gives a term of order $1/r^2$, which can be neglected. The particle velocity components transverse to r are similarly negligible. Thus, the particle velocity in the radiation field is equal to its radial component which is

$$v_r \simeq \left(\frac{1}{a_0}\right)\frac{\partial\phi}{\partial t} = \frac{(p - p_0)}{\rho_0 a_0} = a_0 s. \qquad (1.2.16)$$

This simple but valuable result shows that in the radiation field the acoustic pressure and particle velocity are in phase and that the ratio of the pressure to the velocity is $\rho_0 a_0$, the characteristic specific acoustic resistance of the medium, just as in a plane progressive wave.

The result has an additional important consequence, namely that the mean acoustic intensity in the radiation field is equal to its radial component $\bar{J}_r$, which is

$$\bar{J}_r = \overline{(p - p_0)v_r} \simeq \frac{\overline{(p - p_0)^2}}{\rho_0 a_0} \simeq \rho_0 a_0 \overline{v_r^2},$$

or

$$\frac{\bar{J}_r}{2a_0} \simeq \frac{\overline{(p - p_0)^2}}{(2\rho_0 a_0^2)} \simeq \frac{\rho_0 \overline{v_r^2}}{2}. \tag{1.2.17}$$

These relationships show that measurements of the mean square pressure in the radiation field are in effect measurements of mean acoustic intensity and, further, that the mean potential energy density is here equal to the mean kinetic energy density and also to $1/2$ the mean intensity divided by the speed of sound a_0. Thus microphone measurements of the pressure in the radiation field can be used to find the total power output of the source.

The total mean power output of the source, $\bar{W}$, can be expressed as the integral of $\bar{J}$ over the surface of any sphere, $r = a$, which contains all sources,

$$\bar{W} = \int_0^{2\pi} \int_0^{\pi} \bar{J}_r(a, \theta, \varphi) a^2 \sin\theta \, d\theta \, d\varphi.$$

The output must be independent of the radius of the sphere. When the sphere is large enough so that its surface lies wholly in the radiation field

$$\bar{W} = \int_0^{2\pi} \int_0^{\pi} \left(\frac{1}{\rho_0 a_0}\right) \overline{(p - p_0)^2} \, r^2 \sin\theta \, d\theta \, d\varphi, \tag{1.2.18}$$

which is, using the notation of (1.2.14) and the fact that $p - p_0 = \rho_0 a_0^2 s$,

$$\bar{W} = \frac{\rho_0 a_0^3}{16\pi^2} \int_0^{2\pi} \int_0^{\pi} \overline{f^2(\theta, \varphi, t - r/a_0)} \sin\theta \, d\theta \, d\varphi. \tag{1.2.19}$$

If the fluctuations have simple harmonic time dependence the velocity potential in the radiation field can be written as

$$\phi \simeq \frac{1}{i\omega\rho_0} P(\theta, \varphi) \frac{1}{4\pi r} \exp i(\omega t - kr)$$

so that

$$p - p_0 = i\omega\rho_0\phi = P(\theta, \varphi) \frac{1}{4\pi r} \exp i(\omega t - kr) = \rho_0 a_0^2 s \simeq \rho_0 a_0 v_r,$$

$P(\theta, \varphi)$ being the complex amplitude of $4\pi r(p - p_0)$. The mean intensity is

$$\bar{J}_r \simeq \frac{\frac{1}{2}|P(\theta, \varphi)|^2}{(4\pi r)^2 \rho_0 a_0}, \tag{1.2.20}$$

and the source power output is

$$\bar{W} = \left\{\frac{1}{32\pi^2 \rho_0 a_0}\right\} \int_0^{2\pi} \int_0^{\pi} |P(\theta, \varphi)|^2 \sin\theta \, d\theta \, d\varphi, \tag{1.2.21}$$

showing that the power depends on the square of the modulus of the pressure directivity function, $P(\theta, \varphi)$, averaged over all angles.

When complex numbers of the form $\exp A(i\omega t)$ are being used to represent physical quantities and mean products of the form

$$\overline{\text{Re}\{A \exp (i\omega t)\} \, \text{Re} \{B \exp (i\omega t)\}}$$

are required, it is usually convenient to use the complex product $AB^*/2$, the star denoting a complex conjugate, because

$$\text{Re} \left(\frac{AB^*}{2} \right) = \overline{\text{Re} \{A \exp (i\omega t)\} \, \text{Re} \{B \exp (i\omega t)\}}.$$

This result can be verified by writing out real and imaginary parts of the complex numbers involved. Thus for simple harmonic fluctuations a complex intensity vector, $\tilde{\mathbf{J}}$, is often used, where

$$\tilde{\mathbf{J}} = \frac{(p - p_0)\mathbf{v}^*}{2} \text{ so that } \mathbf{J} = \text{Re } \tilde{\mathbf{J}},$$

$\overline{\mathbf{J}}$ being the real mean acoustic intensity vector. This representation is an exact one, valid for the near field as well as for the radiation field.

1.3 Simple sound radiators

As outlined in the previous sections the acoustic field of a sound radiator is in general a superposition of a reactive near field and a radiative (active) far field, both of which are sensitive to the details of the vibration pattern and the shape of the radiator. Many of the features of acoustic radiation caused by radiators of various degrees of complication, however, are present in the radiation from two of the simplest kinds of radiator, namely a spherical surface which is either uniformly pulsating, or uniformly oscillating as a rigid body. From these simple examples indications of wider applicability can be obtained about the quantitative dependence of the near and far fields, and of the radiated energy, upon such variables as the frequency of oscillation, size and shape of the radiator, etc. (see, e.g., references 2 and 3).

1.3.1 *Pulsating sphere*

Consider first the case of a uniformly pulsating sphere, surrounded by a uniform material of infinite extent and of sound speed a_0 and ambient mass density ρ_0. Because of the symmetry, the field must only depend upon the radial space co-ordinate, and the time. The wave equation 1.2.5 can thus be written as

$$\frac{1}{r^2} \frac{\partial (r^2 \partial \phi / \partial r)}{\partial r} - \frac{1}{a_0^2} \frac{\partial^2 \phi}{\partial t^2} = 0. \qquad (1.3.1)$$

It can be verified by direct substitution that this equation is satisfied by the so-called 'primitive' solutions:

$$\phi = \frac{1}{4\pi r} Q_+ \left(t - \frac{r}{a_0}\right) + \frac{1}{4\pi r} Q_- \left(t + \frac{r}{a_0}\right), \qquad (1.3.2)$$

where Q_+ and Q_- are suitably continuous functions, but otherwise of arbitrary form. $Q_+/4\pi r$ represents a wave travelling radially outwards and $Q_-/4\pi r$ represents a wave travelling radially inwards. In both cases the

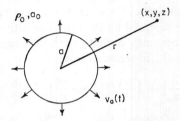

Fig. 1.6   Uniformly pulsating sphere.

amplitude of the wave is proportional to the radial distance, r, but the relative shape of the waves (determined respectively by the Q_+ and Q_- factors) remains unchanged.

Suppose the radial velocity of the sphere (of equilibrium radius $r = a$) has some known time dependence, $v_a(t)$ say, at $r = a$ (for small displacements). Then the radial gradient of the velocity potential at $r = a$ must satisfy

$$-\frac{\partial \phi}{\partial r} (a, t) = v_a(t). \qquad (1.3.3)$$

This relationship, however, is sufficient to determine only one of the arbitrary functions Q_+ and Q_-. To determine both functions a second boundary condition is needed. When the surrounding material is effectively infinite in extent—i.e., there are no reflecting surfaces in the neighbourhood of the pulsating sphere and no other sources of acoustic disturbances outside the surrounding material and directing acoustic waves into it (this is called a *free field* situation)—then the disturbances from the sphere itself must propagate outwards from it, and not inwards towards it, so that the time sequence of cause–effect events as between source and observer is preserved. Mathematically, one can say that at large distances from the source the waves must be radially outgoing ones and this boundary condition, which is in effect applied at the boundary $r \to \infty$, is called the *radiation condition*.

Application of the radiation condition obviously gives the result in this case that $(1/4\pi r)Q_-(t + r/a_0)$ must be identically zero. Then equation 1.3.3 can be used to determine the form of the outgoing-wave function, Q_+:

$$-\frac{\partial \phi}{\partial r}(a, t) = \frac{Q_+(t - a/a_0)}{4\pi a^2} + \frac{1}{a_0}\frac{\partial Q_+(t - a/a_0)/\partial t}{4\pi a} = v_a(t),$$

since $\partial Q_+(t - r/a_0)/\partial r = -(1/a_0)\partial Q_+(t - r/a_0)/\partial t$. Also, since $Q_+(t - r/a_0)$ becomes a function only of t when $r = a$, the partial derivative with respect to time can be replaced by the total derivative. Thus the boundary condition at $r = a$ becomes a first-order, ordinary, linear differential equation determining Q_+ in terms of $v_a(t)$:

$$\frac{dQ_+}{dt}\left(t - \frac{a}{a_0}\right) + \left(\frac{a_0}{a}\right)Q_+\left(t - \frac{a}{a_0}\right) = \left(\frac{a_0}{a}\right)4\pi a^2 v_a(t). \tag{1.3.4}$$

(It should be noted, of course, that only the particular solution of this equation is required here; the complementary function is available to describe the velocity potential due to any previous excitation of the surface.)

In the general case where $v_a(t)$ may have arbitrary time dependence, $\exp\{(a_0/a)t\}$ is an integrating factor for equation 1.3.4 and the particular solution is

$$Q_+\left(t - \frac{a}{a_0}\right) = \int_{-\infty}^{t} \exp\left\{-\left(\frac{a_0}{a}\right)(t - \tau)\right\}4\pi a^2 v_a(\tau)\left(\frac{a_0}{a}\right)dx. \tag{1.3.5}$$

Two limiting cases are of special interest. When the rate of change of the velocity $v_a(t)$ is relatively slow—i.e. $(\omega a/a_0) \ll 1$, where ω is a radian frequency representative of the rate of change—then the first term on the left-hand side of equation 1.3.4 is negligible and one has

$$Q_+\left(t - \frac{a}{a_0}\right) = 4\pi a^2 v_a(t). \tag{1.3.6}$$

The quantity $4\pi a^2 v_a(t)$ is the total *volume velocity* of the pulsating sphere.

At the other extreme is the case where $(\omega a/a_0) \gg 1$—i.e. the rate of change is very rapid. Then the second term on the left hand side of equation 1.3.4 is relatively negligible and Q_+ becomes of order $4\pi a^2 v_a(t)/(\omega a/a_0)$. The quantity $(\omega a/a_0)$ is, for simple harmonic waves, just ka, where k is the wavenumber, $k = \omega/a_0$. Evidently ka is the ratio of the circumference, $2\pi a$, of the sphere to the wavelength, $\lambda = 2\pi/k = 2\pi a_0/\omega$, of the acoustic waves.

The importance of this non-dimensional parameter, ka, in determining the acoustic radiation from a pulsating surface can be seen by considering the case where the surface velocity has simple harmonic time dependence, $v_a(t) = v_0 \exp(i\omega t)$. Then equation 1.3.4 gives

$$Q_+\left(t - \frac{a}{a_0}\right) = \frac{\exp(ika)}{1 + ika} 4\pi a^2 v_0 \exp\left\{i\omega\left(t - \frac{a}{a_0}\right)\right\}, \qquad (1.3.7)$$

and hence the velocity potential $\phi(r, t)$ is

$$\phi(r, t) = \frac{1}{1 + ika} 4\pi a^2 v_0 \frac{\exp i\omega\{t - (r - a)/a_0\}}{4\pi r}. \qquad (1.3.8)$$

(This is the complex velocity potential, of course; the real part of this is the physically significant quantity.) From equations 1.2.4, and 1.3.8, the pressure and the radial particle velocity are, respectively,

$$p - p_0 = \rho_0 a_0^2 \frac{ika}{1 + ika} \left(\frac{v_0}{a_0}\right) \frac{\exp i\omega\{t - (r - a)/a_0\}}{(r/a)} \qquad (1.3.9)$$

and

$$v_r = \frac{(p - p_0)}{\rho_0 a_0 ikr/(1 + ikr)}. \qquad (1.3.10)$$

Note that only at very large distances from the origin ($kr \gg 1$) is the particle velocity strictly in phase with the pressure and is the specific acoustic impedance, $(p - p_0)/v_r$, equal to the characteristic specific acoustic impedance, $\rho_0 a_0$, of the material.

The mean square pressure (the quantity usually measured) is

$$\overline{(p - p_0)^2} = \frac{1}{2} (\rho_0 a_0^2)^2 \frac{(ka)^2}{1 + (ka)^2} \left(\frac{v_0}{a_0}\right)^2 \left(\frac{a}{r}\right)^2. \qquad (1.3.11)$$

This is always proportional to the inverse square of the radial distance, r, and to the square of the surface velocity Mach number, v_0/a_0, but at lower frequencies ($ka \ll 1$) it is also proportional to $(ka)^2$. For the same surface velocity amplitude and at the same number of diameters distant (a/r constant), a large pulsating sphere is therefore a much more effective radiator of acoustic energy than a small one:

$$\overline{(p - p_0)^2} = \frac{1}{2} (\rho_0 a_0^2)^2 \left(\frac{v_0}{a_0}\right)^2 \left(\frac{a}{r}\right)^2 \left\{\begin{matrix} (ka)^2, & ka \ll 1 \\ 1, & ka \gg 1 \end{matrix}\right\}. \qquad (1.3.12)$$

Similarly, from equation 1.3.10, the specific acoustic impedance at the surface of the sphere is $\rho_0 a_0 ika$ for a small sphere and $\rho_0 a_0$ for a large one.

2

If equation 1.3.10 is written as

$$p - p_0 = \rho_0 a_0 \left\{ \frac{1}{1 + 1/(kr)^2} - \frac{1/ikr}{1 + 1/(kr)^2} \right\} v_r \qquad (1.3.13)$$

it is clear that at $kr = 1$ the reactive and active components of the pressure are equal in magnitude. Only for $(kr)^2 \gg 1$, however, does the active component take on its radiation field value. Thus $kr \simeq 1$ may be regarded in this case as a boundary between the active field, $kr \simeq 1$, and the reactive field, $kr < 1$.

When the sphere is very small, $ka \ll 1$, so that from equations 1.3.6 and 1.3.2

$$\phi(r, t) \simeq \frac{4\pi a^2 v_0(t - r/a_0)}{4\pi r} \equiv \frac{Q(t - r/a_0)}{4\pi r}, \qquad (1.3.14)$$

where $Q(t)$ is the volume velocity, the sphere is said to be a *point-monopole radiator*, or 'simple source'. The acoustic pressure is then given by

$$p - p_0 = \frac{\partial M(t - r/a_0)/\partial t}{4\pi r} \qquad (1.3.15)$$

where $M(t)$ is the rate of addition of mass from the neighbourhood of the source ($r = 0$) to the surroundings [or the 'mass velocity', here defined as $M(t) = \rho_0 Q(t)$]. The pressure is thus proportional to the rate of change of this rate of mass addition.

1.3.2 *Oscillating sphere*

Suppose that instead of pulsating uniformly the surface of the sphere (again of radius a) is oscillating back and forth as a rigid body, in the

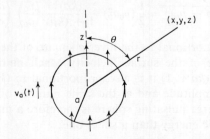

Fig. 1.7   Uniformly oscillating sphere.

z-direction, the normal particle velocity at the surface thus being $v_a(t)$ $\cos\theta$, where θ is the polar angle.

When the velocity potential depends upon θ as well as upon r and t the wave equation 1.2.5 becomes, with the appropriate form of the Laplacian operator, ∇^2, for this case (see reference 2, p. 314, e.g.)

$$\frac{1}{r^2}\frac{\partial}{\partial r}\left(r^2\frac{\partial\phi}{\partial r}\right) + \frac{1}{r^2\sin\theta}\frac{\partial}{\partial\theta}\left(\sin\theta\frac{\partial\phi}{\partial\theta}\right) - \frac{1}{a_0^2}\frac{\partial^2\phi}{\partial t^2} = 0. \qquad (1.3.16)$$

It may be verified by direct substitution that

$$\phi_\pm = -\frac{\partial}{\partial z}\left\{\frac{aQ_\pm(t\mp r/a_0)}{4\pi r}\right\} = -\cos\theta\frac{\partial}{\partial r}\left\{\frac{aQ_\pm(t\mp r/a_0)}{4\pi r}\right\} \qquad (1.3.17)$$

are solutions of this form of the wave equation. It is obvious, in any case, that these are solutions of the wave equation (except at $r = 0$, where, like $aQ_\pm(t\mp r/a_0)/(4\pi r)$, they are singular), since they are simply partial derivatives with respect to rectangular Cartesian co-ordinates of functions which themselves satisfy the wave equation, and hence they will also satisfy the wave equation if these functions are sufficiently continuous to permit interchange of the order of differentiation.

It is also evident that functions of the form given in equation 1.3.17 are suitable for matching normal velocity boundary conditions of the form $-\partial\phi/\partial r = v_a(t)\cos\theta$ on a spherical surface. Thus the velocity potential produced by an oscillating sphere can be found.

Again only the outgoing-wave solution in equation 1.3.17 need be retained, because of the radiation condition at $r \to \infty$. The boundary condition on the normal velocity at $r = a$ then determines the form of Q_+:

$$-\frac{\partial\phi}{\partial r}(a, t) = \cos\theta\frac{\partial^2}{\partial r^2}\left\{\frac{aQ_+(t-r/a_0)}{4\pi r}\right\}_{r=a} = \cos\theta\, v_a(t),$$

or

$$\frac{\partial^2}{\partial r^2}\left\{\frac{aQ_+(t-r/a_0)}{4\pi r}\right\}_{r=a} = v_a(t). \qquad (1.3.18)$$

Performing the differentiations and the substitution indicated in equation 1.3.18, noting again that $\partial\phi_+/\partial r = -(1/a_0)\partial\phi_+/\partial t$, leads directly to the differential equation determining Q_+, namely

$$\left\{\left(\frac{a_0}{a}\right)^2\frac{d^2}{dt^2} + 2\frac{a}{a_0}\frac{d}{dt} + 2\right\}Q_+\left(t - \frac{a}{a_0}\right) = 4\pi a^2 v_a(t), \qquad (1.3.19)$$

which should be compared with the corresponding equation 1.3.4 for the simpler case of the pulsating sphere.

Like equation 1.3.4, equation 1.3.19 can be directly integrated by standard methods to give a general result for $Q_+(t - a/a_0)$ in terms of $v_a(t)$.

This is

$$Q_+\left(t - \frac{a}{a_0}\right) = \int_{-\infty}^{t} \exp\left\{-\left(\frac{a_0}{a}\right)(t - \tau)\right\} \sin\left\{\left(\frac{a_0}{a}\right)(t - \tau)\right\}$$

$$4\pi a^2 v_a(\tau)\left(\frac{a_0}{a}\right)d\tau. \qquad (1.3.20)$$

For the special case of simple harmonic motion, where $v_a(t) = v_0 \exp (i\omega t)$, say, equation 1.3.19 yields directly

$$Q_+\left(t - \frac{a}{a_0}\right) = \frac{\exp (ika)}{(ika)^2 + 2ika + 2} 4\pi a^2 v_0 \exp\left\{i\omega\left(t - \frac{a}{a_0}\right)\right\}, \qquad (1.3.21)$$

which should be compared with equation 1.3.7 for the pulsating sphere. From equations 1.3.21 and 1.3.17 the velocity potential of the oscillating sphere can be written down:

$$\phi(r, \theta, t) = \frac{(ika)}{(ika)^2 + 2\,ika + 2} 4\pi a^2 v_0 \frac{\exp i\omega\{t - (r - a)/a_0\}}{4\pi r}$$

$$\left(1 + \frac{1}{ikr}\right)\cos\theta. \qquad (1.3.22)$$

The acoustic pressure is then (from equations 1.3.22 and 1.2.4)

$$p - p_0 = \rho_0 a_0^2 \frac{(ika)^2}{(ika)^2 + 2\,ika + 2}\left(\frac{v_0}{a_0}\right)\frac{\exp i\omega\{t - (r - a)/a_0\}}{(r/a)}$$

$$\left(1 + \frac{1}{ikr}\right)\cos\theta \qquad (1.3.23)$$

and the radial particle velocity is (from equations 1.3.22 and 1.2.3)

$$v_r = a_0 \frac{(ika)^2}{(ika)^2 + 2\,ika + 2}\left(\frac{v_0}{a_0}\right)\frac{\exp i\omega\{t - (r - a)/a_0\}}{(r/a)}$$

$$\frac{(ikr)^2 + 2\,ikr + 2}{(ikr)^2}\cos\theta. \qquad (1.3.24)$$

(Note that the particle velocity also has a reactive component in the θ-direction).

Thus the radial specific acoustic impedance is

$$\frac{p - p_0}{v_r} = \rho_0 a_0\, ikr \frac{1}{ikr + 1 + 1/(ikr + 1)}. \qquad (1.3.25)$$

Equation 1.3.25 should be compared with equation 1.3.10 for the impedance in the acoustic field of a pulsating sphere. When written out explicitly in terms of real and imaginary parts the two impedances are

$$\frac{p - p_0}{v_r} = \rho_0 a_0 \left\{ \frac{(kr)^2}{(kr)^2 + 1} + ikr \frac{1}{(kr)^2 + 1} \right\}, \qquad \text{(pulsating sphere),} \tag{1.3.10a}$$

$$\frac{p - p_0}{v_r} = \rho_0 a_0 \left\{ \frac{(kr)^4}{(kr)^4 + 4} + ikr \frac{(kr)^2 + 2}{(kr)^4 + 4} \right\}, \qquad \text{(oscillating sphere)} \tag{1.3.25a}$$

These expressions are valid for all positions $r \geq a$, of course. At small distances (compared with the wavelength), $kr \ll 1$, both impedances are predominantly reactive, that for the pulsating sphere approaching $\rho_0 a_0 ikr$ and that for the oscillating sphere approaching $\rho_0 a_0 ikr/2$. Thus, in the neighbourhood of a small pulsating or oscillating sphere (or, by inference, near any small radiator) the acoustic field is largely reactive, the pressure being nearly out of phase with the particle velocity.

The mean radial acoustic intensity, being $\overline{(p - p_0)v_r}$, is proportional to the real part of the specific acoustic impedance. Thus, at low frequencies, the mean intensity (and hence the acoustic power output) from the pulsating sphere is proportional to $(ka)^2$ but that from the oscillating sphere is proportional to $(ka)^4$, a much smaller factor since at low frequencies $(ka) \ll 1$. Thus a small oscillating sphere is, in this sense, a much less efficient radiator than a small pulsating sphere.

At high frequencies (i.e., large distances), however, where (kr), or (ka), is much greater than unity the respective impedances are both predominantly real and equal to $\rho_0 a_0$. Hence, a large oscillating sphere is (again in this sense, without reference to directional effects) just as efficient a radiator as the pulsating sphere at high frequencies. This behaviour is general. Any surface over which the velocity is in phase and fairly uniform in amplitude, and for which the representative dimensions (including the radius of curvature) are large compared with the wavelength, is an efficient radiator, the normal specific acoustic impedance on the surface, $(p - p_0)/v_a$, being everywhere approximately equal to its plane-wave, or radiation-field, value of $\rho_0 a_0$.

The mean square pressures and intensities for the pulsating and oscillating spheres, respectively, are

$$\frac{\bar{J}_r}{\rho_0 a_0^3} = \frac{\overline{(p - p_0)^2}}{(\rho_0 a_0^2)^2} = \frac{1}{2} \left(\frac{v_0}{a_0} \right)^2 \left(\frac{a}{r} \right)^2 \frac{(ka)^2}{(ka)^2 + 1}, \qquad \text{(pulsating sphere).} \tag{1.3.26}$$

$$\frac{\bar{J}_r}{\rho_> a_0^3} = \frac{\overline{(p - p_0)^2}}{(\rho_0 a_0^2)^2} \frac{(kr)^2}{(kr)^2 + 1} = \frac{1}{2} \left(\frac{v_0}{a_0} \right)^2 \cos^2\theta \left(\frac{a}{r} \right)^2 \frac{(ka)^4}{ka^4 + 4}, \qquad \begin{array}{l} \text{(oscillating} \\ \text{sphere)} \end{array} \tag{1.3.27}$$

Note that for the pulsating sphere, but not for the oscillating sphere (or for other radiators, in general), $\bar{J}_r = \overline{(p - p_0)^2}/(\rho_0 a_0)$ even in the reactive near field. The mean square pressure (or intensity) of the oscillating sphere is directional, having a 'figure-eight' pattern determined by the $\cos^2\theta$ factor.

From an oscillating sphere there is no instantaneous net volume flow across the average position of the surface, $r = a$, as there is from a pulsating sphere. It is physically obvious, however, that the oscillating sphere must exert an instantaneous resultant force on the material surrounding it, and that this resultant force is entirely in the direction of the oscillation (provided, of course, that tangential forces such as those due to viscosity, or other shear force mechanisms, are ignored).

The resultant force that the oscillating sphere exerts on the surrounding material is

$$f_z = \int_0^\pi (p - p_0) \cos\theta \, dS$$

$$= \int_0^\pi \rho_0 \frac{\partial}{\partial t} \left\{ - a \frac{\partial}{\partial z} \left[\frac{Q_+(t - r/a_0)}{4\pi r} \right]_{r=a} \cos\theta \right\} 2\pi a \sin\theta \, ad\theta.$$

Since $\partial/\partial z$ of the quantity in the square bracket, evaluated at $r = a$, is $(\partial/\partial a)\{Q_+(t - a/a_0)/4\pi a\} \cos\theta$, the force is

$$f_z = -\rho_0 a \frac{\partial}{\partial a} \left\{ \frac{\partial Q_+(t - a/a_0)/\partial t}{4\pi a} \right\} \int_0^\pi \cos^2\theta \, 2\pi a^2 \sin\theta \, d\theta$$

or

$$f_z = \frac{4}{3} \pi a^3 \rho_0 \left\{ 1 + \left(\frac{a}{a_0} \right) \frac{\partial}{\partial t} \right\} \left\{ \frac{\partial Q_+(t - a/a_0)/\partial t}{4\pi a^2} \right\}. \tag{1.3.28}$$

From the boundary condition equation 1.3.19, which determines Q_+ in terms of the oscillatory velocity, $v_a(t)$,

$$\left\{ 1 + \left(\frac{a}{a_0} \right) \frac{\partial}{\partial t} \right\} Q_+\left(t - \frac{a}{a_0} \right) = \frac{1}{2} \left\{ 4\pi a^2 v_a(t) - \left(\frac{a}{a_0} \right)^2 \frac{\partial^2 Q_+(t - a/a_0)}{at^2} \right\}. \tag{1.3.29}$$

Thus, for spheres of circumference small compared with a representative wavelength—i.e. $(a/a_0)\partial Q_+/\partial t| \ll |Q_+|$—the force is

$$f_z = \frac{4}{3} \pi a^3 \rho_0 \frac{1}{2} \frac{\partial v_a(t)}{\partial t} = \frac{4}{3} \pi a^3 \rho_0 \frac{\partial Q_+(t - a/a_0)\partial t}{4\pi a^2}. \tag{1.3.30}$$

The force in equation 1.3.30 is only that exerted by the sphere on the surrounding material, and therefore is only that necessary to maintain the acoustic field. The *total force* necessary both to move material of mass density ρ_0 inside the sphere with uniform velocity $v_a(t)$, as well as to maintain the field, would be

$$F_z = \frac{4}{3} \pi a^3 \rho_0 \left\{ \frac{\partial v_a(t)}{\partial t} \right\} \left\{ 1 + \frac{1}{2} \right\}$$

or

$$F_z = \frac{3}{2} \frac{4}{3} \pi a^3 \rho_0 \frac{\partial v_a(t)}{\partial t}. \tag{1.3.31}$$

It is clear from this that the acoustic field contributes an inertia, or 'attached mass' (see, e.g., reference 3, pp. 83 and 93) in addition to that of the material being directly accelerated by the force. This attached mass arises only in accelerated motion, as is clear from equation 1.3.30 for the special case of the small oscillating sphere. The attached mass of a pulsating sphere is three times the mass of the surrounding material displaced by the sphere, instead of just one-half of this amount for the oscillating sphere.

The acoustic pressure produced by a small oscillating sphere can be conveniently expressed in terms of the total force F_z. From equations 1.2.4 and 1.3.17, for an oscillating sphere of any size,

$$p - p_0 = \rho_0 \frac{\partial \phi}{\partial t} = -\rho_0 \frac{\partial}{\partial z} \left\{ \frac{a Q_+(t - r/a_0)/\partial t}{4\pi r} \right\}. \tag{1.3.32}$$

For a small oscillating sphere $\rho_0 a \partial Q_+/\partial t = 3f_z = F_z$ (from equations 1.3.30 and 1.3.31) so that

$$p - p_0 = -\frac{\partial}{\partial z} \left\{ \frac{F_z(t - r/a_0)}{4\pi r} \right\}. \tag{1.3.33}$$

For a small sphere, the total force $F_z(t)$ is effectively concentrated at a point—the centre of the sphere, here taken as the origin.

A pressure field of the kind given by equation 1.3.33, produced by a locally applied force, is called a *point-dipole field*. The force component, $F_z(t)$, is here the *dipole moment* of the *point-dipole source*.

In general, the dipole moment, being just the locally applied force vector, is itself a vector quantity.

From comparison of equation 1.3.32 with equation 1.3.33 it is evident that whether an oscillating sphere is small or not it produces a field identical to that of a point-dipole of moment $F_z(t) = a \partial Q_+(t)/\partial t$. However, for a finite size sphere, the simple relationships given in equations 1.3.30 and 1.3.31 do not necessarily apply and it is necessary to use the exact equations

1.3.29 and 1.3.28 to determine $Q_+(t)$ in terms of $v_a(t)$, $f_z(t)$ in terms of $Q_+(t)$, and hence, ultimately, $F_z(t)$ in terms of $v_a(t)$.

The field of two point-monopole sources of mass velocity $\partial M(t)/\partial t$ and $-\partial M(t)/\partial t$, placed respectively at the points $(0, 0, a/2)$ and $(0, 0, -a/2)$, is, at large distances from the origin $(r \gg a)$, exactly that of a point-dipole of moment $F_z(t) = a\partial M(t)/\partial t$: i.e.,

$$p - p_0 = -\frac{\partial}{\partial z}\left\{\frac{a\partial M(t - r/a_0)/\partial t}{4\pi r}\right\}.$$

This expression can easily be verified by writing down the fields of the two point-monopoles, using equation 1.3.15, and then expanding the resultant expression in powers of a/r. The point-dipole source is therefore a limiting case of two point-monopole sources, of opposite signs. In the example given above, the two monopole sources may be very close together but a non-vanishing field will still be produced if the product $aM(t)$ remains finite.

1.4 Acoustic radiation and scattering

Methods of solving radiation and scattering problems, solved problems and experimental results are available extensively in the acoustical literature. Introductions to these subjects can be found in the books by P. M. Morse and S. N. Rschevkin[2,3]. The methods, although generally straightforward, are too lengthy to be described here. In many of the more complicated situations, however, rough estimates of acoustic radiation and scattering can often be made, or facilitated, by the use of the concept of *radiation impedance* and of the *principle of reciprocity*. An understanding of these also gives useful physical insight into the nature of acoustic fields.

1.4.1 *Radiation impedance*

In a number of cases of physical interest it is possible to make order of magnitude estimates of the radiated power from a limited knowledge of conditions on the bounding surface S_R of the source region. In the absence of volume sources, if both the pressure and the normal velocity are known at all points of S_R, the power is determined exactly by equation 1.2.13,

$$\overline{W} = \int_{S_R} \overline{(p - p_0)v_n} \, dS = \int_{S_R} \overline{(p - p_0)v_n} \, dS.$$

In practice these quantities are usually not both known. Results have been worked out in a number of instances where the surface is of a simple shape and the distribution of normal velocity is also a simple one (see, e.g. reference 2, Ch. VII, and other standard acoustics textbooks). Using these results as a guide one can make physically reasonable estimates of what may occur in more complicated practical cases.

Consider as an example a situation in which no volume sources are present, but an acoustic disturbance is being produced by an oscillatory action of a part of the surface of an otherwise rigid body surrounded by an unbounded expanse of material. Suppose that the oscillatory part of the surface has a uniform normal velocity $v_0 \exp(i\omega t)$, where v_0 is a constant, and has an area A.

It is convenient in such problems to recognize explicitly that only the part of $p - p_0$ that is in phase with the normal velocity v_n contributes to the radiated power. The remainder of the fluctuating pressure, $p - p_0$, is out of phase with v_n and hence is associated only with the reactive field. It is therefore useful to introduce the *normal specific acoustic impedance ratio*, ζ, defined by

$$\zeta = \frac{(p - p_0)}{(\rho_0 a_0 v_n)} = \mu + i\chi. \tag{1.4.1}$$

Here μ is the *specific acoustic resistance ratio* and χ is the *specific acoustic reactance ratio*. Both in general are real functions of position on the surface, but not of time.

The expression for the complex intensity then gives, with equation 1.4.1,

$$\bar{J}_n = \overline{(p - p_0)v_n} = \mathrm{Re}\left\{\frac{1}{2}(p - p_0)v_n^*\right\} = \mathrm{Re}\,\frac{1}{2}\left\{\rho_0 a_0(\mu + i\chi)|v_n|^2\right\},$$

or

$$\bar{J}_n = \tfrac{1}{2}\rho_0 a_0 \mu v_0^2,$$

so that

$$\overline{W} = \frac{1}{2}v_0^2 \int_{S_R} \rho_0 a_0 \mu \, dS \equiv \frac{1}{2} v_0^2 R_{\mathrm{rad}},$$

where R_{rad} is called the *radiation resistance* of the oscillating part of the surface. It is equal to $\rho_0 a_0 A \langle \mu \rangle$ where $\langle \mu \rangle$ is the space-average value of the specific resistance ratio over the area of oscillation A. The radiation resistance is proportional to the real part of the average value of the fluctuating pressure.

Morse (see reference 2, pp. 323 ff.) gives some useful curves of radiation resistance and reactance, and directivity patterns, for an example roughly representative of a number of practical situations—a human mouth or ear, a loud-speaker in a baffle box or a microphone diaphragm in its casing, to mention a few. The example is a piston consisting of a radially oscillating segment of the surface of an otherwise rigid sphere.

The effective radius of the segment is $b = 2a \sin(\alpha/2)$ where a is the radius of the sphere and 2α is the angle subtended by the segment at the centre of the sphere.

The general nature of these radiation resistance curves is shown in the sketch below, where the average specific resistance ratio $\langle\mu\rangle$ is plotted against kb (the ratio of the effective piston circumference to the wavelength).

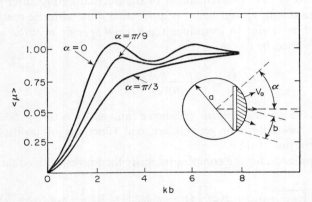

Fig. 1.8   Radiation resistance of a piston set in a spherical baffle.

The top curve, $\alpha = 0$, gives the radiation resistance for a point source (or piston of radius $b \ll a$) on the surface of the sphere. Therefore, it also represents the radiation resistance of a piston in an infinite, plane, rigid baffle when also ka is large.

For $kb \ll 1$, $\langle\mu\rangle$ is proportional to $(kb)^2$, for all the values of α. At high frequencies ($kb \gg 1$) the curves all approach unity, indicating that $p - p_0 \sim \rho_0 a_0 v_n$, as in a plane wave or in the radiation field. The average reactance ratio $\langle\chi\rangle$, which is not shown here, is always a mass reactance representing the attached mass of the surrounding air in the near field which tends to move with the piston. The reactance decreases to zero as $kb \to \infty$, but rather more gradually than the average specific resistance tends towards unity.

This illustrates the following result of practical interest. If the piston surface is relatively large, $kb \gg 1$, then the radiation resistance is approximately $\rho_0 a_0 A$, where A is again the piston area, and hence the total power output is approximately $(1/2)\rho_0 a_0 A v_0^2$, where v_0 is the amplitude of the normal velocity fluctuations. The power radiated has become independent of the size of the sphere itself. Presumably, then, it does not depend on the shape of the body in which the piston is embedded. Even when kb is not large the curves show that the radiated power depends more critically on the piston size than on the size of the sphere itself.

It is to be expected, therefore, that the power radiated by a vibrating portion of a rigid surface depends principally on the size of the vibrating portion. The size and shape of the surface as a whole may critically affect the directionality pattern of the radiation.

If several portions of a surface vibrate and if they are out of phase with one another the remarks above no longer apply. Pressure disturbances from one portion will affect the pressure distribution over the other portions and hence alter the radiation resistance and the total power radiated.

If, for example, the piston in the sphere were split in half, and the two halves vibrated in antiphase, there would obviously be no monopole radiation at all*. The leading term in the multipole expansion would be the dipole term and at low frequencies this means a considerably reduced power output.

At high frequencies the power tends to be concentrated in sharp beams. If the geometry is such that these beams do not overlap and also the near fields of the several pistons do not affect each other, then the total power output would be the sum of the power outputs of the several pistons taken individually as above.

1.4.2 *The principle of reciprocity*

A remarkable and important result of the theory of acoustic waves is the principle of reciprocity, which is due to Helmholtz and can be stated as follows:

> The acoustic field at a point (x, y, z) caused by a point source at (x', y', z') is the same as the field that would be produced at (x', y', z') if the same point source were at (x, y, z).

This result is obvious for a point source in free field from the form of the function representing the field. For example, the pressure from a point source of total rate of mass addition $M(t)$ is†

$$p - p_0 = \frac{\partial M(t - |\mathbf{r} - \mathbf{r}'|/a_0)/\partial t}{4\pi|\mathbf{r} - \mathbf{r}'|}$$

As $|\mathbf{r} - \mathbf{r}'| = \sqrt{(x - x')^2 + (y - y')^2 + (z - z')^2}$, the expression can be

---

* See sections 1.3.2 and 1.5.
† See section 1.3.1, equation 1.3.15.

interpreted as the acoustic pressure at (x, y, z) due to a point source of strength $M(t)$ at (x', y', z') or equally well as the pressure at (x', y', z') due to $M(t)$ at (x, y, z).

The principle also applies when reflecting objects are present. Consider a point source in air above an infinite, plane, air-water interface. The boundary condition at the interface is that the normal particle velocity must vanish, approximately, since the water is much more dense than the air. This is equivalent to $\partial w/\partial t = 0$, if the interface is the plane $z = 0$, and this in turn is equivalent to $\partial(p - p_0)/\partial z = 0$, if no other disturbances are present (see equation 1.2.2). On the other hand, if the source is in the water the boundary condition at the interface, as seen from the water, is that $p - p_0$ must vanish, approximately (the interface being nearly a pressure release surface because the air is so much less dense than the water).

Both these cases are included in the result

$$p - p_0 = \left\{ \frac{\partial M(t - |\mathbf{r} - \mathbf{r'}|/a_0)/\partial t}{4\pi|\mathbf{r} - \mathbf{r'}|} \pm \frac{\partial M(t - |\mathbf{r} - \mathbf{r''}|/a_0)/\partial t}{4\pi|\mathbf{r} - \mathbf{r''}|} \right\}$$

$$(1.4.2)$$

where $|\mathbf{r} - \mathbf{r''}|$ is the image of $|\mathbf{r} - \mathbf{r'}|$ in the plane $z = 0$,

$$|\mathbf{r} - \mathbf{r'}| = \sqrt{(x - x')^2 + (y - y')^2 + (z - z')^2}; \; |\mathbf{r} - \mathbf{r''}|$$

$$= \sqrt{(x - x')^2 + (y - y')^2 + (z + z')^2}. \qquad (1.4.3)$$

For air-water (x, y, z) is an observation point in air, ρ_0 and a_0 have values appropriate to air, and the plus sign is used. For water-air (x, y, z) is an observation point in the water, (x', y', z') is the source point in the water, ρ_0 and a_0 refer to the water, and the minus sign is used. In either case the total pressure is made up of that coming direct from the source together with a contribution from a point source (or sink) at the image point.

It is evident from equation 1.4.3 that equation 1.4.2 is invariant if x and x', y and y', and z and z', respectively, are interchanged, which verifies the principle of reciprocity for these cases. As these two situations of either $p - p_0 = 0$, or $\partial(p - p_0)/\partial n = 0$ on the surface represent extreme conditions, it is plausible to suppose that the principle is valid regardless of the properties or shapes of any reflecting surfaces that may be present. This supposition is correct. A formal proof is given by Lord Rayleigh[4].

A practical use of the principle is in converting a solution of a radiation problem involving a point source in the presence of various reflectors into the solution of the reciprocal problem of the scattering of an incident wave by these same reflectors. This can be illustrated by referring to the problem of the radiation from a piston in a sphere, discussed in the preceding section.

The case $\alpha = 0$ in that problem is of the radiation due to a simple harmonic point source on the surface of the sphere. Let the co-ordinates of the point source be taken to be $r = a$, $\theta = 0$, and $\varphi = 0$ in spherical polar co-ordinates. Then the pressure at $(r, 0, 0)$ due to the source at $(a, 0, 0)$ is the same as would be produced at $(a, 0, 0)$ were the source at $(r, 0, 0)$. Again, the pressure at (r, θ, φ) due to the source at $(a, 0, 0)$ is the same as

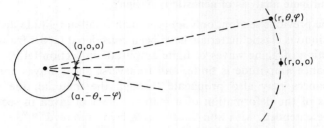

Fig. 1.9   Reciprocity for radiation from a point source on a sphere and scattering of a point source field by a sphere.

would be produced at $(a, 0, 0)$ were the source at $(r, 0, 0)$. Again, the pressure at (r, θ, φ) due to the source at $(a, 0, 0)$ is the same as would be produced at $(a, 0, 0)$ were the source at (r, θ, φ), or, what is the same thing because of the symmetry, it is the same as would be produced at $(a, - \theta, - \varphi)$ were the source at $(r, 0, 0)$.

This with the notation $(p - p_0)$ $(r, \theta, \varphi; M; a, 0, 0)$ denoting the pressure at (r, θ, φ) due to a point simple source of strength M at $(a, 0, 0)$,

$$(p - p_0)(r, \theta, \varphi; M; a, 0, 0) = (p - p_0)(a, - \theta, - \varphi; M; r, 0, 0),$$

indicating the one-to-one correspondence between the pressure at (r, θ, φ) on the surface of a sphere of radius r due to the simple source M at $(a, 0, 0)$ and the pressure at $(a, - \theta, - \varphi)$ on the surface of the sphere of radius a due to the simple source M at $(r, 0, 0)$. The solution for the acoustic pressure caused by the point source at $(a, 0, 0)$ on the surface of the sphere $r = a$ thus can also be used to give the acoustic pressure on this surface when the point source is located at some other (arbitrary) point.

If the point $(r, 0, 0)$ is very far from the sphere $r = a$, then the waves from it arriving at the sphere are essentially plane so the pressure distribution on the sphere due to an incident plane wave has also been obtained. This pressure distribution is obviously just the function describing the distribution in angle of the radiation field due to the source at $(a, 0, 0)$, apart, of course, from a suitable constant of proportionality and the changes of sign in θ and φ. It is evident from this that reciprocity is closely connected with the fact that the radiation field is the spatial Fourier transform of the source distribution causing it*.

By employing the principle of superposition together with the principle of reciprocity one can build up solutions of many reciprocal problems of practical interest in scattering and radiation that involve distributions of sources, either discrete or continuous.

1.5 Multipole analysis of acoustic radiation†

In the previous section, only longitudinal, small-amplitude disturbances in continuous elastic materials have been considered (apart from the brief description of plane waves of finite amplitude in section 1.1.2). When the disturbance amplitude is finite, and transverse shear waves and thermal disturbances may also propagate through the material, the simplified analysis of the deformation of a material element given in section 1.2.1 must be extended, with non-linear terms being retained[5, 6, 7].

It then appears, as Lighthill[8] was the first to point out, that non-acoustic, macroscopic disturbances, such as turbulence in fluids or thermal fluctuations, can be regarded as producing equivalent volume-source distributions which generate acoustic waves. Such volume-source distributions are basically continuous distributions of monopole sources over a region of finite size, not being necessarily confined to a surface. In a similar way, intense acoustic waves can generate turbulence or thermal disturbances. Chu and Kovasznay[9] have classified the various possible second-order, cross-coupling mechanisms occurring in fluids. In the general case, however, acoustic disturbances cannot be unequivocally separated from other types of disturbances—superposition not applying—and considerable difficulties arise in defining the acoustic part of the disturbance, even to only second-order accuracy[10].

Because of these difficulties—practical, conceptual and mathematical—current theories of internally generated sound in continuous materials are almost inevitably constrained to descriptions in terms of equivalent acoustic waves in a fictitious material, the properties of which approximate to those of the actual material[11]. It is an approach of this type, which has had considerable initial success, which is presented in the discussion of aerodynamically generated sound in Chapter 5.

Descriptions of this kind are based on an *inhomogeneous acoustic wave equation*, which can be derived for the general case as follows[12]. If $\rho(x_k, t)$ is the mass density of any distributed component of the material, $v_i(x_k, t)$ is its local barycentric velocity and $\rho\zeta$ is the local rate of production of the

* See section 1.5.

† Cartesian tensor notation is used in this section for brevity; a repeated suffix indicates summation.

component, then the equation of mass transport of the component is

$$\frac{\partial \rho}{\partial t} + \frac{\partial (\rho v_i)}{\partial x_i} = \rho \zeta. \tag{1.5.1}$$

The equation of linear momentum transport of the component is

$$\frac{\partial (\rho v_i)}{\partial t} + \frac{\partial}{\partial x_j} (\rho v_i v_j + p_{ij}) = \rho f_i + \rho \zeta v_i, \tag{1.5.2}$$

where $f_i(x_k, t)$ is the external force per unit mass on the component and $p_{ij}(x_k, t)$ is the stress tensor for internal forces acting on the component.

Elimination of terms depending only on ρv_i between equations 1.5.1 and 1.5.2, together with some rearrangement, results in the exact equation

$$\frac{\partial^2 \rho}{\partial x_j^2} - c^{-2} \frac{\partial^2 \rho}{\partial t^2} = -c^{-2} \left\{ \frac{\partial (\rho \zeta)}{\partial t} - \frac{\partial (\rho f_i + \rho \zeta v_i)}{\partial x_i} + \right.$$
$$\left. + \frac{\partial^2 (\rho v_i v_j + p_{ij} - c^2 \rho \delta_{ij})}{\partial x_i \partial x_j} \right\}. \tag{1.5.3}$$

where c is an arbitrary function of time only. It is then customary to assign to the arbitrary function of time, c, the value, a_0, of the actual ambient speed of sound in the component (space-averaged, of course, over the region concerned). The equation 1.5.3 can then be regarded as an *inhomogeneous scalar wave equation* for the mass density, referred to the fictitious material of uniform sound speed a_0. If, as is often the case, the sound speed in the actual component is in fact reasonably constant over the region concerned, then this description has some physical validity, as well as mathematical convenience.

From this point of view, then, *all* mass density variations in a material can be regarded as being built up of equivalent acoustic waves (including those of zero frequency) in the fictitious material of constant sound speed a_0, these waves being caused by the presence of a distribution of equivalent acoustic sources—i.e., the right-hand side of equation 1.5.3. Such an equivalent source distribution includes both 'true' sources, such as that due to the external force field, and fictitious sources (such as those, e.g., equivalent in effect to the convection and refraction of the actual acoustic waves in the material by the mean motion.)

The description is useless, both mathematically and physically, unless the source distribution is known to sufficient accuracy from theoretical or experimental evidence, independently of the mass density.

In addition to these problems of the applicability of this description and of its physical interpretation, there is the interrelated problem of the dependence of the mass density, as calculated from equation 1.5.3, on the detailed character of a given, or assumed, source distribution. This problem is interrelated because the accuracy to which the source distribution must be determined to provide reasonable mass density predictions in any given case largely depends upon the sensitivity of the mass density to the detailed behaviour of the source distribution. As the exact source distribution usually cannot be determined with any high degree of precision, either theoretically or experimentally, this is of practical importance.

Also, the dependence of the mass density on the functional character of the source distribution is of fundamental theoretical importance. Knowledge of this dependence can be used to provide insights into the relationship of the equivalent acoustic waves to the real acoustic waves, and to identify the true sources of internally generated sound (i.e., those which create actual acoustic radiation, as distinct from those which are equivalents for such processes as convection or refraction).

Multipole analysis is a general technique for solving the inhomogeneous scalar wave equation, in which the dependences referred to in the two preceding paragraphs are particularly emphasized. It originated about a century ago with Maxwell and has been widely used for Poisson's equation in electromagnetism but tended to be applied only in a fragmentary way for acoustic and electromagnetic radiation.

In the various formulations of the theory of internally generated sound the dependent variable in the relevant inhomogeneous wave equation may be the mass density, the pressure or the velocity potential. Therefore the analysis is outlined here for a general wave quantity, $\phi(x_k, t)$. From equation 1.5.3 it is evident that the different terms of the source distribution will fall into three functional categories: (i) a simple scalar $c^{-2}\partial(\rho\zeta)/\partial t$, which in the general analysis is denoted by $q(x_k, t)$; (ii) the divergence of a vector, $-c^{-2}\partial(\rho f_i + \rho\zeta v_i)/\partial x_i$, which is denoted by $-\partial q_i(x_k, t)/\partial x_i$; (iii) the double divergence of a tensor, $c^{-2}\partial^2(\rho v_i v_j + p_{ij} - c^2\rho\delta_{ij})/\partial x_i\partial x_j$, denoted by $\partial^2 q_{ij}(x_k, t)/\partial x_i\partial x_j$. The entire source distribution function, the sum of these three terms, is called $\hat{q}(x_k, t)$. In this notation the inhomogeneous scalar wave equation is

$$\frac{\partial^2\phi}{\partial x_i^2} - a_0^{-2}\frac{\partial^2\phi}{\partial t^2} = -\hat{q}, \tag{1.5.4}$$

where c has been given the value of the appropriate, space-averaged speed of sound, a_0.

For simplicity, equation 1.5.4 is assumed to be valid everywhere in a region of infinite extent; scattering objects are not considered here. Extension of the analysis to include such objects is straightforward in principle; the objects can be replaced by equivalent, generalized image source distributions to which the multipole analysis may be applied in turn. Taking the wave speed as a space-averaged, constant, ambient sound speed, a_0, of course restricts the applicability of the results to physical situations in which for all sufficiently large distances from the source distribution the real material is quiescent except for the presence of small, superposed acoustic, or other, disturbances, the speed of actual acoustic waves in this region being a_0.

The source strength density, or monopole moment density, $\hat{q}(x_k, t)$, is assumed to vanish and have vanishing space derivatives of all orders on and outside a surface S_R enclosing a region R. It then follows that the exact and unique solution of equation 1.5.4 for ϕ is

$$\phi(x_k, t) = \int_R \frac{\hat{q}(y_k, t - |x_k - y_k|/a_0)}{4\pi|x_k - y_k|} \, dy_1 \, dy_2 \, dy_3. \tag{1.5.5}$$

Equation 1.5.5 shows clearly that each volume element, $dy_1 \, dy_2 \, dy_3$, emits elementary spherical acoustic waves, as from a point-monopole, the elementary waves from all volume elements being superposed to give the resultant field. If $\hat{q}$ is zero over any sub-region of R, that sub-region contributes nothing whatever to the radiation. For example, a solenoidal vector field $q_i(x_k, t)$ would contribute no radiation even if q_i itself were non-vanishing in R.

The integrand in equation 1.5.5 can be expanded in a form which separates out termwise the effects on ϕ of the spatial dependence of $\hat{q}$ and of its time dependence, respectively. Term by term integration then gives ϕ, again exactly and uniquely, as a superposition of waves from an infinite sequence of point-multipole sources all located at the origin of co-ordinates, the multipole moments of these sources being instantaneous, normalized, spatial moments of the source distribution with respect to the origin:

$$\phi(_k, t) = \sum_{l, m, n = 0}^{\infty} D_{lmn}\left(x, \frac{\partial}{\partial x_k}\right) Q_{lmn}\left(t - \frac{x}{c}\right), \tag{1.5.6}$$

where

$$D_{lmn}\left(x, \frac{\partial}{\partial x_k}\right) \equiv -\frac{\partial^{l+m+n}}{\partial x_1^l \partial x_2^m \partial x_3^n}\left\{\frac{1}{4\pi x} \cdots \right\}, \tag{1.5.7}$$

$$Q_{lmn}(t) \equiv \int_R \frac{y_1^l y_2^m y_3^n}{l! m! n!} \hat{q}(y_k, t) \, dy_1 \, dy_2 \, dy_3. \tag{1.5.8}$$

The quantities $Q_{lmn}(t)$ are the *overall, instantaneous multipole moments*. D_{lmn} are the *multipole operators*, and $x \equiv |x_k|$. The normalization ensures that each multipole term is also the leading term at large distances of the

radiation from an appropriate cubical lattice of equal positive and negative point-monopole sources, and of unit lattice spacing.

By combining terms having equal values of $l + m + n$, tensor multipole
moments and operators of all orders can be defined:

$$Q \equiv \int_R \hat{q}\, dy_1\, dy_2\, dy_3, \quad Q_i \equiv \int_R y_i \hat{q}\, dy_1\, dy_2\, dy_3, \quad Q_{ij} \equiv \frac{1}{2} \int_R y_i y_j \hat{q}\, dy_1\, dy_2$$

$$dy_3 \ldots;$$

$$D \equiv \frac{1}{4\pi x}, \quad D_i \equiv \frac{\partial}{\partial x_i}\left\{\frac{1}{4\pi x}, \ldots\right\}, \quad D_{ij} \equiv \frac{\partial^2}{\partial x_i \partial x_j}\left\{\frac{1}{4\pi x}, \ldots\right\}, \ldots.$$

The series in equation 1.5.6 can then be written alternatively as $\phi = DQ$
$+ D_i Q_i + D_{ij} Q_{ij} + \cdots$. Q is the overall monopole moment of the source
distribution, Q_i is the overall dipole moment, Q_{ij} is the overall *quadrupole
moment*, and so on.

The following result can then be proved (for convenience of reference
it is called the equivalence theorem). If $\hat{q}(x_k, t)$ is a regular bounded function on a bounded region R, which vanishes with all its space derivatives
on the surface S_R of R, and if $\hat{q}$ is of the functional form

$$\hat{q}(x_k, t) = \{(-\partial)^N/(\partial x_{i_1} \partial x_{i_2} \ldots \partial x_{i_N})\} q_{i_1 i_2} \ldots {}_{i_N}(x_k, t),$$

then

$$\phi(x_k, t) = \frac{(-\partial)_N}{\partial x_{i_1} \partial x_{i_2} \cdots \partial x_{i_N}} \int_R \frac{q_{i_1 i_2} \cdots {}_{iN}(y_k, t - |x_k - y_k|/a_0)}{4\pi|x_k - y_k|} dy_1\, dy_2,\, dy_3,$$

and the leading term in the multipole expansion for ϕ is $D_{i_1 i_2} \cdots {}_{i_N} Q_{i_1 i_2} \cdots {}_{i_N}$.

The theorem shows that a source (i.e., monopole) distribution can be
said to be of multipole order 2^N if it is functionally expressible as the Nth
order divergence of a suitably regular and bounded N^{th}-order tensor field.
For example, $-\partial q_i/\partial x_i$ is a source distribution of dipole order, or a dipole
source-distribution (but *not* a *dipole-source* distribution). It is evident on
mathematical grounds alone that all source distributions which are diver-
gences of regular tensor functions are fundamentally monopole distributions. In any physical theory based on the conception of matter and
motion as continuous fields, therefore, all source distributions must also
be conceived of as fundamentally monopole in nature. That is to say, each
small microscopic volume element in the source region emits not higher
order radiation but monopole radiation; the local monopole moment
density is not zero, for if it were, the element would emit no radiation at
all.

The source distribution as a whole, however, because of phase cancellation, may not emit any net monopole radiation (or dipole, or quadrupole, for that matter, depending upon the degree of phase cancellation).

The radiation field occupies the region where the distance, x, from the source distribution is large compared with both the representative wavelength, $\lambda = 2\pi/k = 2\pi a_0/\omega$, where ω is a radian frequency representative of the moment density rates of change, and with the representative linear dimension of the source region, R. In the radiation field region, terms in $D_{lmn}Q_{lmn}$ of order higher than the first power in $1/x$ can be neglected and consequently ϕ becomes

$$\phi(x_k, t) \simeq \sum_{l, m, n = 0}^{\infty} \frac{(x_1/x)^l (x_2/x)^m (x_3/x)^n}{4\pi x} \left(\frac{1}{a_0} \frac{\partial}{\partial t}\right)^{l+m+n} Q_{lmn}\left(t - \frac{x}{c}\right).$$

(1.5.9)

Examination of this expression (e.g., for any distribution having symmetry about the origin) shows that although each multipole term contributes radiation of a distinctive directional pattern, nevertheless a source distribution producing a radiation field directional pattern characteristic of a certain point-multipole is not necessarily a distribution of that multipole order. This can happen because the direction cosines are not linearly independent.

Equation 1.5.9 shows that the radiation field of any source distribution is of the form

$$\frac{1}{4\pi r} f\left(\theta, \varphi, t - \frac{r}{a_0}\right)$$

where (r, θ, φ) are spherical polar co-ordinates; for a source distribution of simple harmonic time dependence this is

$$\frac{1}{4\pi r} A(\theta, \varphi) \exp i\{\omega t - kr - \alpha(\theta, \varphi)\}.$$

Taking the Fourier transform of equation 1.5.9 with respect to time gives

$$\phi(x_k, \omega) \simeq \frac{\exp(-ikx)}{4\pi x} \sum_{l, m, n = 0}^{\infty} \left(\frac{ikx_1}{x}\right)^l \left(\frac{ikx_2}{x}\right)^m \left(\frac{ikx_3}{x}\right)^n Q_{lmn}(\omega),$$

or

$$\phi(x_k, \omega) \simeq \frac{\exp(-ikx)}{4\pi x} \int\int\int_{-\infty}^{\infty} \hat{q}(y_k, \omega) \exp i\left\{\frac{kx_j}{x} y_j\right\} dy_1\, dy_2\, dy_3. \quad (1.5.10)$$

This is the familiar result that the directivity pattern of the spectral density of the radiation field and the source distribution spectral density are reciprocal functions of a triple Fourier transform, where the pairs of transform variables are the position of the source element, y_i, and the products of the observer's direction cosines by the wavenumber, kx_i/x. The source distribution spectral density thus can be uniquely determined from observation of the directivity pattern, as well as *vice versa*.

Finally, Fourier inversion of equation 1.5.10 with respect to ω yields $\phi(x_k, t)$ in the form of a fourfold Fourier transform,

$$\phi(x_k, t) \simeq \frac{1}{4\pi x} \int\int\int_{-\infty}^{\infty}\int \hat{q}(y_k, \omega) \exp i\left\{\omega t - kx + \frac{kxj}{x} y_j\right\} d\omega \, dy_1 \, dy_2 \, dy_3.$$

(1.5.11)

The multipole expansion in the radiation field is thus seen to be simply the term by term expansion of this familiar Fourier representation. Either the Fourier representation, the multipole expansion in terms of overall multipole moments Q_{lmn} or direct integration over an appropriate equivalent multipole distribution can be used to calculate radiation fields of source distributions. Calculation of radiation from idealized model source distributions by the Fourier transform method is particularly easy and from such calculations the effects on the radiation of the size, shape and frequency of a distribution can be seen.

Consider a source region R of representative dimensions (a, b, d) in the x_1-, x_2- and x_3- directions, respectively, and containing a source distribution of multipole order $2^{l+m+n}(l + m + n = N)$. This source distribution then has an equivalent, multipole-source distribution of this order. For each overall multipole moment of the spectral density of the distribution, $Q_{lmn}(\omega)$, a space-averaged multipole moment density, $q_{lmn}(\omega) = Q_{lmn}(\omega)/V$, can be defined, where V is the volume of the region. The radiation field of such a uniform multipole moment density (i.e., one which is both coherent and of constant amplitude over the region) can be written as

$$\phi_{lmn}(x_k, \omega) \simeq \left(\frac{-ikx_1}{x}\right)^l \left(\frac{-ikx_2}{x}\right)^m \left(\frac{-ikx_3}{x}\right)^n \frac{\exp(-ikx)}{4\pi x}$$

$$\times q_{lmn}(\omega) V F\left(\frac{kax_1}{x}, \frac{kbx_2}{x}, \frac{kdx_3}{x}, \ldots\right).$$

The factor $F(\ldots)$ depends on the size and shape of the region (the correlation region) and on the frequency, but *not* on the multipole order of the distribution. As indicated, it is primarily a function of the quantities $kax_1/x, \ldots$.

The directionality of the radiation thus depends both on the multipole directionality factors, kx_i/x, and on the shape and size of the correlation region. For a correlation region in the form of a rectangular parallelepiped the factor is exactly a triple product of the familiar $(\sin z)/z$ type of functions, namely

$$\frac{\sin (kax_1/x)}{(kax_1/x)} \frac{\sin (kbx_2/x)}{(kbx_2/x)} \frac{\sin (kdx_3/x)}{(kdx_3/x)}.$$

If the region is spherical, and of radius a, the factor is (again exactly) just

$$3 (\sin ka - ka \cos ka)/(ka)^3.$$

Thus, at low frequencies (ka, kb, kd all less than unity), the radiation from the region is independent of the shape, and is equal to

$$\left| \phi_{lmn}(x_k, \omega) \right| \simeq \left| \left(\frac{x_1}{x} \right)^l \left(\frac{x_2}{x} \right)^m \left(\frac{x_3}{x} \right)^n \right| \cdot \left| \frac{1}{4\pi x} k^N \right| \cdot \left| q_{lmn} \right| \cdot V.$$

At high frequencies, however, the results for the parallelepiped and the sphere are respectively,

$$\left| \phi_{lmn}(x_k, \omega) \right| \simeq \left| \left(\frac{x_1}{x} \right)^{l-1} \left(\frac{x_2}{x} \right)^{m-1} \left(\frac{x_3}{x} \right)^{n-1} \right| \cdot \left| \frac{1}{4\pi x} k^{N-3} \right| \cdot \left| q_{lmn} \right| \times$$

$$\times \left| 8 \sin \left(\frac{kax_1}{x} \right) \sin \left(\frac{kbx_2}{x} \right) \sin \left(\frac{kdx_3}{x} \right) \right|,$$

$$\left| \phi_{lmn}(x_k, \omega) \right| \simeq \left| \left(\frac{x_1}{x} \right)^l \left(\frac{x_2}{x} \right)^m \left(\frac{x_3}{x} \right)^n \right| \cdot \left| \frac{1}{4\pi x} k^{N-3} \right| \cdot \left| q_{lmn} \right| \cdot \left| 4 \pi ka \cos ka \right|,$$

showing significantly different behaviour for the two shapes, as functions of frequency.

References

1. M. J. Lighthill. In *Surveys in Mechanics* (ed. G. K. Batchelor), Cambridge University Press (1956): see pp. 250–351.
2. P. M. Morse. *Vibration and Sound* (second edition), McGraw-Hill Book Co., Inc., New York (1948).
3. S. N. Rschevkin. *A Course of Lectures on the Theory of Sound*, Pergamon Press, Oxford (1963).
4. Lord Rayleigh. *Theory of Sound* (second edition, 1945 re-issue), Dover Publications, New York (1945): see Volume II, p. 145, paragraph 294.
5. L. D. Landau and E. M. Lifshitz. *Fluid Mechanics*, Pergamon Press, Oxford (1959).
6. L. D. Landau and E. M. Lifshitz. *Theory of Elasticity*, Pergamon Press, Oxford (1959).
7. R. N. Thurston. In *Physical Acoustics* (ed. W. P. Mason), Academic Press, New York (1964): see chapter 1, Volume 1, Part A.
8. M. J. Lighthill. *Proc. Roy. Soc. A.*, **211**, 564 (1952).
9. B.-T. Chu and L. S. G. Kovasznay. *J. Fluid Mech.*, 3, 494 (1958).

10. C. L. Morfey. *Proc. 5th International Congress on Acoustics*, Liége, paper K 15 (1965).
11. P. E. Doak. *J. Sound Vib.*, **2**, 53 (1965).
12. P. E. Doak. *Proc. 5th International Congress on Acoustics*, Liége, paper K 56 (1965).

CHAPTER 2

Elements of Sound Propagation

2.0 Introduction

In chapter 1 the radiation of sound from sources in the absence of reflecting or absorbing surfaces has been considered and in chapter 3 the reflection, transmission and absorption of sound at such surfaces will be discussed. In this chapter we shall be concerned with the effects of fairly rigid reflecting surfaces on the propagation of sound waves. Because air is so light most solid bodies reflect a considerable amount of the acoustic energy impinging on them and thus tend principally to redirect or guide the flow of acoustic energy. The main effects of this can be studied by treating the solid bodies as though they were rigid or nearly so, and usually we shall make this assumption in what follows.

We shall also briefly outline the effects of the principal factors influencing the propagation of sound waves in the atmosphere; namely, the attenuation of the waves due to shear viscosity, thermal conduction and relaxation processes, and the refraction caused by winds and temperature gradients.

2.1 Sound propagation in ducts

Waves from a source of sound in a duct will be reflected from the walls of the duct. For waves of a given frequency, interference among these multiple reflections results in the establishment of certain definite patterns of variation of the amplitude over the cross section of the duct—the *modes* of the duct. An infinite (but denumerable) number of such modes are possible, but at a given frequency only a finite number of them can be propagated down the duct. The amplitude of the others decays exponentially with distance along the duct.

If we have a certain distribution of simple harmonic sources over some cross section of the duct only those elements of the pattern of the distribution that are identical to the pattern of one or another of the propagating modes of the duct will produce radiation of sound along the duct.

43

2.1.1 *Modes of a rectangular duct*

To see more precisely how this comes about we consider a rectangular duct of very great length and cross-section dimensions a and b, the interior of the duct being the region $0 < x < a$, $0 < y < b$, $-\infty < z < \infty$. The walls are assumed to be rigid. The acoustic velocity potential in the duct must satisfy the wave equation and the condition that the normal particle velocity vanishes at the rigid walls. Assuming simple harmonic motion, and writing the velocity potential as $\phi(x, y, z)\, e^{i\omega t}$, we must have

$$\nabla^2\phi + k^2\phi = 0, \qquad (k = \omega/a_0),$$

where a_0 is the speed of sound in the fluid contained in the duct. We solve this by the method of separation of variables, obtaining solutions of the form $\phi = X(x)\,Y(y)\,Z(z)$, where $X(x) = A_x \cos k_x x + B_x \sin k_x x$, $Y(y) = A_y \cos k_y y + B_y \sin k_y y$ and $Z(z) = A_z\, e^{i k_z z} + B_z\, e^{-i k_z z}$. Here the A's and B's are arbitrary constants and the wave numbers must satisfy the condition

$$k_x^2 + k_y^2 + k_z^2 = k^2. \tag{2.1.1}$$

The exponential form is used for $Z(z)$ as it turns out to be more convenient.

The normal particle velocity will be zero at the walls $x = 0$ and $x = a$ if $-\partial\phi/\partial x = 0$ there. Thus we must have also

$$k_x A_x \sin k_x x - k_x B_x \cos k_x x = 0, \qquad \text{for } x = 0, a.$$

One solution is $k_x = 0$. If $k_x \neq 0$, the condition will be satisfied at $x = 0$ only if $B_x = 0$, and then at $x = a$ only if $\sin k_x a = 0$ (apart from the trivial case $A_x = 0$). Both the solutions for $k_x = 0$ and the others are included if

$$k_x = m\pi/a, \qquad m = 0, 1, 2, \ldots .$$

Thus we have found a sequence of possibilities for $X(x)$,

$$X(x) = A_{xm} \cos m\pi x/a, \qquad m = 0, 1, 2, \ldots .$$

Similarly, we find for $Y(y)$

$$Y(y) = A_{yn} \cos n\pi y/b, \qquad n = 0, 1, 2, \ldots .$$

From equation 2.1.1 we see that for a given pair of the indices m, n, we must have

$$k_z = \pm k_{mn} = \pm\{k^2 - (m\pi/a)^2 - (n\pi/b)^2\}^{1/2}. \tag{2.1.2}$$

Because the system is linear the complete solution is the superposition of all the linearly independent solutions

$$\phi(x, y, z)\, e^{i\omega t} = \sum_{m,n=0}^{\infty} \cos{(m\pi x/a)}\cos{(n\pi y/b)}(A_{mn}\, e^{i\omega t - ik_{mn}z}$$
$$+ B_{mn}\, e^{i\omega t + ik_{mn}z}), \qquad (2.1.3)$$

where the A_{mn}'s and B_{mn}'s are arbitrary constants.

The terms in equation 2.1.3 can be interpreted as follows. The term $\cos{(m\pi x/a)}\cos{(n\pi y/b)}A_{mn}\exp{(i\omega t - ik_{mn}z)}$, which we shall call the positive-going part of the (m, n)th *mode*, represents a wave whose amplitude $A_{mn}\cos{(m\pi x/a)}\cos{(n\pi y/b)}$ varies over its front (i.e., the wave front is space modulated). This wave has frequency ω and is travelling in the $(+z)$ direction at speed $c_{mn} = \omega/k_{mn}$, this speed being greater than the speed of sound a_0 for $m, n \neq 0$. When m and n are equal to zero the speed is equal to the speed of sound and the wave is a plane acoustic wave, $A\exp{(i\omega t - ikz)}$. Similarly, the negative-going part of the (m, n)th mode is $\cos{(m\pi x/a)}\cos{(n\pi y/b)}B_{mn}\exp{(i\omega t + ik_{mn}z)}$ and is a front-modulated wave of frequency ω travelling in the $(-z)$ direction at speed c_{mn}.

Of course, each of these waves is itself a superposition of interfering acoustic waves that travel at the speed of sound. For example, if $n = 0$ we can write the positive-going part of the $(m, 0)$ mode as

$$(A_{m0}/2)\,\{\exp{(i\omega t + im\pi x/a - ik_{mn}z)} + \exp{(i\omega t - im\pi x/a - ik_{mn}z)}\}.$$

The first exponential is a plane acoustic wave travelling in a direction whose direction cosines with respect to the x and z axes are, respectively, $\cos{\alpha_m} = -(m\pi/a)/k$, $\cos{\gamma_m} = k_{mn}/k$. The second exponential similarly is a plane acoustic wave in the direction defined by $\cos{\alpha_m} = (m\pi/a)k$, $\cos{\gamma_m} = k_{mn}/k$. Each of these waves is simply the reflection of the other at the plane $x = 0$, or at the plane $x = a$. Thus each mode arises from the interference among plane acoustic waves travelling at different angles and being reflected back and forth between the duct walls. It is basically a propagating interference pattern.

Not all the modes that may be excited at a given frequency actually propagate as waves, however. Only a finite number do. We can see from equation 2.1.2 that when m, n are sufficiently large, k_{mn} becomes a pure imaginary quantity, $\pm i|k_{mn}|$. For a given mode (m, n) this occurs if the frequency is less than the *mode cut-off frequency*, ω_{mn}, that is,

$$\omega < \omega_{mn} = a_0\{(m\pi/a)^2 + (n\pi/b)^2\}^{1/2}. \qquad (2.1.4)$$

If the frequency is lower than the cut-off frequency the propagation factor for the positive-going part of the mode becomes $\exp{(i\omega t - |k_{mn}|z)}$ and that for the negative-going part $\exp{(i\omega t + |k_{mn}|z)}$. Thus the amplitude of these modes decays exponentially with the axial distance from the source of excitation of the mode.

We have seen that the effect of the duct walls is to permit a simple harmonic source to establish only certain definite types of acoustic fluctuations, each with a characteristic amplitude pattern over the cross section, depending only on the cross section shape and size and each with a characteristic speed of propagation in the axial direction. At a given frequency only a finite number of these modes actually propagate. The speed of propagation of the others becomes purely imaginary and they decay exponentially with axial distance from the source.

2.1.2 *Excitation of duct modes*

To complete the picture we now turn to the question of which modes a given source would excite. Let us suppose that a vibrating panel is producing over the cross section $z = 0$ a certain velocity $U(x, y)\, e^{i\omega t}$ in the z-direction, this velocity distribution being maintained by some external force. We need consider only the region $z > 0$ in the duct because the acoustic field produced in the other half would be the same but with z replaced by $-z$ and U by $-U$. We suppose also that there are no reflections from the end of the duct. This means that only positive-going waves can be present and so the B_{mn}'s in equation 2.1.3 are zero in this case.

The boundary condition to be satisfied at the panel is that the z-component of the acoustic velocity field must be equal to the panel velocity. Expressing the acoustic velocity potential by means of equation 2.1.3 and differentiating it with respect to $-z$ to obtain the z-component of the acoustic particle velocity, we can write this condition as

$$U(x, y) = \sum_{m,n=0}^{\infty} \cos (m\pi x/a) \cos (n\pi y/b)\, i k_{mn} A_{mn}. \qquad (2.1.5)$$

The factor $e^{i\omega t}$ common to both sides has been suppressed.

To determine the coefficients A_{mn} in equation 2.1.5 we first express $U(x, y)$ as a double Fourier cosine series,

$$U(x, y) = \sum_{m,n=0}^{\infty} u_{mn}(2 - \delta_{0m})(2 - \delta_{0n}) \cos (m\pi x/a) \cos (n\pi y/b),$$

$$(2.1.6)$$

where δ_{ij} is the Kronecker delta symbol, defined by

$$\delta_{ij} = \begin{cases} 0, & i \neq j \\ 1, & i = j. \end{cases}$$

The coefficients u_{mn} are obtained in the usual manner by multiplying both sides of equation 2.1.6 by $\cos (m'\pi x/a) \cos (n'\pi y/b)$, and integrating over the area of cross section. Because the characteristic functions are *ortho-gonal*, i.e., because

$$\int_0^a \int_0^b \cos(m'\pi x/a)\cos(n'\pi y/b)\cos(m\pi x/a)\cos(n\pi y/b)\,dx\,dy$$

$$= ab\,\frac{\delta_{m'm}\,\delta_{n'n}}{(2-\delta_{m0})(2-\delta_{n0})},$$

we obtain

$$u_{mn} = (1/ab)\int_0^a \int_0^b \cos(m\pi x/a)\cos(n\pi y/b)\,U(x,y)\,dx\,dy, \quad (2.1.7)$$

thus determining u_{mn}. The coefficients u_{mn} are weighted averages of $U(x,y)$ over the surface, the weighting factor being the mode amplitude factor $\cos(m\pi x/a)\cos(n\pi y/b)$. To find the A_{mn}'s in terms of the u_{mn}'s we equate to each other the expressions for $U(x,y)$ given in equations 2.1.5 and 2.1.6, obtaining

$$\sum_{m,n=0}^{\infty} \{ik_{mn}A_{mn} - (2-\delta_{0m})(2-\delta_{0n})u_{mn}\}\cos(m\pi x/a)\cos(n\pi y/b) = 0. \quad (2.1.8)$$

Again, because of the orthogonality of the characteristic functions $\cos(m\pi x/a)\cos(n\pi y/b)$ equation 2.1.8 implies that the coefficient of each term in the expansion must be equal to zero so that we have

$$A_{mn} = (2-\delta_{0m})(2-\delta_{0n})\,u_{mn}/ik_{mn}. \quad (2.1.9)$$

The velocity potential caused by the panel is then

$$\phi(x,y,z)\,e^{i\omega t} = \sum_{m,n=0}^{\infty} \cos(m\pi x/a)\cos(n\pi y/b)A_{mn}\,e^{i\omega t - ik_{mn}z}. \quad (2.1.10)$$

In ascertaining the physical meaning of this result it will be helpful to think of a more tangible physical quantity like the pressure, rather than the velocity potential. We denote the fluctuating acoustic pressure (which is defined as the total pressure P less the mean pressure p_0) by $p(x,y,z)$ $e^{i\omega t}$. In terms of the velocity potential it is equal to ρ_0 times the time derivative of the velocity potential (see chapter 1), where ρ_0 is the mean fluid mass density, or, in this case, $p(x,y,z)\,e^{i\omega t} = i\omega\rho_0\phi(x,y,z)\,e^{i\omega t}$. Combining this with equations 2.1.10 and 2.1.9 and writing k_{mn} as ω/c_{mn}, where again c_{mn} is the mode propagation speed, we can write the acoustic pressure as

$$p(x,y,z)\,e^{i\omega t} = \sum_{m,n=0}^{\infty} \rho_0 c_{mn}u_{mn}(2-\delta_{0m})(2-\delta_{0n})\cos(m\pi x/a)$$
$$\cos(n\pi y/b)\exp(i\omega t - ik_{mn}z). \quad (2.1.11)$$

Comparing this result with equation 2.1.6 we see that the complex amplitude of each term in the expansion for the acoustic pressure is equal to $\rho_0 c_{mn}$ times the corresponding term in the expansion (equation 2.1.6) for the panel velocity amplitude.

To put this in a concise form we can write, from equation 2.1.6, for the velocity component $U_{mn}(x, y)$ in the (m, n)th mode

$$U_{mn}(x, y) = u_{mn}(2 - \delta_{0m})(2 - \delta_{0n}) \cos (m\pi x/a) \cos (n\pi y/b), \qquad (2.1.12)$$

so that

$$U(x, y) = \sum_{m,n=0}^{\infty} U_{mn}(x, y),$$

and

$$p(x, y, z)\, e^{i\omega t} = \sum_{m,n=0}^{\infty} \rho_0 c_{mn} U_{mn}(x, y) \exp i\omega(t - z/c_{mn}). \qquad (2.1.13)$$

For propagating modes, $\omega_{mn} < \omega$ (see equation 2.1.4). The mode speed c_{mn} is real and equal to its absolute value $|c_{mn}|$ given by

$$|c_{mn}| = a_0/|1 - (m\pi/k_a)^2 - (n\pi/k_b)^2|^{1/2}, \qquad (2.1.14)$$

and for non-propagating modes, $\omega_{mn} > \omega$, c_{mn} in equation 2.1.13 is to be taken equal to $i|c_{mn}|$.

The quantity $\rho_0 c_{mn}$ is the *characteristic impedance of the mode*. It is the ratio of the acoustic pressure in the mode to the z-component of the acoustic particle velocity in the mode.

The pressure observed in the duct thus depends on two distinct factors. In the first place, only components of the panel velocity exactly proportional to $\cos (m\pi x/a) \cos (n\pi y/b)$ will produce any disturbance at all in the fluid in the duct. The Fourier coefficients u_{mn} are the measures of these components. If all modes propagated this would not necessarily result in the acoustic disturbance having a different amplitude pattern over the cross section from that of the panel velocity. But not all modes will propagate and so only a finite number of the panel velocity components $U_{mn}(x, y)$ will appear in the acoustic disturbance at reasonable axial distances from the panel. The others will be negligible because of their exponential decay with distance. Those that do appear are those having the lower values of m and n. In general, therefore, the variation of the amplitude of the acoustic disturbance over the cross section is different from that of the panel causing the disturbance.

The relationship between the panel velocity pattern and the space modulation pattern of the acoustic wave fronts is rather like that between the input and the output signals of a low-pass electrical filter. The filter rejects high frequency components of the input signal and accepts the low frequency components. The output waveform is thus smoother than the input. Similarly, the duct rejects short wavelength components of the panel velocity pattern—because they couple with modes of large (m, n) and hence are not propagated—and accepts long wavelength components, which couple with modes of (m, n) values below cut-off. The wavefront pattern therefore is smoother than the panel velocity pattern.

2.1.3 *Some examples*

(i) *Uniform excitation.* If the velocity distribution over the panel is uniform and equal to $U_0 \, e^{i\omega t}$, all the u'_{mn}s except u_{00} are zero, and $u_{00} = U_0 = U_{00}$, as can be seen from equations 2.1.7 and 2.1.12. Only the plane wave mode $(0, 0)$ is excited at all and the pressure is, from equation 2.1.13,

$$p(x, y, z) \, e^{i\omega t} = \rho_0 a_0 U_0 \exp i\omega(t - z/a).$$

(ii) *Point excitation.* Suppose we have a rigid wall at $z = 0$ containing a small piston of area S' at the point (x', y'). The velocity is then zero everywhere except on the piston and there we suppose that it is uniform and equal to $U' e^{i\omega t}$. We denote by U_0 a velocity amplitude equal to $(S'/ab)U'$, so that U_0 is the uniform velocity amplitude that the whole cross section would have to have to produce the same total volume velocity as the small piston. Then from equation 2.1.7 we obtain in the limit of very small S'

$$u_{mn} = U_0 \cos (m\pi x'/a) \cos (n\pi y'/b),$$

and from equation 2.1.12,

$$
\begin{aligned}
U_{mn}(x, y) = (2 - \delta_{0m})(2 - \delta_{0n})U_0 &\cos (m\pi x'/a) \cos (n\pi y'/b) \\
&\times \cos (m\pi x/a) \cos (n\pi y/b). \quad (2.1.15)
\end{aligned}
$$

The pressure $p(x, y, z) \, e^{i\omega t}$ is obtained by substituting this expression into equation 2.1.13.

Suppose that the dimension b of the cross section is less than a. Then the frequency ω will be less than the lowest cut-off frequency ω_{mn}, if $\omega < a_0\pi/a$, as can be seen from equation 2.1.4. This condition is equivalent to $ka < \pi$ which is $2\pi a/\lambda < \pi$ or $a < \lambda/2$, where λ is the acoustic wavelength (of a plane acoustic wave of frequency ω). If this condition is satisfied, only the plane-wave mode will be propagated. All the other modes are excited, but being all beyond cut-off, they decay with axial distance like $\exp (- |k_{mn}|z)$, the decay constant k_{mn} being larger the higher the order (m, n) of the mode. The pressure in the duct would then be (except very close to the piston at $z = 0$)

$$p(x, y, z) \, e^{i\omega t} = \rho_0 a_0 U_0 \exp i\omega(t - z/a_0).$$

This is exactly what it would have been for uniform excitation at the same volume velocity.

Suppose now that, as before, $b < a$, but that the frequency is greater than $a_0\pi/a$ and less than the cut-off frequencies ω_{mn} of all the other modes. Then only the plane wave mode $(0, 0)$ and the mode $(1, 0)$ will propagate. The mode $(1, 0)$ being proportional to $\cos(\pi x/a)$, has one line of zero pressure, at $x = a/2$. The propagated pressure now has two terms and is

$$p(x, y, z)\, e^{i\omega t} = \rho_0 a_0 U_0 \exp i\omega(t - z/a_0)$$

$$+ \rho_0\, \frac{a_0}{|1 - (\pi/ka)^2|^{1/2}}\, U_0 \cos(\pi x'/a) \cos(\pi x/a)$$

$$\exp i\omega\left(t - \frac{z|1 - (\pi/ka)^2|^{1/2}}{a_0}\right). \qquad (2.1.16)$$

It is necessary to include all of this rather bulky expression because all these factors in the second term, which represents the contribution of the $(1, 0)$ mode, critically affect the pressure. The quantity $a_0/|1 - (\pi/ka)^2|^{1/2}$ is the mode speed c_{10}, obtained from equation 2.1.14.

If we assume that $b < a/2$, then our assumption that only the $(0, 0)$ and $(1, 0)$ modes are propagated implies that the frequencies we are dealing with lie in the range $a_0\pi/a < \omega < a_0 2\pi/a$, and the mode speed c_{10} lies in the range $\infty > c_{10} > 2a_0/\sqrt{3}$. Thus as ω approaches the $(1, 0)$ mode cut-off frequency $a_0\pi/a$ from above equation 2.1.16 predicts an infinite amplitude for the $(1, 0)$ mode term, which also becomes nearly independent of z (the effective wavelength in the z-direction tending to infinity with the mode speed).

We have here a clear-cut resonance effect. It is perhaps unexpected because one might have thought that as energy seemingly can travel away unimpeded to infinity in the z-direction, this leak in the system should have excluded such a possibility. In fact, as cut-off is approached, the $(1, 0)$ mode pressure becomes a pure standing wave across the tube, being proportional to $\cos(\pi x/a) \exp(i\omega t)$. No energy is required to maintain this motion. On the other hand, the mode particle velocity component in the z direction is $1/\rho_0 c_{10}$ times the mode pressure and remains finite. It is in phase with the mode pressure so that their product, which represents the *mode intensity* (the flow of energy carried by the mode in the z direction), is infinite. As the pressure is infinite it is clear that to produce this mode at cut-off an infinite force (and infinite energy) would have to be supplied to the piston. In practice the pressure amplitude would approach a limiting value determined by one or more of several possible factors, of which the finite impedance of any actual piston (or other source) is one. In a finite duct a reflected negative-going $(1, 0)$ mode component would be present and by its reaction back on the source could prevent the pressure from going to infinity, and ensure that instead it approached a finite value consistent with the actual finite amount of energy being radiated from the end of the duct. Of course, energy dissipation by viscous friction and thermal conduction would also play a part, but usually in cases of this kind it is the *radiation damping* (i.e., the energy loss by sound radiation) that limits the pressure build-up in a mode as the cut-off frequency is approached. Another related effect is that the cut-off is not sharp for a finite duct (see section 10.3.1).

We have thus seen that because of a resonance effect a given mode may tend to dominate the propagated pressure when the frequency is close to its cut-off frequency. Going back to the electrical filter analogy for a moment, we see now that the analogous filter is not a broad band low-pass filter, but rather a low-pass filter consisting of a number of narrow band-pass filters in parallel.

The second critical factor is the piston location which is expressed through the factor cos $(\pi x'/a)$ in equation 2.1.16. This is zero when $x = a/2$ (i.e., when the piston is on the line of zero pressure for the mode) and it is of maximum magnitude (unity) if $x = 0$ or $x = a$. If we had been dealing with the (1, 1) mode the factor would have been cos $(\pi x'/a)$ cos $(\pi y'/b)$ and would have had its maximum magnitude at the corners. This illustrates a principle of general validity, namely that in the presence of reflecting surfaces a source is most effective in a corner or as near to as many reflecting surfaces as possible.

Finally, it is of interest to note from equation 2.1.16 that the pressure at (x, z) when the piston is at $(x', 0)$ is the same as the pressure at x', z when the piston is at $(x, 0)$. This is an instance of the celebrated principle of reciprocity.

(iii) *General remarks.* The complexity of the situation even when only two of the simplest modes are involved, as in (ii) above, is a true indication of the behaviour of real ducts. The example shows all the principal factors affecting the excitation and propagation. The general problem is more complicated only in degree, not in kind.

2.1.4 *Ducts of arbitrary cross section*

There is no essential difference between the behaviour of a duct of any constant cross section and that of a rectangular duct. Instead of the functions cos $(m\pi x/a)$ cos $(n\pi y/b)$ we would obtain a different set of orthogonal functions $\psi_{mn}(x, y)$, say. As before, we would have

$$p(x, y, z) \, e^{i\omega t} = \sum_{m,n=0}^{\infty} \rho_0 c_{mn} U_{mn}(x, y) \exp i\omega(t - z/c_{mn}). \qquad (2.1.17)$$

The mode speeds c_{mn} would be defined in the same way in terms of the wave-numbers k_{mn} characteristic of the set of orthogonal functions ψ_{mn}

(x, y), as would the cut-off frequencies ω_{mn}. For roughly equivalent shapes and sizes of cross section there would be roughly the same number of cut-off frequencies (and hence the same number of available modes) in a given frequency band.

The velocity components $U_{mn}(x, y)$ would be

$$U_{mn}(x, y) = u_{mn}\psi_{mn}(x, y) \tag{2.1.18}$$

and if S is the area of cross section we would have

$$u_{mn} = \int\int_S \psi_{mn}(x, y)U(x, y)\, dx\, dy \tag{2.1.19}$$

provided we adjust the values of the ψ_{mn}'s by suitable constant multipliers (the normalization constants) so that

$$\int\int_S \psi_{mn}^*\psi_{\mu\nu}\, dx\, dy = \begin{cases} 0, & (m, n) \neq (\mu, \nu) \\ 1, & (m, n) = (\mu, \nu), \end{cases} \tag{2.1.20}$$

i.e., provided the ψ_{mn}'s are the normalized orthogonal functions.

In the case of a circular duct of radius a the ψ_{mn}'s are proportional to $J_m(K_{mn}r) \exp(-im\theta)$ where r and θ are polar coordinates, $J_m(w)$ is the Bessel function of order m, K_{mn} is the nth root of $J_m'(w) = 0$, $m = 0, \pm 1, \pm 2, \ldots$ and $n = 0, 1, 2, \ldots$. The cut-off frequencies here are $\omega_{mn} = a_0 K_{mn}$.

2.1.5 *Intensity and power*

For simple harmonic waves the mean acoustic intensity J_n in a given direction, n, can be written

$$J_n = \overline{\mathrm{Re}\,\{p(x, y, z)\exp(i\omega t)\}\,\mathrm{Re}\,\{u_n(x, y, z)\exp(i\omega t)\}}$$

where $u_n(x, y, z)$ is the complex amplitude of the particle velocity component in the n-direction, $\mathrm{Re}(\ )$ denotes 'real part of' and the bar denotes the time average (over a period, in this case). It can be shown simply by writing out real and imaginary parts that if A and B are two complex numbers then

$$\overline{\mathrm{Re}\,\{A\exp(i\omega t)\}\,\mathrm{Re}\,\{B\exp(i\omega t)\}} = \tfrac{1}{2}\,\mathrm{Re}\,(AB^*),$$

where the asterisk denotes the complex conjugate.

Thus J_n can be conveniently written as

$$J_n(x, y, z) = (1/2)\,\mathrm{Re}\,\{p(x, y, z)u_n^*(x, y, z)\}, \tag{2.1.21}$$

where $p(x, y, z)$ and $u_n(x, y, z)$ are the complex amplitudes of the pressure and of the particle velocity in the n-direction, respectively.

The pressure in the duct is (in the absence of reflections)

$$p(x, y, z)\, e^{i\omega t} = \sum_{m,n=0}^{\infty} \rho_0 c_{mn} U_{mn}(x, y) \exp i\omega(t - z/c_{mn})$$

and, as the particle velocity in the z-direction, u_z, is equal to $-(1/i\omega\rho_0)$ $\partial p/\partial z$, we have

$$u_z(x, y, z)\, e^{i\omega t} = \sum_{m,n=0}^{\infty} U_{mn}(x, y) \exp i\omega(t - z/c_{mn}).$$

Further, from equation 2.1.18 we have $U_{mn}(x, y) = u_{mn}\psi_{mn}(x, y)$, in general, so that the mean acoustic intensity in the z-direction is

$$J_z(x, y) = (1/2)\, \mathrm{Re} \left\{ \sum_{m,n=0}^{\infty} \sum_{\mu,\nu=0}^{\infty} \rho_0 c_{mn} u_{mn} \psi_{mn}(x, y)\, u_{\mu\nu}^* \psi_{\mu\nu}^*(x, y) \right\}.$$
$$(2.1.22)$$

The total power radiated down the duct is

$$W = \iint_S J_z(x, y)\, dx\, dy. \qquad (2.1.23)$$

Inserting the series 2.1.22 for $J_z(x, y)$ into equation 2.1.23 and integrating we get, by virtue of the orthogonality properties of the ψ_{mn}'s (see equation 2.1.20),

$$W = (1/2)\, \mathrm{Re} \left\{ \sum_{m,n=0}^{\infty} \rho_0 c_{mn} |u_{mn}|^2 \right\} \qquad (2.1.24)$$

This expression shows that non-propagating modes, for which c_{mn} is purely imaginary, carry no energy. They produce a purely reactive field in which the pressure is out of phase with the particle velocity. Thus in practice the sum in equation 2.1.24 contains only a finite number of terms. The presence of the mode impedance factor $\rho_0 c_{mn}$ in each term indicates again that at a frequency near the cut-off frequency ω_{mn} we expect the (m, n)th mode to dominate the radiation as c_{mn} becomes very large. This is provided, of course, that the degree to which it is excited by the source distribution, represented here by the factor $|u_{mn}|^2$, is comparable to the degree of excitation of other modes.

Note that the coefficients u_{mn} in this and the preceding section are not quite the same as those defined in equation 2.1.7 and used in sections 2.1.2 and 2.1.3. The difference is that the u_{mn}'s in equation 2.1.7 are defined in terms of un-normalized orthogonal functions, and therefore should be modified by the appropriate constants if substituted into equation 2.1.24.

3

2.1.6 *Influence of mean flow and wall absorption*

When a mean flow is present both volume sources of sound (such as the quadrupole sources associated with turbulence) and the radiated sound itself are convected by the flow. Here we shall consider only the convection of the sound itself by a uniform mean flow in a duct.

We suppose that the uniform mean flow is of speed V_0 in the z direction. The equations governing acoustic motion in the presence of a mean flow are, for an ideal compressible fluid,

$$(1/\rho_0)D_0\rho/Dt + \partial v_i/\partial x_i = 0 \quad \text{(mass balance)}$$
$$\rho_0 D_0 v_i/Dt + \partial p/\partial x_i \quad = 0 \quad \text{(linear momentum balance)}$$
$$D_0(p - a_0^2\rho)/Dt \quad = 0 \quad \text{(energy balance, or isentropy condition).}$$
$$(2.1.25)$$

Here D_0/Dt represents the time rate of change moving with the mean flow

$$D_0/Dt = \partial/\partial t + V_0\partial/\partial z.$$

The notation $\partial v_i/\partial x_i$ means $\partial u/\partial x + \partial v/\partial y + \partial w/\partial z$. In the momentum equation v_i is the velocity vector (u, v, w) and $\partial p/\partial x_i$ is the pressure gradient.

Combining the mass and energy equations gives

$$(1/a_0^2)D_0p/Dt + \rho_0\partial v_i/\partial x_i = 0.$$

Eliminating v_i between this equation and the momentum equation gives

$$\nabla^2 p - (1/a_0^2)D_0^2p/Dt^2 = 0. \qquad (2.1.26)$$

This is identical to the usual wave equation, but with D_0/Dt replacing $\partial/\partial t$.

For the rectangular duct discussed in section 2.1.1 the pressure again may be taken to be $p(x, y, z)\,e^{i\omega t}$ and to satisfy boundary conditions on the side-walls we must again have modes of the form

$$p_{mn}(x, y, z) = \cos(m\pi x/a)\cos(n\pi y/b)Z(z).$$

Substituting $p_{mn}(x, y, z)\,e^{i\omega t}$ into equation 2.1.26 as a trial solution we find that $Z(z)$ must satisfy

$$\left\{(1 - M_0^2)\frac{d^2}{dz^2} - 2ikM_0\frac{d}{dz} + (k^2 - K_{mn}^2)\right\}Z(z) = 0,$$

where M_0 is the mean flow Mach number V_0/a_0, $K_{mn}^2 = (m\pi/a)^2 + (n\pi/b)^2$, and $k = \omega/a_0$ is the usual wave number. This equation has solutions of the form $\exp(ik_{mn}z)$ provided that

$$k_{mn} = (kM_0 \pm \{k^2 - (1 - M_0^2)K_{mn}^2\}^{1/2})/(1 - M_0^2). \qquad (2.1.27)$$

Although it is a straightforward matter to obtain this result its physical interpretation is more difficult to see. The term $kM_0/(1 - M_0^2)$ indicates that waves travelling downstream do so at a greater speed than those travelling upstream.

Downstream propagation occurs if $+ \{k^2 - (1 - M_0^2)K_{mn}^2\}^{1/2}$ is real and is greater than kM_0. When these conditions are satisfied the k_{mn}'s with the minus signs in equation 2.1.27 are the downstream wave numbers. The conditions are

$$k^2 > (1 - M_0^2)K_{mn}^2; \quad k^2 - (1 - M_0^2)K_{mn}^2 > k^2M_0^2.$$

It is seen that the second condition is more restrictive and is equivalent in fact to $k^2 > K_{mn}^2$, which is the cut-off condition in the absence of mean flow. Downstream modes, therefore, have cut-off frequencies equal to those in the absence of mean flow.

Cut-off for upstream modes, however, can occur only if the argument of the square root is negative, that is if $k^2 < (1 - M_0^2)K_{mn}$. Upstream mode cut-off frequencies therefore are a factor $(1 - M_0^2)^{1/2}$ less than those in the absence of mean flow.

The characteristic impedances of the modes are also affected. The upstream mode speed is less than that in the absence of flow so that the upstream mode impedance is reduced and hence less intensity is radiated upstream from a given velocity distribution. Similarly, the downstream mode speed is greater; the downstream mode impedance is therefore greater and more intensity is radiated downstream.

We now turn briefly to the effects of absorption at the walls of the duct. This subject has been studied in considerable detail in connection with the practical problems of reducing noise propagated along heating and ventilating ducts and in mufflers and silencers for aircraft engines and other noisy machines.

Propagation in the absence of mean flow has been discussed in some detail by Morse[1]. This discussion shows that all modes attenuate exponentially with distance when the walls are at all absorbent. Each mode contains a factor $\exp(- \alpha_{mn}z)$ where α_{mn} is real and depends on the absorptivity of the walls. The absorption affects the mode wave numbers and hence the mode impedances and cut-off frequencies as well.

The absorptive properties of a wall can be characterized, to a first approximation, by the normal specific acoustic impedance $Z = p/v_n$ where v_n is the particle velocity normal to the wall (from the fluid to the wall). This complex quantity, in general a function of frequency, is approximately independent of the sound field under a wide range of conditions and hence is a measurable property of the material itself. The specific acoustic admittance ratio, η, is defined as $\eta = \rho_0 a_0/Z$ and is expressed in terms of its real and imaginary parts as $\eta = \kappa + i\sigma$ where κ is called the acoustic conductance ratio and σ the acoustic susceptance ratio.

Morse finds that the attenuation of the (nearly) plane wave mode in a duct with uniformly absorbent walls is approximately $4.34\,LK/S$ dB per ft, where L is the duct perimeter in feet and S is the area of cross section in square feet. For modes of the type $(m, 0)$ in a rectangular duct of dimensions (a, b) as above the attenuation is $4.34(L - b)K/\gamma_0(m, n)S$ dB per ft, where $\gamma_0(m, n)$ is the ratio of the sound speed a_0 to the mode speed c_{mn} in the absence of absorption, $\gamma_0(m, n) = a_0/c_{mn} = \{1 - (m\pi/ka)^2 - (n\pi/kb)^2\}^{1/2}$. For modes (m, n), both m and n greater than zero, the attenuation is $8.68\,LK/\gamma_0(m, n)S$ dB per ft. An exposition of the principles and design methods for ventilating duct absorption is given by R. W. Leonard, (chapter 27 reference 2).

Typical attenuations obtained in lined ventilating ducts of this kind are of the order of several dB per ft. The influence of the mean flow on the impedance of a porous material has not received much attention. There is some recent evidence that porous materials are less absorptive when a mean flow is present[3].

2.1.7 *Rotating modes*

Recent work[4,5] has shown that rotating modes in circular or annular ducts can play an important role in the propagation of aircraft engine compressor noise.

A rotating propeller or rotor stage of a compressor produces a rotating pressure pattern. Such a pressure pattern in a duct produces spiralling sound waves. Assuming the rotation to be in the direction of positive θ, (r, θ) being cylindrical polar coordinates, the modes excited and propagating in the positive z-direction will be of the form

$$\tilde{C}_{mn}(r) \exp i(i\omega t - m\theta - ik_{mn}z)$$

where m, instead of taking on all values $m = 0, \pm 1, \pm 2, \ldots$ as in the similar expression given in section 2.1.4, is restricted to the values $m = 0, 1, 2, \ldots$. The functions $\tilde{C}_{mn}(r)$ are appropriate cylinder functions (such that the normal acoustic particle velocity vanishes at the walls). For a circular duct $\tilde{C}_{mn}(r)$ would be $J_m(K_{mn}r)$, a Bessel function of order m, as was mentioned in section 2.1.4.

The dependence on θ and z can be written in the form $\exp i\{\omega t - (m/a)\,a\theta - k_{mn}z\}$. From this it is seen that the direction of propagation of the mode has direction cosines $(m/a)K_{mn}$ and k_{mn}/K_{mn}, respectively, relative to the tangential and axial directions. Here, $K_{mn} = \{(m/a)^2 + k_{mn}^2\}^{1/2}$. This means that an element of a wavefront moves along a spiral path.

Unlike higher order modes in a rectangular duct these modes always travel parallel to the duct wall. Their cut-off frequencies are similar to those for the corresponding modes of a squarish rectangular duct of approximately equal area of cross section. A particular feature of compressor noise is that because of the large number of blades, modes having many diametral pressure nodes are excited. The index m may well be of the order of tens rather than units. When stators are present to superpose a stationary pattern on the rotating one, modes rotating in both directions may be excited.

Mode theory can be applied to three fundamental aspects of the generation and propagation of compressor noise: (i) the amplitude of excitation of the various modes by rotor and rotor–stator stages; (ii) the attenuation of the modes by absorbent duct linings or absorbent vanes; (iii) reflection of the modes by other stator or rotor rows and radiation from the intake, including the effects of intake flare or other alterations in intake shape and size. Further reference will be made in chapter 10 to work reported on these problems.

2.2 Sound propagation in enclosures

When a volume of fluid is completely enclosed by boundaries that are acoustically partially reflecting and partially absorbent and a fluctuating force is applied to some movable portion of the boundary (a piston or panel, say) then the acoustic field set up in the enclosure will depend even more on the shape, size and acoustic impedance of the boundaries than it did in the duct, where there was always the possibility that energy could be radiated away unimpeded along the axis. In the enclosure there is no direction in which energy can be radiated away unimpeded.

Pressure waves sent out from the vibrating panel will be reflected back to it from the boundaries and will alter the pressure distribution on it. This reaction may either oppose or assist the force being applied to move the panel. Either more or less energy may thus be supplied to the enclosure than would be supplied to the same volume by the vibrating panel in the absence of the boundaries.

When the boundaries are not too absorbent (and this is the case more often than not) the acoustical dynamical behaviour of the fluid in it in response to the forcing motion of the panel is almost exactly like that of a linear mechanical system of small damping when a force is applied to it, the system having a very large number of degrees of freedom. Such a system is characterized by a sequence of normal modes of vibration, each having a characteristic normal frequency and a characteristic decay time, the number of normal modes being equal to the number of degrees of freedom of the system. All the remarks in the preceding sentence apply equally well to the fluid in the enclosure. The only difference is that the fluid in the enclosure has an infinite number of acoustic degrees of freedom, and hence an infinite number of normal modes. It is thus a limiting case of the mechanical system.

2.2.1 *Free oscillations of fluid in a rectangular enclosure*

To see why these normal modes occur we consider, as the simplest representative case, the acoustic motion of fluid in a rectangular enclosure with hard walls at $x = 0,a$, $y = 0,b$ and $z = 0,d$.

When the motion is simple harmonic the acoustic pressure $p(x, y, z)$ exp $(i\omega t)$ satisfies $(\nabla^2 + k^2)p = 0$, in the absence of sources. As we found in section 2.1.1 for the rectangular duct, the condition that the normal particle velocity vanishes at $x = 0,a$ and $y = 0,b$ requires that $p(x, y, z)$ be of the form

$$\cos k_x x \cos k_y y (A \cos k_z z + B \sin k_z z)$$

where the separation constants must satisfy $k_x^2 + k_y^2 + k_z^2 = k^2 = \omega^2/a_0^2$, and the vanishing of the normal particle velocity on the x and y walls requires that $k_x = l\pi/a$, $l = 0,1,2,\dots$ and $k_y = m\pi/b$, $m = 0,1,2,\dots$. But here we must also have the normal particle velocity vanishing on both z walls: i.e., $v_z = -(1/i\omega\rho_0)\, \partial p/\partial z = 0$ on $z = 0,d$. Hence $B = 0$ to satisfy the condition at $z = 0$ and $k_z = n\pi/d$, $n = 0,1,2,\dots$ to satisfy the condition at $z = d$. Thus $p(x, y, z)$ is

$$A \cos l\pi x/a \cos m\pi y/b \cos n\pi z/d; \qquad l,m,n = 0,1,2,\dots.$$

But now the condition $k_x^2 + k_y^2 + k_z^2 = \omega^2/a_0^2$ implies that for each (l, m, n) the frequency ω cannot be chosen arbitrarily but must have the characteristic value

$$\omega_{lmn} = a_0\{(l\pi/a)^2 + (m\pi/b)^2 + (n\pi/d)^2\}^{1/2}. \qquad (2.2.1)$$

These are the *normal frequencies* of the enclosure.

Thus for free oscillations the acoustic pressure is

$$p(x, y, z, t) = \sum_{l,m,n=0}^{\infty} a_{lmn}\psi_{lmn}(x, y, z) \exp(i\omega_{lmn}t), \qquad (2.2.2)$$

a sum of a triply infinite set of *normal modes of vibration*, namely ψ_{lmn} $(x, y, z) \exp(i\omega_{lmn}t)$ (the constants a_{lmn} being determined by the initial conditions), where the functions $\psi_{lmn}(x, y, z)$ are the *normal functions* (or eigenfunctions) of the enclosure

$$\psi_{lmn}(x, y, z) = \{\epsilon_l \epsilon_m \epsilon_n/\sqrt{(abd)}\} \cos l\pi x/a \cos m\pi y/b \cos n\pi z/d. \qquad (2.2.3)$$

Here the numbers ϵ_j are equal to unity if $j = 0$ and to $\sqrt{2}$ if $j = 1, 2, \ldots$, so that the functions ψ_{lmn} are *normalized orthogonal functions*, i.e.

$$\int_0^a \int_0^b \int_0^d \psi_{lmn}^* \psi_{\lambda\mu\nu} \, dx \, dy \, dz = \begin{cases} 0, & (\lambda, \mu, \nu) \neq (l, m, n) \\ 1, & (\lambda, \mu, \nu) = (l, m, n). \end{cases} \qquad (2.2.4)$$

The normal modes described above are pure standing waves. This is because the walls have been assumed to be perfectly rigid. As in the case of the duct, each of these standing waves can be analysed as a super-position of travelling waves.

If the walls are neither perfectly rigid nor too absorbent (their average impedance must be somewhat greater than the characteristic impedance of the fluid $\rho_0 a_0$), then it can be shown (reference 1, para. 33, pp. 401 ff) that ω_{lmn} above becomes complex, having an imaginary part representing the damping present. Equations 2.2.2 to 2.2.4 above still apply, to a first approximation, except that ω_{lmn} in equation 2.2.2 should be replaced by $\omega_{lmn} + i\mu_{lmn}$ where μ_{lmn} is the damping constant for the (l, m, n) mode,

$$\mu_{lmn} \approx (a_0/8abd)(\epsilon_l^2 a_x/2 + \epsilon_m^2 a_y/2 + \epsilon_n^2 a_z/2). \qquad (2.2.5)$$

Here, a_x, a_y, a_z represent the total acoustic absorption of the x, y and z walls, respectively. If there are N_x patches of absorbent material on the x walls, the ith patch having an acoustic conductance ratio κ_i and an area A_i, the x wall absorption is

$$a_x \approx \sum_{i=1}^{N_x} (8\kappa_i) A_i, \qquad (2.2.6)$$

and similarly for a_y and a_z. To the same order of approximation the quantity $8\kappa_i$ is equal to the *acoustic absorption coefficient*, α_i of the ith patch of material. The absorption coefficient is the quantity usually used to characterize the acoustical absorbing properties of a material, and its values for various materials at various frequencies are given extensively in the literature[2].

Thus when damping is present the amplitude of the contribution of the (l, m, n) mode to the total pressure in the free oscillations decays exponentially with time like $\exp(-\mu_{lmn}t)$.

The *reverberation time*, T_{lmn}, for this mode is defined as the time it takes for the energy in the mode to decay to 10^{-6} its initial value (the energy, of course, being proportional to the mean square pressure in the mode). This time is,

$$T_{lmn} \approx 6.91/\mu_{lmn}. \qquad (2.2.7)$$

We can classify the normal modes as *axial, tangential* and *oblique*. Axial modes are those for which two of the indices (l, m, n) are zero, tangential modes are those for which one of the indices (l, m, n) is zero, and for oblique modes none of the indices are zero. Axial modes represent motion parallel to two of the three pairs of walls. Tangential modes represent motion parallel to one pair of walls. The motion in oblique modes is not parallel to any of the walls.

We see from equation 2.2.5 that if the absorption is uniformly distributed on the walls the decay constants μ_{lmn} for axial, tangential and oblique modes are in the ratio 4, 5, 6, respectively. Thus oblique modes decay 3/2 times as fast as axial modes. Assuming all modes to be equally excited when the source is turned off, the decay goes through three distinct phases. First the oblique modes decay, then the tangential, and finally we are left with the axial modes. As the decay proceeds, the rate of decay becomes slower, approaching the axial decay rate as a limit. In an irregularly shaped room we would expect all modes to behave more or less like oblique modes and hence that the decay rate would be more nearly constant and equal, approximately, to $\mu \simeq a_0 a/8V$, where a is the total absorption,

$$a = \sum_{\text{all walls}} (8\kappa_i)A_i,$$

and V is the room volume.

When a source is first turned on, the pressure in each mode is proportional to the factor $1 - \exp(-\mu_{lmn}t)$. It reaches $(1 - 1/e)$ its final value —assuming the source to have a steady amplitude—in a time $t = 1/\mu_{lmn}$ or $t \simeq T_{lmn}/7$: i.e., one-seventh of the mode reverberation time.

The mean square pressure during decay is, approximately, $\overline{p^2}(x, y, z, t) \simeq p(x, y, z, t)p^*(x, y, z, t)/2$. Because of the orthogonality of the normal functions the average over the enclosure of the mean square pressure is, using equation 2.2.2 and integrating over the volume V,

$$\langle \overline{p^2} \rangle_V \simeq \sum_{l,m,n=0}^{\infty} (|a_{lmn}|^2/2) \exp(-2\mu_{lmn}t).$$

Although the mean square pressure at a point depends upon products of the pressure in one mode with that in another, its average over the enclosure is the sum of the mean square pressures in the various normal modes.

2.2.2 *Forced oscillations of fluid in a rectangular enclosure*

We can now describe the forced oscillations of fluid in a rectangular enclosure. We assume that the sound is caused by a volume distribution of velocity potential sources of strength $q(x, y, z) \exp(i\omega t)$ per unit volume. The source strength density $q(x, y, z) \exp(i\omega t)$ is the volume flow of fluid out of a point, per unit volume. The acoustic pressure $p(x, y, z) \exp(i\omega t)$ then satisfies (see, e.g., chapter 1),

$$\{\nabla^2 - (1/a_0^2)\partial^2/\partial t^2\}p(x, y, z) \exp(i\omega t) = -\rho_0 \partial\{q(x, y, z) e^{i\omega t}\}/\partial t.$$

Thus

$$(a_0^2\nabla^2 + \omega^2)p(x, y, z) = -i\omega\rho_0 a_0^2 q(x, y, z). \qquad (2.2.8)$$

To solve equation 2.2.8 we write both $p(x, y, x)$ and $q(x, y, z)$ as series expansions in the normal functions $\psi_{lmn}(x, y, z)$. We have

$$q(x, y, z) = \sum_{l,m,n=0}^{\infty} Q_{lmn}\psi_{lmn}(x, y, z). \qquad (2.2.9)$$

The functions ψ_{lmn} are normalized orthogonal functions so that

$$Q_{lmn} = \int\int\int_V q(x, y, z)\psi_{lmn}^*(x, y, z) \, dx \, dy \, dz. \qquad (2.2.10)$$

Also we put

$$p(x, y, z) = \sum_{l,m,n=0}^{\infty} a_{lmn}\psi_{lmn}(x, y, z). \qquad (2.2.11)$$

When the walls are absorbent the normal functions ψ_{lmn} satisfy

$$a_0^2\nabla^2\psi_{lmn}(x, y, z) = (i\omega_{lmn} - \mu_{lmn})^2\psi_{lmn}(x, y, z), \qquad (2.2.12)$$

where ω_{lmn} and μ_{lmn} are as defined in the preceding section. Substituting these expressions into equation 2.2.8 we have (because the orthogonality implies that the two sides must be equal term by term)

$$a_{lmn}\{(i\omega_{lmn} - \mu_{lmn})^2 + \omega^2\} = i\omega\rho_0 a_0^2 Q_{lmn},$$

or, as we may assume $\mu_{lmn}/\omega_{lmn} \ll 1$,

$$a_{lmn} = \rho_0 a_0^2\omega Q_{lmn}/\{2\omega_{lmn}\mu_{lmn} + i(\omega^2 - \omega_{lmn}^2)\}. \qquad (2.2.13)$$

If the source is a small piston of area S_p on the wall $z = 0$ at the point $(x', y', 0)$, say, and oscillating with normal velocity amplitude V_0, then $Q_{lmn} \approx S_p V_0\psi_{lmn}^*(x', y', 0)$ and the pressure caused by it is

$$p(x, y, z) \exp(i\omega t) \approx \rho_0 a_0^2\omega S_p V_0 \exp(i\omega t) \sum_{l,m,n=0}^{\infty} \frac{\psi_{lmn}^*(x', y', 0)\psi_{lmn}(x, y, z)}{2\omega_{lmn}\mu_{lmn} + i(\omega^2 - \omega_{lmn}^2)}. \qquad (2.2.14)$$

The pressure in each mode is proportional to the mode excitation factor $S_p V_0\psi_{lmn}^*(x', y', 0)$, to the mode amplitude at the observation point, and to a resonant factor (the denominator of each term in equation 2.2.14).

For relatively non-absorbent walls, we may assume that the normal functions $\psi_{lmn}(x, y, z)$ are given to a first approximation by equation 2.2.3. We see that if the piston is in the corner $x' = y' = z' = 0$. Then the excitation factor is the same for all modes. But if the piston is at the mid-point of the wall, none of the modes for which l or m is odd will be excited. If the room shape is irregular we would expect the excitation factor to depend less critically on position of the source. Similar remarks apply to the influence of the observation point.

The pressure amplitude in each mode (assuming uniform excitation) is a maximum when the driving frequency is equal to the normal frequency of the mode, $\omega = \omega_{lmn}$. For example, the pressure amplitude in the mode at the piston itself is then

$$\rho_0^2 a_0^2 S_p V_0 |\psi_{lmn}(x', y', 0)|^2 / 2\mu_{lmn}.$$

The mode pressure is thus in phase with the source velocity at mode resonance. The form of the resonance factor shows that only those modes having frequencies near that of the driving frequency can have an appreciable amplitude. Therefore, the overall dependence of the pressure on frequency must depend particularly on the number of normal modes having such frequencies.

Using equation 2.2.1 we can easily calculate that the number δN of the normal modes having frequencies lying between ω and $\omega + \delta\omega$ is, approximately,

$$\delta N \simeq \{ V\omega^2 / 2\pi^2 a_0^3 + A\omega / 8\pi a_0^2 + L/16\pi a_0 \} \delta\omega,$$

where V is the volume, A is the total surface area, and L is the sum of the lengths of all the edges of the rectangular enclosure. The first term is the number of oblique modes, the second term is the number of tangential modes, and the third is the number of axial modes.

Just as we had three phases of the decay of free oscillations, so we have three phases for forced oscillations as the forcing frequency varies. At low frequencies, axial modes predominate. At intermediate frequencies tangential modes are most numerous and at high frequencies, oblique modes.

Again, as for the decay of free oscillations, the local mean square pressure of the forced oscillations, being $(1/2)p(x, y, z) \times p^*(x, y, z)$, where p is given by equation 2.2.11, depends upon products of the pressure in one mode by those in others, but the average of the mean square pressure over the enclosure is independent of these products, because of the orthogonality of the normal functions. It is

$$\langle \overline{p^2} \rangle_V = \sum_{l,m,n=0}^{\infty} |a_{lmn}|^2 / 2,$$

i.e., one half the sum of the squares of the amplitudes of the mode pressures, these amplitudes a_{lmn} being given by equation 2.2.13.

2.2.3 *Average behaviour at high frequencies*

We see that at high frequencies, where the oblique modes predominate, the pressure is the sum of contributions from a very large number of normal modes, as a large number of modes have frequencies near any driving frequency. Each normal mode in turn can be regarded as the superposition of several plane waves travelling in several distinct directions. Thus at high frequencies we approach the situation where the pressure can be regarded as being due to a very large number of plane waves travelling in all different directions, and we can write

$$p(x, y, z) \exp(i\omega t) \approx \int_0^{2\pi} \int_0^{\pi} A(\theta, \phi) \exp(i\omega t - i\mathbf{k} \cdot \mathbf{r}) \sin \theta \, d\theta \, d\varphi.$$

In this $A(\theta, \varphi)$ is the amplitude and phase of the plane wave component at (x, y, z) travelling in the direction (θ, φ) and

$$\mathbf{k} \cdot \mathbf{r} = kx \sin \theta \cos \varphi + ky \sin \theta \sin \varphi + kz \cos \theta = k(x^2 + y^2 + z^2)^{1/2} \cos \chi.$$

χ is the angle between the direction of propagation of the plane wave component in the (θ, φ) direction and the position vector of (x, y, z).

The mean energy density in each plane wave component $A(\theta, \varphi) \times \exp(i\omega t - i\mathbf{k} \cdot \mathbf{r})$ is, since for a plane wave the mean energy density w is $w = \overline{p_2}/\rho_0 a_0^2$,

$$\delta w(x, y, z) = |A(\theta, \varphi)|^2/(2\rho_0 a_0) \sin \theta \delta \theta \delta \varphi.$$

Hence, if we assume that the phases of the plane waves are uncorrelated, we find the total mean energy density to be the sum of the mean energy densities of the components,

$$w(x, y, z) \approx \int_0^{2\pi} \int_0^{\pi} |A(\theta, \varphi)|^2/(2\rho_0 a_0^2) \sin \theta \, d\theta \, d\varphi.$$

The scalar mean acoustic intensity, $\hat{J}$, is defined here as the mean flow of *energy density* across unit area. The wave incident at angle Φ to the normal of the area element thus contributes an amount $\delta \hat{J}(x, y, z) \approx a_0 \delta w \cos \Phi$ to the total intensity, which is, therefore (again assuming the phases uncorrelated)

$$\hat{J}(x, y, z) \approx \int_0^{2\pi} \int_0^{\pi} |A(\theta, \phi)|^2/(2\rho_0 a_0) \cos \Phi \sin \Phi \, d\Phi \, d\varphi.$$

The scalar mean acoustic 'intensity' used here is somewhat different from the vector acoustic intensity used elsewhere in this and the previous chapter, as will be explained subsequently.

In the limiting case of high frequencies in enclosures with even moderately irregular wall shapes, it can be assumed that the amplitude $A(\theta, \varphi)$ is substantially independent of angle and, on average, independent of position (x, y, z). Then the energy density and scalar intensity are, respectively,

$$w = 2\pi|A|^2/\rho_0 a_0^2; \qquad \hat{J} = \pi|A|^2/2\rho_0 a_0 = a_0 w/4.$$

When the energy density is thus uniformly distributed the power lost at the walls is $\hat{J}A_w$, where A_w is the total wall absorption, $A_w = \sum_i \alpha_i S_i$, α_i being the absorption coefficient of the ith patch of absorbing material of area S_i, as before. If $W(t)$ is the power supplied by the source and it is assumed that the energy density responds instantaneously to changes in source power, then, as the total energy is $wV = 4\hat{J}V/a_0$ where V is the volume of the enclosure, the energy balance is

$$d(4\hat{J}V/a_0)/dt = W(t) - \hat{J}A_w.$$

This equation has the solution

$$\hat{J} = (a_0/4V) \exp(-A_w a_0 t/4V) \int_{-\infty}^{t} \exp(A_w a_0 t'/4V) W(t')\, dt'.$$

The solution indicates that the scalar intensity depends primarily on the power output for the previous $4V/A_w a_0$ seconds. If the power output changes slowly in this period the solution is

$$\hat{J} \approx W(t)/A_w$$

and in decibels we have

(scalar intensity level) $= 10 \log_{10} \hat{J} \simeq 10 \log_{10}(W/A_w) + 130$ dB.

W is in watts and A_w is in square feet.

If the source is suddenly shut off the scalar intensity decreases exponentially,

$$\hat{J} \approx \hat{J}_0 \exp(-A_w a_0 t/4V),$$

and decays to 10^{-6} its initial value in a time T, called the *reverberation time*, which becomes

$$T = 0.049\,(V/A_w)$$

when V is in cubic feet and A_w in square feet.

These results should be compared with the corresponding ones in section 2.2.1.

The scalar intensity $\hat{J}$ defined in this section is an intensity for a sound field in which energy flows uniformly *in all directions*. A microphone measures pressure rather than intensity. The mean square pressure is always related to the mean energy density by $\overline{p^2} = \rho_0 a_0^2 w$. Thus the measurable quantity, the mean square pressure, is related to the scalar intensity $\hat{J}$ used here by $\overline{p^2} = 4\rho_0 a_0 \hat{J}$. (Note that when the field is a plane wave field or the radiation field of a source, where the energy flow is in one direction only, the relationship between mean square pressure and the magnitude of the true, vector intensity (directed energy flow) is $\overline{p^2} = \rho_0 a_0 \hat{J}$.)

Near the source the pressure falls off like $1/r$, where r is the distance from the source, as it would were the source in free field, the mean square pressure there being $\overline{p^2} \simeq \rho_0 a_0 W / 4\pi r^2$, assuming a source radiating uniformly in all directions. Since far from the source the multiply reflected waves making up the uniform reverberant field in the enclosure predominate and the mean square pressure there is, as above, $\overline{p^2} \simeq 4\rho_0 a_0 W / A_w$, we find by equating the two formulae that the first is accurate for $r^2 < A_w/50$ and the second for $r^2 > A_w/50$. Thus the pressure level in the enclosure can be written approximately as

$$\text{Pressure level} = 20 \log_{10} |\overline{p^2}|^{1/2}$$
$$\approx \begin{cases} 10 \log_{10} W - 20 \log_{10} r + 49 \text{ dB}, \, r^2 < A_w/50 \\ 10 \log_{10} W - 10 \log_{10} A_w + 66 \text{ dB}, \, r^2 > A_w/50 \end{cases},$$

where W is in watts, r is in feet and A_w in square feet.

2.2.4 *Deviations from average behaviour*

It is an easily verifiable physical fact that no matter what the shape and size of an enclosure the sound pressure level is not uniform at a given position as the source frequency varies, nor at a given frequency as the source position varies.

It has been shown[6] that when the source frequency is high enough so that the spacing between normal frequencies is small compared with the half-power breadth of an individual normal mode resonance (this condition being quantitatively $f > 24 \times 10^3 (T/V)$ c/s where f is the frequency, T is the reverberation time in seconds and V is the volume in cubic feet) then the r.m.s. deviation of the pressure level from its average value is the same for variation in the source or receiver position as it is for variation of the source frequency, namely, 5.5 dB. Near to the source, or at lower frequencies, the size and shape of a typical fluctuation as source or receiver position varies depend upon the source directionality, the reverberation time, the room volume and the type of normal functions (axial, tangential, or oblique) predominant among those with normal frequencies near the driving frequency. Approximate formulas taking some of these factors into account are given in reference 6.

2.3 Low-frequency approximations and impedances

When the dimensions of an enclosure are small compared with the acoustic wavelength the pressure is nearly uniform over the entire volume, and the reaction of the fluid in the enclosure to an externally applied pressure is exactly like that of a spring.

Consider, for example, fluid in a tube rigidly closed at $z = l$, with rigid side-walls. The pressure amplitude for simple harmonic motion is, if the tube perimeter is small compared with the wavelength,

$$p(z) = p_l \cos k(l - z).$$

The velocity amplitude is

$$v = -(1/i\omega\rho_0)\partial p(z)/\partial z = i(pl/\rho_0 a_0) \sin k(l - z).$$

The acoustic impedance, p/v, at $z = 0$ is then

$$p(0)/v(0) = \rho_0 a_0/i \tan kl.$$

When kl is small we have $\tan kl \sim kl$ and the impedance becomes

$$p(0)/v(0) \approx 1/(i\omega l/\rho_0 a_0^2).$$

The *analogous impedance* is defined to be the ratio of the pressure to the volume flow Sv, where S is the cross sectional area. The analogous impedance of the small volume is then

$$p(0)/Sv(0) \approx 1/(i\omega V/\rho_0 a_0^2)$$

where $V = Sl$ is the volume of the tube section.

If we have a constriction of small dimensions in a tube the fluid in the constriction will tend to move back and forth incompressibly*. The pressure gradient in the constriction will be nearly constant and equal to $(p_2 - p_1)/l_e$ where p_2 is the pressure at the downstream end of the constriction, p_1 that at the upstream end, and l_e is the effective length of the constriction (somewhat greater than the actual length because some of the fluid just outside the ends of the constriction will tend to move together with that in the constriction itself). The fluid velocity v in the constriction is given by $i\omega\rho_0 v = -\partial p/\partial z$ and so is

$$v \approx (p_1 - p_2)/i\omega\rho_0 l_e.$$

The impedance looking into the constriction from the upstream end is, taking p_2 as a reference level of zero,

$$p_1/v \approx p_1/(p_1/i\omega_0 l_e) = i\omega\rho_0 l_e.$$

* There can be exceptions to this behaviour if the tube itself is terminated in an impedance which creates a large impedance at the mouth of the constriction. In general, fluid in a length of tube possesses both inertance and stiffness.

This is an *inertance* (i.e. a mass reactance), as is to be expected because the inertia of the fluid in the constriction causes the velocity to lag behind the pressure by $\pi/2$. The analogous impedance of the constriction is

$$p_1/S_c v \simeq i\omega\rho_0 l_e/S_c$$

where S_c is the cross sectional area of the constriction.

If we consider conditions at a discontinuity in cross sectional area in a tube we see that the pressure must be continuous across the area discontinuity (neglecting terms of order of the square of the velocity) and also the mass flow $\rho S v$ must be continuous. For acoustic disturbances the density ρ can be replaced by its reference value ρ_0 in the mass flow condition and so this becomes $S v = $ constant. These two conditions imply that the analagous impedance is continuous across the area discontinuity.

Consider a bottle. The analogous impedance at the cavity end of the neck is $1/(i\omega V/\rho_0 a_0^2)$, and since the analogous impedance is continuous this is also the impedance just inside the cavity end of the neck. Thus for the neck

$$p_2/S_c v = 1/(i\omega V/\rho_0 a_0^2).$$

At the open end of the neck we have

$$(p_1 - p_2)/S_c v = i\omega\rho_0 l_e/S_c,$$

and so

$$p_1/S_c v = (p_2/S_c v) + i\omega\rho_0 l_e/S_c$$
$$= i\omega(\rho_0 l_e/S_c) + 1/i\omega(V/\rho_0 a_0^2).$$

The bottle obviously acts like a resonator of resonant frequency

$$\omega_0 = a_0(S_c/l_e V)^{1/2}.$$

When loss of energy by radiation or by friction is taken into account an active (resistive) term will also appear in the impedance.

A system of constrictions and cavities obviously will behave like a mechanical system of point masses, springs and dash pots (or like an electrical circuit of inductances, capacitances and resistances). At lower audio frequencies in particular the losses (mechanical resistances) of these acoustical circuits can be very small. A microphone in the cavity of a bottle, for example, can serve as a very good narrow-band analyser.

More detailed analysis of acoustical circuits is given in standard acoustics text-books. In particular, references 7 and 8 contain extensive discussions of the subject and list many typical arrangements and their impedances.

2.4 Propagation of sound in the atmosphere

The principal factors causing propagation of sound in the atmosphere to be somewhat different from that in an ideal fluid at rest are, in the order of their usual importance: (i) winds and other effective mean flows due to large scale intense atmospheric turbulence; (ii) temperature gradients; (iii) irreversible mechanisms for the internal exchange of energy such as viscosity, heat conduction, relaxation processes and electromagnetic radiation.

2.4.1 *Effects of mean flows and temperature gradients*

In assessing the effects of mean flows (or, for that matter, turbulent flows) in acoustic motion it is essential to bear in mind that the basic property of an acoustic disturbance is that it is propagated from particle to particle at the local sound speed $a = \sqrt{(\gamma p/\rho)}$, which is a function of temperature only and, in a moving medium, is not the apparent propagation speed measured by a fixed observer. Being a property of the medium an acoustic disturbance is convected by a mean flow just as a fluid particle is. If an element of an acoustic wavefront is propagating through a fluid particle at speed a relative to the particle, and if the particle speed is U then the speed of propagation of the wavefront element as seen by an observer fixed in space is $a + U \cos \psi$, where ψ is the angle between the direction of propagation of the wave and the particle velocity. Thus a wavefront element travels upstream at speed $a - U$, downstream at speed $a + U$, and normal to the stream at speed a. The mean flow affects both the magnitude and direction of the velocity of propagation of the wave front, as seen by a fixed observer.

As a mean flow alters the apparent sound speed, so a velocity gradient, as well as a temperature gradient, must result in reflection and refraction of an incident sound wave. The sketch below shows the manner in which sound is reflected and refracted by a mean velocity or temperature gradient. It is assumed for simplicity that the gradient is concentrated in a very thin layer.

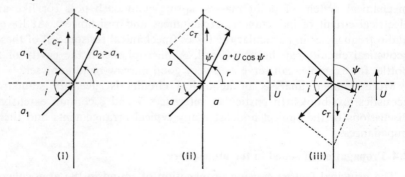

Fig. 2.1   Sketch illustrating reflection and refraction of sound
rays by a mean velocity or temperature gradient.

In (i) the wave is incident at an angle i on a plane interface between two regions of different temperature, the region to the left containing the incident wave and being at the lower temperature. The condition governing the angles of reflection and refraction is that the trace of the incident wavefront on the interface must move at the same speed as the trace of the transmitted wavefront. This speed, c_T, will be

$$c_T = \frac{a_1}{\sin i} = \frac{a_2}{\sin r} \qquad (2.4.1)$$

where r is here the angle of refraction. Hence

$$\sin r = (a_2/a_1) \sin i. \qquad (2.4.2)$$

The refracted wave in this case bends away from the normal to the interface ($r > i$). The reflected and refracted waves obey Snell's laws as in optics.

Refraction by wind is essentially different. As it depends upon the convection of the wave by the wind, the speed of the wave as well as its direction of propagation in the moving medium depends on the angle the incident wave makes with the wind direction. In (ii) a wave is incident in fluid at rest on fluid in motion with uniform speed U, the temperature being the same so that the speed of sound from particle to particle remains the same. As $\cos \psi = \sin r$ the condition on the trace speed is

$$\frac{a}{\sin i} = \frac{a}{\sin r} + U \qquad (2.4.3)$$

so that the angle of refraction is not given by Snell's law but by

$$\sin r = \sin i \Big/ \Big(1 - \frac{U}{a} \sin i\Big). \qquad (2.4.4)$$

The result is again that the refracted wave is bent away from the normal, but only provided that the direction of the incident wave makes an acute angle with the wind direction as in (ii). Total reflection occurs when $\sin r = 1$, or $\cos i = 1/(1 + U/a)$. Inserting some numbers shows that a wind of 36 m.p.h. produces total reflection of all sound rays having the same azimuth as the wind and being inclined at less than about 18° to the horizontal.

For an incident wave of azimuth opposite to the wind as in (iii) the situation is reversed. The refracted waves are bent towards the normal, the angle of refraction being

$$\sin r = \sin i \Big/ \Big(1 + \frac{U}{a} \sin i\Big). \qquad (2.4.5)$$

Total reflection cannot occur. On the contrary, at a wind speed of 36 m.p.h. even rays at grazing incidence (and azimuth opposite to that of the wind, of course) are refracted upwards at an inclination to the horizontal of about 18°.

These numbers show that the effects of wind on sound transmission in the atmosphere are appreciable.

Downwind transmission is reinforced by the rays which initially were inclined upwards but have been refracted down again. Sound can penetrate down behind hills or other obstacles in this manner. Upwind transmission is impaired because initially horizontal rays are bent upwards and will pass over the head of an observer.

The theory of reflection and refraction by wind given above is valid for an individual ray, but is only approximately valid for the wavefronts themselves. This is because the wavefront formed in the moving fluid in general is not normal to the ray. Consider an element of wavefront incident on the interface in Figure 2.1 for example. The part of the wavefront reaching the interface later falls on an element of moving fluid that was a certain distance upstream when the first part of the wavefront reached the interface. This fluid element is closer therefore to the element disturbed by the first part of the wavefront than would be the case in the absence of motion. This means that the wavefront in the moving fluid will form along a surface that is not normal to the vector but rather is normal to a vector at a somewhat larger angle of refraction.

Further analysis confirms this picture and shows that the angle of refraction of the *normal to the wavefront* in the moving medium is given by

$$\sin r = \sin i \Big/ \left(1 - 2\frac{U}{a}\sin i\right)^{1/2}$$

rather than by equation 2.4.4. The difference between this result and equation 2.4.4 evidently is small when the Mach number is small compared with unity. This question has recently been discussed by Warren[9].

On average, over long periods, the temperature decrease with altitude is about 8°C per mile up to a height of about eight miles. As the temperature decreases with height in this region initially horizontal rays tend to become refracted upwards (bent towards the normal) and so the overall effect of the temperature gradient is to reduce the sound heard along the horizontal from a source on the ground. A general exception to this rule is that very loud sounds propagating over great distances may be brought to earth again by the high altitude temperature inversion. Using the figure of 8°C in Snell's law suggests that an initially almost horizontal ray will be inclined at an angle of about 10° to the horizontal when it has reached a height of one mile, which would occur, roughly at about ten miles from the source. As we have just seen, however, moderate winds can produce deflections of this order over much shorter distances. The deflection due to the vertical temperature gradient is independent of the azimuth of the ray.

Although the effects of the mean atmospheric temperature gradient are not usually comparable with those of wind and atmospheric turbulence, local temperature gradients—like those occurring over lakes, for example —may produce significant effects on occasion.

Reflection, refraction, and the general scattering of sound by mean flows, temperature gradients and turbulence, and indeed all variations in the propagation of condensation waves can be regarded as caused by an appropriate volume distribution of quadrupole sources in the fluid.

A recent comprehensive and authoritative discussion of the propagation of waves in various kinds of stratified media will be found in reference 10.

In chapter 5 the effects of the convection of sound sources themselves will be described.

2.4.2 *Effects of irreversible internal energy exchange*

The attenuation and dispersion of sound waves due to viscosity, thermal conduction and relaxation effects is a large subject in itself. The amount of attenuation depends on the molecular structure of the fluid and often is very strongly influenced by impurities. A recent survey of the subject[11] gives a connected account from the point of view of an aerodynamicist.

The results can be summarized briefly by saying that the acoustic intensity in a simple harmonic plane progressive wave, or in an outwardly radiating wave at large distances from the source, is exponentially attenuated with distance according to the factor $\exp(-\delta\omega^2 z/a_0^3)$, where z is distance in the direction of propagation, ω is the radian frequency, a_0 the speed of sound and δ is the 'diffusivity' of sound, which can be written as

$$\rho\delta = (4/3)\mu + (\gamma - 1)(k/C_p) + \mu',$$

where μ is the coefficient of shear viscosity, γ is the ratio of specific heats, k is the coefficient of thermal conduction, C_p is the specific heat at constant pressure, and μ' is an effective bulk viscosity.

It turns out that, to first order, relaxation effects are equivalent to an effective bulk viscosity that is a function of frequency.

In atmospheric air the dominant factor in attenuation is often the humidity. If h is the ratio of the number of water vapour molecules per unit volume to the total number of molecules per unit volume of the atmosphere (the absolute humidity times the ratio of the representative mass of an atmospheric molecule to the mass of the water vapour molecule), then, except in very dry air, we can represent the attenuation to a good approximation by taking μ' as

$$\mu' \sim 1.4 \times 10^{-3}\mu h^{-2},$$

for temperatures in the region of 20°C. For representative values of h, μ' is of the order of ten to one hundred times μ. In air the thermal conduction term in the acoustic diffusivity δ is of the order of 0.4 times the shear viscosity term.

Even assuming μ' to be 100μ we would have for the attenuation constant $\delta\omega^2/a_0^3$

$$\delta\omega^2/a_0^3 = O\left\{\frac{30 \times 10^2 \times 1.5 \times 10^{-1}}{(3.44 \times 10^4)^3}\,\omega^2/\text{ft}\right\}$$

or $\delta\omega^2/a_0^3 \sim 10^{-11}\omega^2/\text{ft}$, so that only at the higher audio frequencies does the attenuation become appreciable except over long distances.

References

1. P. M. Morse. *Vibration and sound*, McGraw-Hill, New York (1945). 2nd ed. See para. 31, pp. 368–376 and as indicated in subsequent citations.
2. C. M. Harris. (Ed.). *Handbook of noise control*, McGraw-Hill, New York (1957).
3. F. Mechel and W. Schilz. *Acustica*, **14**, 325 (1964).
4. J. M. Tyler and T. G. Sofrin. *Trans. S.A.E.*, **70**, 309 (1962).
5. C. L. Morfey. *J. Sound Vib.*, **1**, 60 (1964).
6. P. E. Doak. *Acustica*, **9**, 1 (1959).
7. L. L. Beranek. *Acoustics*, McGraw-Hill, New York (1954).
8. S. N. Rschevkin. *A course of lectures on the theory of sound*, Pergamon Press, Oxford (1963).
9. C. H. E. Warren. *J. Sound Vib.*, **1**, 175 (1964).
10. L. M. Brekhovskikh. *Waves in layered media*, Academic Press, New York (1960).
11. M. J. Lighthill. Viscosity in waves of finite amplitude, in *Surveys in mechanics* (G. K. Batchelor, Ed.), Cambridge Univ. Press (1956). See especially pp. 251–280.

Supplementary General References

12. L. E. Kinsler and A. R. Frey. *Fundamentals of acoustics*, Wiley, New York (1950).
13. A. Wood. *Acoustics*. Blackie, London (1940).
14. Lord Rayleigh. *Theory of sound* (two volumes), Dover, New York (1945).
15. M. Redwood. *Mechanical wave guides*, Pergamon Press, Oxford (1960).

CHAPTER 3

The Elements of Sound Transmission

3.1 Acoustic impedance

The term 'acoustic impedance' has already been mentioned in chapters 1 and 2. Before dealing with some elementary sound transmission problems, we shall consider it once again in some detail.

The *specific acoustic impedance* of a medium is the ratio of the harmonic acoustic pressure in the medium to the associated particle velocity. It is not only dependent upon the characteristics of the medium, but also upon the characteristics of the wave, i.e. whether it is a travelling wave moving to the right or left, or whether a curved wave, or standing wave. For a plane harmonic wave travelling to the right, we may express the pressure in the form

$$p = A \, e^{i(\omega t - kx)}$$

where $k \, (= 2\pi/\lambda = \omega/a)$ is the wave number. The velocity, u, is related to the pressure by

$$\rho \frac{\partial u}{\partial t} = -\frac{\partial p}{\partial x}$$

from which we find

$$\frac{p}{u} = +\rho a = z.$$

For a wave travelling to the left ($p = A \, e^{i(\omega t + kx)}$), $p/u = -\rho a$. The specific acoustic impedance is therefore the harmonic pressure required to produce unit harmonic particle velocity. For a curved or standing wave, it will be found to be a complex quantity,

$$\frac{p}{u} = R + iX.$$

This shows that, in general, the pressure and particle velocity are not exactly in phase as for the plane travelling wave. R is called the *specific*

73

acoustic resistance and X is called the *specific acoustic reactance* of the medium for the particular wave motion being considered.

The product ρa is a characteristic of the medium alone, and obviously does not necessarily apply to any particular wave motion. Its importance is probably greater (for acousticians) than that of ρ or a alone, and for this reason is often referred to as the 'characteristic impedance' of the medium, or the 'characteristic resistance'.

3.2 Transmission from one medium to another (normal incidence)

We consider here the transmission of sound between two media, the boundary between which is a plane surface (00′). The plane wavefronts are parallel to this surface. (See Fig. 3.1.)

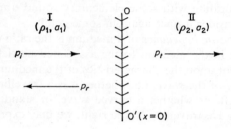

Fig. 3.1   Waves at normal incidence impinging on an interface.

The characteristic impedances of the two media are $\rho_1 a_1$ and $\rho_2 a_2$. The pressure of the wave incident upon the boundary is p_i, and of that transmitted is p_t. A wave p_r is reflected at the boundary back into medium I.

The waves travelling to the right may be represented by

$$p_i = A_1 \, e^{i(\omega t - k_1 x)} \tag{3.1a}$$

and

$$p_t = A_2 \, e^{i(\omega t - k_2 x)} \tag{3.1b}$$

and the reflected wave travelling to the left by

$$p_r = B_1 \, e^{i(\omega t + k_1 x)}. \tag{3.1c}$$

The frequencies of all the waves must be the same, but the wavelengths in the two media (embodied in the wave numbers k_1 and k_2) are different.

At the boundary ($x = 0$) both the pressure and particle velocity in each medium must be identical, for continuity. The pressure equality yields:

$$A_1 + B_1 = A_2. \tag{3.2}$$

The particle velocity associated with each of these waves is $\pm p/\rho a$ (+ve for p_i and p_t, $-$ve for p_r). Continuity of particle velocity therefore yields:

$$\frac{A_1 - B_1}{\rho_1 a_1} = \frac{A_2}{\rho_2 a_2}. \tag{3.3}$$

Eliminating A_2 from equations 3.2 and 3.3, and writing $r_{12} = \rho_2 a_2/\rho_1 a_1$, we find

$$B_1 = A_1 \frac{r_{12} - 1}{r_{12} + 1}. \tag{3.4}$$

r_{12} is the 'relative characteristic impedance' of medium II with respect to medium I.

Since B_1/A_1 is evidently a real number, it follows that at the boundary, the reflected wave is exactly in phase, or in counterphase, with the incident pressure. If $r_{12} > 1$, i.e. the characteristic impedance of medium II is greater than that of medium I (as when a plane wave in air impinges on an air–water interface), then a positive excess pressure is reflected as a positive excess pressure, and a rarefaction as a rarefaction. On the other hand if $r_{12} < 1$ (e.g. a plane wave in water incident upon a water–air interface) a positive excess pressure is reflected as a rarefaction, and a rarefaction as a positive pressure. If $r_{12} = 1$ (i.e. the characteristic impedances of the two media are equal) there is no reflection. If $r_{12} \to \infty$ (i.e. the second medium is rigid) there is total reflection, $B_1 = A_1$, and the total pressure on the interface is *double* that of the incident wave.

Now eliminate B_1 from equations 3.2 and 3.3. We find

$$A_2 = A_1 \frac{2r_{12}}{r_{12} + 1}. \tag{3.5}$$

At the boundary, therefore, the pressures of the incident and transmitted waves are always in phase. For large r_{12} (as for air–water) $A_2 \approx 2A_1$; for small r_{12} (water–air) A_2 is very small compared with A_1.

The acoustic intensity in a plane wave (average energy flux across unit area parallel to wavefront in unit time) is equal to (pressure amplitude)^2 $\div 2\rho a$. For the *incident* wave, therefore, this is $(A_1^2/2\rho_1 a_1) = I_1$, and for the transmitted wave it is $(A_2^2/2\rho_2 a_2) = I_2$. Hence

$$\frac{I_2}{I_1} = \alpha_t = \frac{4r_{12}}{(r_{12} + 1)^2}. \tag{3.6}$$

α_t is called the 'sound transmission coefficient'. Notice that whether $r_{12} \gg 1$ or $r_{12} \ll 1$, α_t is very small. Furthermore, α_t is independent of the direction of the wave motion (from or to medium II. This is shown by replacing r_{12} by $1/r_{12}$ in equation 3.6). This demonstrates the important 'reciprocal property' of transmission.

The value of r_{12} for air–water is 3560, hence

$$\alpha_t = 0.00112$$

i.e. the intensity of a plane wave passing at normal incidence between air and water (or vice versa) is reduced by approximately 30 dB.

3.3 Transmission from one medium to another. Oblique incidence

The incident wave is now inclined to the interface, such that the normal to the wavefronts makes an angle θ_i with the normal to the interface. (See Figure 3.2.) The reflected and transmitted waves will be inclined likewise by angles θ_r and θ_t:

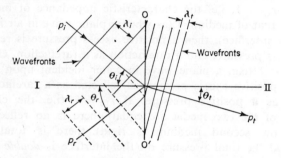

Fig. 3.2   Waves at oblique incidence impinging on an interface.

Now the frequencies of the incident, reflected, and transmitted waves must all be the same. The wavelengths of the incident and reflected waves must be the same ($\lambda_i = \lambda_r$), but the wavelength of the transmitted wave is given by

$$\lambda_t = \frac{2\pi a_2}{\omega}.$$

But

$$\lambda_i = \frac{2\pi a_1}{\omega}.$$

Hence

$$\lambda_t = \lambda_i \frac{a_2}{a_1}. \tag{3.7}$$

For continuity of the component of particle velocity normal to the interface, it is evident that, firstly, the intercepts of the wavefronts of the three waves on the interface must be equal, i.e.

$$\lambda_i \operatorname{cosec} \theta_i = \lambda_r \operatorname{cosec} \theta_r = \lambda_t \operatorname{cosec} \theta_t.$$

Hence $\theta_i = \theta_r$ (angle of incidence = angle of reflection) and

$$\frac{\sin \theta_t}{\sin \theta_i} = \frac{\lambda_t}{\lambda_i} = \frac{a_2}{a_1} = \text{constant.} \tag{3.8}$$

This is Snell's law of refraction.

Equating now the total components of particle velocity in the x-direction on either side of the interface, we find

$$\frac{A_1}{\rho_1 a_1} \cos \theta_i - \frac{B_1}{\rho_1 a_1} \cos \theta_r = \frac{A_2}{\rho_2 a_2} \cos \theta_t. \tag{3.9}$$

The equation for pressure continuity across the interface is the same as before, i.e.

$$A_1 + B_1 = A_2. \tag{3.10}$$

We then find that

$$A_2 = A_1 \frac{2\rho_2 a_2 \cos \theta_i}{\rho_2 a_2 \cos \theta_i + \rho_1 a_1 \cos \theta_t}, \tag{3.11}$$

and

$$B_1 = A_1 \frac{\rho_2 a_2 \cos \theta_i - \rho_1 a_1 \cos \theta_t}{\rho_2 a_2 \cos \theta_i + \rho_1 a_1 \cos \theta_t}. \tag{3.12}$$

Clearly, when $\rho_2 a_2 \cos \theta_i = \rho_1 a_1 \cos \theta_t$, $B_1 = 0$. That is there is no reflection, and all the acoustic power is *transmitted*. Using equation 3.8 we then find the value of θ_i for this condition to be

$$\cot^2 \theta_i = \frac{(a_1/a_2)^2 - 1}{(\rho_2/\rho_1)^2 - (a_1/a_2)^2}. \tag{3.13}$$

For this angle to be real, we must have $\rho_2/\rho_1 > a_1/a_2 > 1$, or $\rho_2/\rho_1 < a_1/a_2 < 1$.

If $a_2 > a_1$, there is a critical angle of incidence, θ_c, for which the refracted wave is parallel to the interface, i.e. $\sin \theta_t = 1$. From equation 3.8, therefore,

$$\sin \theta_c = \frac{a_1}{a_2}. \tag{3.14}$$

If $\theta_i < \theta_c$, no acoustic energy is transmitted into the second medium, all of it being reflected.

3.4 Transmission of plane waves through a massive wall

Figure 3.3 shows two media, I and II, separated by a plane wall of mass per unit area, m. The plate is free to move normal to its own plane with no elastic restraint. For simplicity, suppose that media I and II are the same, or have the same characteristic impedances. Let the displacement of the wall be uniform and equal to ξ normal to its plane.

The resultant pressure on the wall in the x-direction is $\{p_i + p_r - p_t\}_{x=0}$. This causes an acceleration of the wall, $d^2\xi/dt^2$. Hence

$$\{p_i + p_r - p_t\}_{x=0} = m\,\frac{d^2\xi}{dt^2}. \tag{3.15}$$

This equation is our new equation for 'continuity' of pressure, but because of the inertia reaction of the wall, the total pressures at $x = 0$ in each medium are no longer equal.

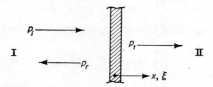

Fig. 3.3   Waves at normal incidence to a massive wall.

We shall assume that the wall is undeformable in the x direction, so that the velocities of its two faces are equal. The particle velocity in the medium adjacent to the left-hand face is then equal to that adjacent to the right-hand face. As before, therefore, the condition for velocity continuity gives:

$$\left\{\frac{p_i}{\rho a} - \frac{p_r}{\rho a}\right\}_{x=0} = \left\{\frac{p_t}{\rho a}\right\}_{x=0} \tag{3.16}$$

Now

$$m\,\frac{d^2\xi}{dt^2} = m\,\frac{\partial}{\partial t}\,(\text{particle velocity at } x = 0)$$

$$= -\frac{m}{\rho}\left\{\frac{\partial}{\partial x}\,p_t\right\}_{x=0}$$

$$= i\left(\frac{\omega m}{\rho a}\right)A_2\,e^{i\omega t}. \tag{3.17}$$

$i\omega m/\rho a$ is the ratio of the plate velocity impedance to the characteristic impedance of the medium. Denote this ratio by χ. Now replace each of the p's in equations 3.15 and 3.16 by the appropriate expression 3.1a to 3.1c and eliminate B_1 and A_2 in turn. We then find

$$\frac{A_2}{A_1} = \frac{1}{1 + \chi/2} \tag{3.18}$$

and

$$\frac{B_1}{A_1} = \frac{\chi/2}{1 + \chi/2}. \tag{3.19}$$

Notice that A_2/A_1 is now a complex quantity, indicating the existence of a phase difference between the incident and transmitted pressures at the wall. Furthermore, when $m\omega/\rho a$ is very much greater than unity, we have the approximation

$$A_2 \approx A_1 \frac{2\rho a}{i\omega m},$$ (3.20)

i.e. the transmitted pressure is inversely proportional to both the frequency and the mass of wall. This demonstrates the 'mass law' of sound transmission through walls.

It is very important now to notice an alternative method of analysing this problem, which greatly simplifies the analysis of more complicated problems.

Suppose, firstly, that the wall is prevented from moving. All the incident pressure is reflected, with the result that the net pressure on the left-hand side of the wall is $\{2p_i\}_{x=0}$. 'Pressure doubling' has occurred.

Next, suppose that the wall oscillates uniformly with a displacement ξ, due to some unspecified means. Due to the positive velocity $d\xi/dt$ there will exist on the right-hand side of the plate a positive acoustic pressure of $\rho a \, d\xi/dt$, and on the left-hand side a negative acoustic pressure of $\rho a \, d\xi/dt$. The total acoustic pressure acting in the positive x-wise direction is therefore $-2\rho a \, d\xi/dt$. Also acting in this direction is the inertia force per unit area of wall, $-m \, d^2\xi/dt^2$. For equilibrium of the element of the wall, there must exist another force (the exciting pressure, p_{ex}) such that

$$p_{ex} - m\frac{d^2\xi}{dt^2} - 2\rho a \frac{d\xi}{dt} = 0$$ (3.21)

or

$$m\frac{d^2\xi}{dt^2} + 2\rho a \frac{d\xi}{dt} = p_{ex}.$$ (3.22)

We may now equate this exciting pressure to the net pressure on the wall when held rigid, i.e. put $p_{ex} = \{2p_i\}_{x=0}$.
Then

$$m\frac{d^2\xi}{dt^2} + 2\rho a \frac{d\xi}{dt} = 2A_1 \, e^{i\omega t}.$$ (3.23)

This gives the equation of motion of the plate. Putting $d\xi/dt = \dot{\xi}_0 \, e^{i\omega t}$ we have

$$\dot{\xi}_0 = \frac{2A_1}{i\omega m + 2\rho a}$$

and the transmitted pressure amplitude, $A_2 \, (=\rho a \dot{\xi}_0)$ becomes

$$A_2 = \frac{A_1}{1 + i\omega m/2\rho a}$$ (3.24)

which is the same as equation 3.18.

With these ideas in mind, we can now distinguish between three or four different acoustic waves acting on the wall:

1. The incident wave.

2. The reflected wave, which derives from the presence of a rigid wall, and which results in the doubling of the effective exciting pressure on the wall.

3. The re-radiated wave: (a) to the left. This is radiated back into the incident field by virtue of the motion of the wall.

                               (b) to the right. This is again radiated by virtue of the motion of the wall, and is the *transmitted* wave.

Notice that the re-radiated waves appear in the equation of motion of the wall as a *damping* term, proportional to velocity. This is the 'acoustic damping' which plays an important part in vibrating structures.

Notice, too, that acoustic effects are not involved in the inertia (acceleration) term of the equation of motion. This is purely due to the simplicity of the system we have chosen, and is not a general characteristic of all systems.

3.5 Transmission of plane waves through a sprung wall

The wall of the last paragraph is now restrained from moving by a uniform distribution of elastic 'springs', such that the displacement ξ of a unit area perpendicular to the wall is resisted by an elastic force of $\bar{k}\xi$ per unit area. The treatment of the problem is most simply carried out using the second method of the last section. The elastic force introduces a further term into equation 3.22, since to the inertia and acoustic reactions, $(-m\, d^2\xi/dt^2 - 2\rho a\, d\xi/dt)$, must be added the elastic reaction, $-\bar{k}\xi$. The resultant equation of motion (corresponding to equation 3.23) is now

$$m\frac{d^2\xi}{dt^2} + 2\rho a\frac{d\xi}{dt} + \bar{k}\xi = 2A_1\,e^{i\omega t}. \qquad (3.25)$$

Putting $\xi = \xi_0\,e^{i\omega t}$, we have

$$\xi_0 = \frac{2A_1}{(\bar{k} - \omega^2 m) + 2i\omega\rho a} \qquad (3.26)$$

or, putting $d\xi/dt = \dot{\xi}_0\,e^{i\omega t}$ we find

$$\dot{\xi}_0 = \frac{2A_1}{i(\omega m - \bar{k}/\omega) + 2\rho a}. \qquad (3.27)$$

The transmitted pressure amplitude, A_2 ($= \rho a \dot{\xi}_0$), is now

$$A_2 = \frac{A_1}{1 + i\frac{(\omega m - \bar{k}/\omega)}{2\rho a}}. \tag{3.28}$$

As with the 'massive' wall, the complex form of the denominator here implies a phase difference between the incident and transmitted pressures. *At high frequencies* ($\omega m/2\rho a \gg \bar{k}/\omega 2\rho a$ and $\omega m/2\rho a \gg 1$) the equation degenerates into

$$A_2 \approx \frac{A_1 2\rho a}{i\omega m}. \tag{3.29}$$

The transmitted pressure is now governed by the *mass* of the wall, and is inversely proportional to the mass and frequency.

At low frequencies ($\bar{k}/\omega \gg \omega m$ and $\bar{k}/\omega \gg 2\rho a$) the approximation becomes

$$A_2 \approx -A_1\frac{2\rho a \omega}{i\bar{k}}. \tag{3.30}$$

The transmitted pressure is now governed by the *stiffness* of the springs, and is inversely proportional to the stiffness and proportional to the frequency.

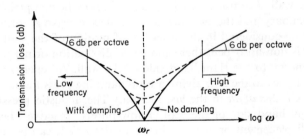

Fig. 3.4 Transmission loss through a spring wall.

When $\omega m = \bar{k}/\omega$ (i.e. at the resonant frequency of the wall, $\omega_r = (\bar{k}/m)^{1/2}$) we have $A_2 = A_1$. The incident wave 'passes through' the wall as though it were not there. The velocity amplitude of the plate (from equation 3.28) is now $A_1/\rho a$, which is the undisturbed particle velocity of the incident wave. The wave re-radiated to the left now exactly cancels out the wave reflected by assuming the wall to be rigid, further demonstrating that the incident wave ignores the presence of the wall at its resonant frequency.

The 'transmission loss' of the wall is defined by

$$-10 \log_{10} \frac{|A_2|^2}{|A_1|^2} = -20 \log_{10} \frac{|A_2|}{|A_1|} \text{ dB.} \qquad (3.31)$$

This is based on the ratio of the intensities of the incident and transmitted waves. Using the equations 3.29 and 3.30, it is evident that at low frequencies the transmission loss decreases by 6 dB per octave, whereas at high frequencies it increases by 6 dB per octave. This is indicated in Fig. 3.4. Owing to the damping inherent in the restraining springs, the ideal condition of zero transmission loss ($A_1 = A_2$) at resonance will not be realized. The effect of this damping may be introduced by replacing the $\bar{k}$ in equation 3.28 by $\bar{k}(1 + i\eta)$. At the frequency ω_r we then have

$$\frac{A_2}{A_1} = \frac{1}{1 + \bar{k}\eta/2\rho a}. \qquad (3.32)$$

The transmission loss curves between the high and low frequency portions follow the lines indicated on Fig. 3.4.

3.6 The effect of a rigid boundary behind the wall

A rigid, non-absorbing boundary is now introduced at a distance d behind the wall. The transmitted wave p_t will now be completely reflected by the boundary and the pressure of the reflected wave p_{tr} on the sprung wall must be considered in the equation of motion of the wall. Since the boundary is non-absorbent there can be no dissipation or net radiation of acoustic energy to the right-hand side of the wall. All energy re-radiated by the wall goes to the left, giving a total damping pressure on the wall of $\rho a \, d\xi/dt$. (Cf. $2\rho a \, d\xi/dt$ in the last section.) The acoustic pressure acting on the right-hand side of the wall is now purely 'reactive', i.e. in phase with the displacement or acceleration, ξ or $d^2\xi/dt^2$. It can be shown[1] that the equation of motion may be written:

$$m \frac{d^2\xi}{dt^2} + \rho a \frac{d\xi}{dt} + \left(\bar{k} + \rho a \omega \cot \frac{\omega d}{a}\right)\xi = 2A_1 \, e^{i\omega t}. \qquad (3.33)$$

The reactive term $\xi \rho a \omega \cot \omega d/a$ exists on account of the standing wave set up in the enclosure behind the wall. At certain frequencies the term becomes zero (when $\omega d/a = 2\pi d/\lambda = \pi/2, 3\pi/2, \ldots, n\pi/2$) and the wall moves as though there was no medium to the right at all, i.e. the acoustic pressure from the right-hand side is zero *at the wall*. The quarter wavelength of the incident sound is then equal to the distance between the wall and the rigid boundary. It may be shown that the pressure amplitude at any point within the enclosure, at a distance x from the wall, is

$$|p| = \rho a \omega \, \frac{\cos 2\pi\left(\dfrac{x-d}{\lambda}\right)}{\sin 2\pi d/\lambda} \, \xi_0 \qquad (3.34)$$

where ξ_0 is the amplitude of the wall displacement. $|p|$ has its maximum value at $x = d$, or at any point distant $n\lambda/2$ from the rigid boundary (within the enclosure). Hence

$$|p_{\max}| = \frac{\rho a \omega}{\sin 2\pi d/\lambda} \, \xi_0. \qquad (3.35)$$

Now the displacement amplitude, ξ_0, is found to be

$$\xi_0 = \frac{2A_1}{|k - \omega^2 m + \rho a \omega \cot \omega d/a + i\rho a \omega|}. \qquad (3.36)$$

When $\cot \omega d/a = 0$ then $\sin 2\pi d/\lambda = \pm 1$. The maximum pressure within the enclosure is therefore

$$|p_{\max}| = \frac{\rho a \omega 2A_1}{\{(k - \omega^2 m)^2 + (\rho a \omega)^2\}^{1/2}}. \qquad (3.37)$$

At certain other frequencies the reactive term in the equation becomes infinite, for $\cot \omega d/a \to \infty$ as $\omega d/a \, (=2\pi d/\lambda) \to n\pi$, i.e. when $\lambda/2 \to d/n$. The acoustic impedance of the enclosure at the wall has now become infinite, and the wall 'hardens up'. It requires an infinite force to cause a finite displacement. There are now an exact number of half sine waves in the standing wave within the enclosure, and the wall is at a velocity node (a pressure anti-node).

To find the maximum pressure within the enclosure at this particular frequency, it is necessary to examine the limiting value of $\rho a \xi_0 / \sin 2\pi d/\lambda$ as the frequency is approached. ξ_0 approaches the value $2A_1/\rho a \omega \cot \omega d/a$ since the cotangent term tends to dominate the denominator. As the frequency is approached

$$\cot \frac{\omega d}{a} \to \frac{a}{\omega d} \quad \text{and} \quad \sin \frac{2\pi d}{\lambda} \to \frac{2\pi d}{\lambda} = \frac{\omega d}{a}.$$

Hence

$$|p_{\max}| = \frac{\rho a \xi_0}{\sin 2\pi d/\lambda} \to 2A_1. \qquad (3.38)$$

The maximum pressure within the enclosure (either at $x = 0$ or $x = d$, in this case) is equal to the incident + reflected pressure on the left-hand side of the wall. Since the panel does not move, there is no re-radiated pressure on its left-hand side, nor is there any inertia force on the panel. The acoustic pressures on each side must obviously therefore be self balancing.

Now consider the term

$$Z = \bar{k} - \omega^2 m + \rho a \omega \cot \frac{\omega d}{a} + i \rho a \omega,$$

which is the 'obstructance' per unit area of the wall i.e. the complex force required to produce unit harmonic displacement. (See also section 13.1). The real part of this vanishes whenever $\rho a \omega \cot \omega d/a = \omega^2 m - \bar{k}$ and since $\cot \omega d/a$ is periodic there are an infinite number of such vanishing points. This is illustrated by Fig. 3.5 which shows $\rho a \omega \cot \omega d/a$ and $(\omega^2 m - \bar{k})$ plotted against ω.

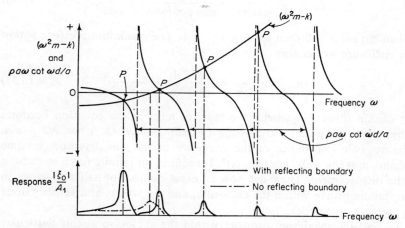

Fig. 3.5   Diagrams illustrating the resonant frequencies and response of a sprung wall backed by a rigid boundary.

The intersections P of the curves in the top figure correspond with the vanishing points. It follows that ξ_0 will exhibit a peak at each of these frequencies, as shown in the figure beneath. Whereas ξ_0 had only one peak when there was no rigid boundary, an infinite set of peaks occurs when the boundary exists. The wall has so coupled with the standing waves within the cavity that multiple resonances occur.

Although in practical structures these effects of the acoustic reactance are not likely to be so pronounced, there have been cases where the effect has been most important.

3.7 Transmission of sound through an infinite, flexible wall; the coincidence effect

Suppose now that the wall of section 3.4 can bend in the direction normal to its own plane and has a flexural stiffness of D per unit width. Let plane harmonic waves be incident upon it and inclined at the angle θ_i as before. There is no boundary now to the right of the wall.

The loading imposed on the wall by the incident wave now varies sinusoidally in space. The wavelength of this distribution is equal to the intercept of the wavefronts on the wall and is called the 'trace-wavelength', λ_t. Evidently $\lambda_t = \lambda_i \operatorname{cosec} \theta_i$. (See Fig. 3.6.)

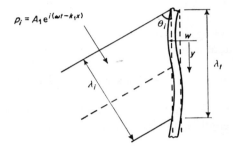

$$p_i = A_1 e^{i(\omega t - k_1 x)}$$

Fig. 3.6   Plane waves incident upon a flexible wall at oblique incidence.

As in section 3.4, we may say that the total pressure on the left-hand surface of the wall is the sum of the incident pressure, the reflected pressure and the pressure re-radiated by the plate motion. On the right-hand surface, the pressure is solely that which is transmitted (i.e. radiated) to the right-hand side.

The incident and reflected waves exert a total pressure on the wall of magnitude $2p_i$. The pressure at any point, y, on the wall due to these waves is

$$2A_1 e^{i(\omega t - 2\pi y/\lambda_t)}. \tag{3.39}$$

This pressure excites in the wall a harmonic wave of wavelength λ_t, frequency ω and local (transverse) velocity $\partial w/\partial t$, where w is the flexural displacement of the wall normal to its surface.

Now the re-radiated and transmitted waves have the same trace wavelength on the wall and the same frequency as the incident wave. The transmitted wavefronts must therefore be parallel with the incident wavefronts and the particle velocity in the transmitted wave has a component normal to the wall which is equal to the local wall velocity. The particle

4

velocity perpendicular to the wavefronts must therefore be sec $\theta_i \, \partial w/\partial t$. The pressure corresponding to this is $\rho a \sec \theta_i \, \partial w/\partial t$ which is therefore the pressure on the right-hand surface of the wall, due to the wall motion. On the other side of the wall the re-radiated pressure is $-\rho a \sec \theta_i \, \partial w/\partial t$, giving a resultant loading on the wall due to re-radiation and transmission of

$$2\rho a \sec \theta_i \frac{\partial w}{\partial t}. \tag{3.40}$$

This loading is always opposing the velocity.

The total acoustic loading on the wall is therefore

$$2A_1 \, e^{i(\omega t - 2\pi y/\lambda_t)} - 2\rho a \sec \theta_i \frac{\partial w}{\partial t}. \tag{3.41}$$

The inertia loading on a unit area of the wall is $-m \, \partial^2 w/\partial t^2$ so that the equation of flexural motion of the wall may be written:

$$D \frac{\partial^4 w}{\partial y^4} = -m \frac{\partial^2 w}{\partial t^2} - 2\rho a \sec \theta_i \frac{\partial w}{\partial t} + 2A_1 \, e^{i(\omega t - 2\pi y/\lambda_t)}. \tag{3.42}$$

Since the forced wave motion in the wall occurs at the same frequency and trace wavelength as the incident wave, we put

$$w = \bar{w} \, e^{i(\omega t - 2\pi y/\lambda_t - \epsilon)} \tag{3.43}$$

in which ϵ has been introduced to allow for a phase difference between the wall motion and the incident pressure fluctuation. Substituting w from equation 3.43 into the equation of motion, equating the real and imaginary parts on both sides and eliminating ϵ, it is found that the wall velocity amplitude $\omega \bar{w}$ is given by

$$\omega \bar{w} = \frac{2A_1}{\sqrt{\{[D(2\pi/\lambda_t)^4/\omega] - m\omega\}^2 + \{2\rho a \sec \theta_i\}^2}}. \tag{3.44}$$

The transmitted pressure amplitude is equal to $\rho a \sec \theta_i$ times this velocity, and may be written in the form

$$A_2 = \frac{A_1}{\sqrt{\left\{ \dfrac{[D(2\pi/\lambda_t)^4/\omega] - m\omega}{2\rho a \sec \theta_i} \right\}^2 + 1}}.$$

At low frequencies, but not too low, the transmitted pressure is now inversely proportional to the mass m and to the frequency. At high frequencies it is inversely proportional to the stiffness D, and to the cube of the frequency. Its maximum value occurs when

$$\frac{D(2\pi/\lambda_t)^4}{\omega} - m\omega = 0, \tag{3.45}$$

under which conditions the transmitted pressure amplitude is found to be simply A_1, the same as the incident pressure. The re-radiated pressure amplitude is also A_1, but it is found that the re-radiated pressure and the reflected pressure exactly cancel out. All-in-all, therefore, the incident wave passes through the wall as though the wall were not there. The wall is acoustically 'transparent' at this particular frequency, reflecting nothing and transmitting everything.

Now equation 3.45 is identical to the equation relating the wavelength λ_t and frequency of a *free* undamped flexural wave in the wall. The above phenomenon of acoustic transmission therefore occurs when the frequency of the incident sound wave is the same as the frequency of a free flexural wave in the wall having a wavelength equal to the trace wavelength of the field on the wall. Alternatively, we may say that it occurs when the phase velocity of the free flexural wave is the same as the trace velocity of the incident field. On account of this necessary coincidence of frequency and velocity, the phenomenon is usually known as the 'coincidence effect'.

The trace wavelength and the frequency of the incident field are related by

$$\frac{\omega}{2\pi} \lambda_t \sin \theta_i = a. \tag{3.46}$$

Using this to eliminate λ_t from equation 3.45 gives the relationship between the frequency at which coincidence transmission occurs and the inclination of the incident field, viz.

$$\sin^2 \theta_i = \frac{a^2}{\omega} \sqrt{\frac{m}{D}}. \tag{3.47}$$

If the right-hand side exceeds unity, there is no (real) inclination which will satisfy this equation and coincidence transmission cannot take place at that frequency. The lowest frequency at which it can occur is obviously given by $\sin^2 \theta_i = 1 = \sqrt{(m/D)}a^2/\omega$, i.e. $\omega = a^2\sqrt{(m/D)}$, when the direction of the incident field is parallel with the plane of the wall.

The phenomenon of coincidence transmission is referred to in some of the following chapters, in particular in chapter 23.

3.8 Transmission of sound into a circular cylinder

We have seen in section 3.6 that standing waves can readily couple with the 'mode' of vibration of the wall—in this case a rigid body translation. In a similar way, the standing waves (or acoustic normal modes) within a circular cylinder can readily couple with the normal modes of the vibrating cylinder. The cylinder modes involve cross-section distortion of the form shown in Figure 3.7.

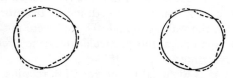

Fig. 3.7 Diagrams illustrating radial displacements of the normal modes of a circular cylinder.

The acoustic normal modes have the same pattern of radial velocity, but the variation across the diameter follows a Bessel's function form. (See Figure 3.8.)

Fig. 3.8 Diagram illustrating radial velocity along a diameter of some acoustic normal modes in a circular cylinder.

There are an infinite set of such diametrical distributions corresponding to the infinite set of circumferential distributions. Each normal mode of the cylinder structure can couple with an infinite number of acoustic normal modes within the cylinder. We therefore obtain the same phenomenon, as before, of each cylinder mode having multiple resonant frequencies. Outside the cylinder, the acoustic medium radiates energy away from the vibrating cylinder, giving rise to an acoustic damping term in the equations of motion. Furthermore, since we are now dealing with curved waves (not plane waves) we find that at the cylinder surface the air *outside* the surface has a reactive component to its impedance, in addition to the solely reactive component of the air inside.

In order to calculate the response of, and sound transmission into, a cylinder, Foxwell and Franklin[2] suggest the following procedure:

(a) Find the generalized force on the cylinder corresponding to each mode, using the pressure distribution on the cylinder corresponding to the incident field and the field reflected from the *rigid* cylinder.

(b) Find the modes of vibration of the cylinder in vacuo, and the corresponding mechanical impedances.

(c) Find the acoustic normal modes within the cylinder corresponding to each cylinder mode and their acoustic reactances. A finite number of these modes are sufficient to give a good approximation to the internal pressure field, but the higher the frequency of excitation, the more acoustic modes must be introduced.

(d) Find the complex acoustic impedance of the external medium corresponding to each mode.

(e) Using the Lagrangian approach, set up the equations of motion of each of the structural modes, the coefficients being complicated functions of ω owing to the acoustic terms.

(f) Solve for the amplitudes of each of the structural and acoustic modes. Add the internal acoustic pressures from as many of the internal modes as necessary.

When a cylinder is excited by plane waves in this way, the distribution of sound pressure inside the cylinder is found to be non-uniform across a diameter, on account of the 'Bessel function' distribution of pressure of the standing waves that are set up. The calculations[2] are carried through, for a particular case, to demonstrate this point.

References

1. J. H. Foxwell and R. E. Franklin. Acoustic effects in the vibrations of structures, *A.R.C.*, **19**, 495 (1957).
2. J. H. Foxwell and R. E. Franklin. The vibrations of a thin-walled stiffened cylinder in an acoustic field, *Aero Qtly.*, **10** (1959).

Many standard works discuss the transmission of sound between two media. E.g. Kinsler and Frey, *Fundamentals of Acoustics*, Wiley (1962).

CHAPTER 4

Random Processes

4.1 Introduction

In many physical phenomena the behaviour of a system cannot be determined completely with 100% confidence. The results themselves may be completely unpredictable or at least partially obscured by random disturbances or noise. The word 'noise' can mean almost anything and so it is important that it should be clear what the word is to mean in the present context. Here, noise will be regarded not just as 'unwanted sound' but as something more explicit, viz. a random disturbance. In a sense this is at once more restrictive and more general: more restrictive because we exclude such phenomena as the audible 50 c/s hum which comes through some radio receivers; more general because we include all processes or experiments in which the results fluctuate irregularly. Thus our definition includes not only the audible noise from a subsonic jet but the tiny voltages which appear across components in electronic circuits and the results of several fatigue tests on apparently identical specimens under identical conditions. In these, the measured quantity (sound pressure, voltage, fatigue life) can be described as noise or said to be partially obscured by noise.

It will quickly be realized that the difficulty has merely been shifted from that of defining 'noise' to that of defining 'random'. Considerations of the examples quoted will reveal that their common feature is that the results are unpredictable and this is what we shall take as our definition of randomness.

Hence, to deal with one problem more fully, if we are conducting research to find the fatigue properties of a particular type of specimen of a certain material, we find that if we carry out a number of tests at one stress level we get a different result for each specimen. The 'true fatigue life' (if we can talk of such a thing) is obscured by random fluctuations in such a way that we can never state precisely what the life of the next specimen will be.

The determination of the fatigue life of a specimen is an experiment the result of which is random; the description of such experiments is the question with which we have to deal. A good general treatment of the engineering applications of random process analysis is given by Bendat and Piersol (1).

4.2 Probability

Suppose we have a random experiment E, the result of which is expected to vary in an unpredictable manner from one repetition of E to another. In general we can assign a definite set S consisting of all results which are possible outcomes of E, e.g.

(i) If E is the observation of the sex of a child, the set S consists of two results: 'boy' and 'girl'.

(ii) If E is a throw with a die, the set S consists of the numbers 1, 2, 3, 4, 5, 6.

(iii) If E is the determination of the fatigue life of a certain specimen under certain conditions, the set S may consist of a large number of ranges such as (0 to N cycles), (N to $2N$ cycles), ($2N$ to $3N$ cycles) and so on.

Suppose now that we are concerned with the event A that the result of E belongs to some specified part of S, i.e. where A is the event that a die turns up a six, or that the fatigue life of a specimen lies between $5N$ and $6N$ cycles.

Let E be performed n times. Then the n experiments are said to constitute an ensemble of experiments E. If the event A occurs on f_A occasions we call f_A the absolute frequency of the event A, and the ratio f_A/n its relative frequency or frequency ratio.

In many cases, if the frequency ratio for many values of n is calculated it is found that when n exceeds a certain value, the frequency ratio is almost constant. If an experiment shows this stability we say that it displays statistical regularity.

To describe the behaviour of the frequency ratio for large n we postulate the existence of a number P_A, an idealization of the ratio f_A/n. This number we call the probability of the event A in the experiment. The frequency ratio is now to be regarded as an experimental determination of the probability P.

Symbolically, we have:

$$P(A, E) = \lim_{n \to \infty} \frac{f_A}{n}.$$

4.3 Presentation of information

In the last section we saw that although we cannot state precisely what the result of a random experiment will be, we can assign certain probabilities to the several possible results. We now look at the methods of presenting such information.

If by $P(X = a)$ we denote 'the probability that X assumes the value a' and by $P(a < X \leq b)$ we denote 'the probability that X assumes a value in the interval $a < X \leq b$', then if we know $P(a < X \leq b)$ for all a and *b* we say that we know the probability distribution of the random variable *X*.

Let *x* be a certain number and consider the probability $P(X \leq x)$. Clearly this is a function of *x*. If we wrote

$$F(x) = P(X \leq x)$$

then $F(x)$ is the distribution function of the variable *X*. It can be shown that distribution functions are monotonically increasing functions.

The distribution function is one way in which statistical information may be presented. Another, and perhaps a clearer way is by plotting a probability diagram. In such diagrams an ordinate equal to the appropriate probability is erected over each value which the variable can assume. Yet another diagram is called the probability density diagram, in which a rectangle is erected over each value which may be assumed by the variable. The width of this rectangle is equal to the interval between successive possible values and its area to the appropriate probability. These three diagrams are shown in Fig. 4.1.

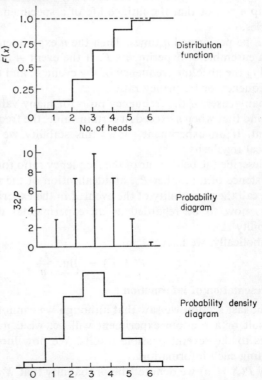

Fig. 4.1   Probability distribution of the number of 'heads' obtained when six coins are tossed.

Description of the results of a random experiment now reduces to a description of the appropriate probability density diagram. The principal pieces of information required are the most probable result, the scatter of the results about the most probable, and whether there is any preference for scatter in a particular direction. These characteristics are described by the mean value (which is the abscissa of the centre of area of the diagram), the second central moment (which is the second moment of area about the ordinate through the mean value), and the third central moment. The actual quantities quoted are the mean, the square root of the second moment (which is known as the standard deviation), and the cube root of the third moment (known as the Kurtosis).

Theoretically all the moments are required to define a distribution completely.

4.4 Examples of probability distributions

In certain experiments conditions of symmetry enable us to calculate the probability of a given result, provided we make assumptions about bias, etc. Thus there are certain theoretical probability distributions.

Those most frequently encountered are the binomial distribution and its two limiting cases the Poisson and Gaussian distribution. The binomial distribution is associated with 'go', 'no-go' situations. As an example, if six coins are tossed simultaneously the probability distribution of X, the number of heads, will tend to follow the binomial law.

In circumstances where the possible range of X is very large but the average value remains finite, its probability distribution tends to the limiting case of the binomial distribution known as Poisson's distribution. An example of this which is often quoted is the number of calls coming into a telephone exchange at a given instant.

On the other hand there are occasions where the average value becomes very large, e.g. the number of molecules of a gas in a certain volume. The binomial distribution then tends to a form known as the Gaussian distribution.

The binomial, Poisson and Gaussian distributions are theoretical and based on certain assumptions. Thus, strictly speaking, they should be checked by experiment when applied to real cases.

In an experiment where no symmetry conditions are apparent, e.g. the fatigue discussed earlier, the probability distribution must be determined by experiment. This involves finding frequency ratios for the various events, examining their stability and determining their limiting values. If these are to be found accurately and the various moments are to be expressed accurately then a large number of experiments have to be carried out. Thus we see that the accuracy of our statements depends on the size of the ensemble.

Where such an ensemble of experiments has been carried out and the result analysed, it may be possible to fit a theoretical distribution to the results.

4.5 Experiments with more than one variable

In many cases the result of an experiment may be expressed by more than one quantity, e.g. the result of a fatigue test on a steel specimen will include the fatigue life, the percentage carbon in the steel and so on.

Suppose for simplicity, that there are two quantities. Then in a similar fashion to that described in section 4.3 we can define a two-dimensional probability distribution and corresponding diagrams. Just as the one-dimensional probability density diagram of section 4.3 can be considered as a distribution of mass along a line, so a two-dimensional diagram may be visualized as a distribution of mass or a varying density of points in a plane. Thus a graph with fatigue life along one axis and percentage carbon along the other will, when the data are plotted, result in a varying density of points. By examining such a distribution, some idea may be gained as to any dependence between the two quantities. If the points are evenly scattered over the plane it is likely that there is no dependence; if they are concentrated about some line there is likely to be a strong dependence (see Fig. 4.2).

If there appears to be some dependence between the variables it is natural to try to express this as a relationship between them, i.e. as a curve in the plane. Assuming that this dependence seems to be linear then the standard technique of fitting a 'best straight line' is to find a line such that the mean square deviation of the points from the line is a minimum. This is the Least Squares Method. We can minimize either the square of the error in the x-direction or that in the y-direction: the resulting lines are said to be respectively, the least-squares regression line of X upon Y and that of Y upon X.

Having found, say, the least-squares regression lines of X on Y, if the mean square error is calculated it is found to be of the form:

$$\sigma_x^2(1 - \rho^2)$$

where

$$\rho = \frac{\overline{(X - m_x)(Y - m_y)}}{\sigma_x \sigma_y}.$$

σ denotes standard deviation, m the mean value, and the bar an average over the ensemble.

ρ is a quantity known as the correlation coefficient and it can be shown that it must lie in the range $-1 \le \rho \le 1$. It can thus be seen that it is a measure of the degree of linear dependence between X and Y for as it approaches its extreme values of 1, so the mean square error of the regression line tends to zero, indicating a complete dependence of X on Y.

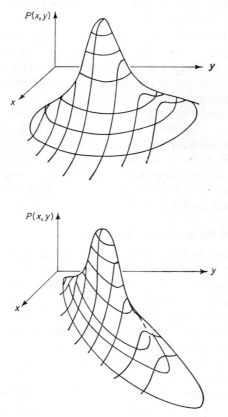

Fig. 4.2   Two-dimensional probability density diagrams.

4.6 Stochastic variables

In the preceding sections we have been dealing with variables in which time has no special significance. The random experiments we have discussed might equally well have been carried out simultaneously as consecutively. In some processes, however, time enters as an independent variable; these processes are becoming known as stochastic processes.

As an example, consider the electronic phenomenon known as shot noise. Because the hot cathode emits electrons in a random manner, the current which flows in a valve fluctuates irregularly with time. If we wish to examine the statistical properties of this fluctuation, then strictly speaking we should collect an ensemble of similar valves operating under identical conditions and, at a given instant, measure the current flowing in them all. By examination of the results we could find the probability distribution of the current, i.e. the quantities $P(I_1 < I \le I_1 + \delta I; t_1)$.

But if we are to describe the statistical properties of the current precisely, this probability distribution is not sufficient, for each of the members of the ensemble is fluctuating with time. In fact we have to determine an infinite set, the first of which we have just defined. The next would be the probability that those valves in the ensemble conducting a current I_1 at time t_1 should conduct a current I_2 at time t_2.

If this second probability distribution is a function only of $(t_2 - t_1)$ and not of either t_2 or t_1, the process is said to be stationary. When this is so (and we shall assume throughout this work that it is), then the ensemble average may be replaced by a time average. Thus instead of an ensemble of similar valves we can work with a single valve, observing the current at different times. Notice now that the conditions are that the current should be observed for a long time.

4.7 Description of the behaviour of a curve

Thus we are now brought to the idea of a quantity which is a random function of time and appears as a continuous fluctuating line when plotted out. In the last section we mentioned the noise current in a valve. Another example is the signal from a strain gauge attached to a panel which is vibrating randomly. Statements about the amplitude of the stress in the panel must be made on a probability basis: the maximum value of the stress during a given period is of little importance compared with a statement of the probable number of times per second that the stress will exceed a given value.

We have stated in the last section too that to describe such a curve we need an infinite set of probability distributions based on the idea: given that I passed through $I_1, I_2, I_3, \ldots I_n$ at times $t_1, t_2, t_3, \ldots t_n$, what is the probability that at time t_{n+1} it will pass through I_{n+1}? In practice, practically all the information we need is given by the first two probability distributions.

The second probability distribution is the two-dimensional distribution of $I(t_1)$ and $I(t_2)$, or (since the actual time is unimportant because of stationarity) of $I(t)$ and $I(t + \tau)$. In examining the distribution we shall look at the correlation coefficient, and this is now (see section 4.5),

$$\rho = \frac{\overline{\{I(t) - \overline{I(t)}\}\{I(t + \tau) - \overline{I(t + \tau)}\}}}{\sigma_{I(t)}\sigma_{I(t+\tau)}}.$$

If the process is stationary, $\sigma_{I(t)} = \sigma_{I(t+\tau)}$, $\overline{I(t)} = \overline{I(t + \tau)}$ and the ensemble average may be replaced by a time average. ρ is a function of τ and if we assume the mean value to be zero we may rewrite this as:

$$\rho(\tau) = \left\{ \lim_{T \to \infty} \frac{1}{T} \int_0^T I(t) \cdot I(t + \tau) \, dt \right\} \Big/ \left\{ \lim_{T \to \infty} \frac{1}{T} \int_0^T I^2(t) \, dt \right\}.$$

This is called the autocorrelation coefficient.

The dimensional form involving only the product term is known as the autocorrelation function $R(\tau)$ where

$$R(\tau) = \lim_{T \to \infty} \frac{1}{T} \int_0^T I(t) I(t + \tau) \, dt.$$

4.8 Physical idea of correlation

Consider first a function which varies with time in an unpredictable way. In actual problems it is most likely that although this variation in time is unpredictable, it will be continuous. Thus we may imagine a pen recorder, where the movement of the pen is proportional to the measured quantity, plotting out a continuous curve on a strip of moving paper. Without being at all rigorous, the idea of continuity is that the pen should move smoothly from one position to the next. Consideration of this leads us to the conclusion that the position of the pen at any instant is dependent to a certain extent on its position earlier; how much earlier is clearly dependent upon the speed at which the quantity whose value is being recorded fluctuates. We feel intuitively that if at one instant the pen is recording a large, increasing positive value, it will not at the next instant be recording a large, increasing negative value. As we increase the interval between the two observed positions of the pen, the dependence between them clearly grows less, and if the interval between the observations is large enough there will be practically no dependence.

To be more precise, what we have been saying (and assuming) in the previous paragraph is that if we have a function $F(t)$ which is a stationary, random function of time, then on the average, if we observe the function at two instants separated by an interval τ the two observations should be correlated to an extent which in general decreases as τ is increased. The phrase 'on the average' must be used because the function is a random one and we must carry out an average over an ensemble, or since the function is stationary, over time.

We are thus describing the measurement:

$$\rho(\tau) = \left\{ \lim_{T \to \infty} \frac{1}{T} \int_0^T F(t) F(t + \tau) \, dt \right\} \Big/ \left\{ \lim_{T \to \infty} \frac{1}{T} \int_0^T F^2(t) \, dt \right\}.$$

defined as the autocorrelation coefficient, and what we have said is, effectively, that all continuous curves have autocorrelation and the interval over which the autocorrelation is non-zero is a measure of the speed of the fluctuations.

The important distinction between a random and a periodic function is that whereas the periodic function repeats itself at regular intervals, the random function never repeats itself. In symbols, if $F(t)$ is periodic with periodic time t_0

$$F(t) = F(t + t_0).$$

If the formula for the autocorrelation coefficient is now referred to, it will be seen that the autocorrelation coefficient for a periodic function will oscillate. In particular, if $F(t) = \sin \omega t$, then:

$$\rho(\tau) = \left\{ \lim_{T \to \infty} \frac{1}{T} \int_0^T \sin \omega t \sin \omega(t + \tau)\, dt \right\} \bigg/ \left\{ \lim_{T \to \infty} \frac{1}{T} \int_0^T \sin^2 \omega t\, dt \right\}$$

$$= \cos \omega \tau$$

As already stated, a random signal never repeats itself, and so the value of its autocorrelation coefficient gradually falls to zero. One of the important applications of autocorrelation can now be seen. If a small periodic signal is buried in noise, the periodic signal may be recovered by an autocorrelation analysis, for at large values of τ the random components will have become incoherent whereas the periodic component remains coherent.

4.9 Harmonic analysis

The statistical analysis of a random function is not always very convenient, for example in the problem of the behaviour of a linear oscillator when excited by noise. The technique of autocorrelation analysis is then useful, but is merely a method of describing the variability of a function and there is another way of doing this.

Confronted with an irregular wave-form, one is tempted to try and resolve it into a Fourier series. In applying this technique to noise certain difficulties arise. In the first place, a noise signal does not repeat itself, i.e. it is not periodic and therefore has no fundamental period. One might choose an arbitrary period t_0 and then let it become infinite, but noise is a statistical quantity and the mean square value is definite (in theory). Thus if

$$\lim_{t_0 \to \infty} \frac{1}{t_0} \int_0^{t_0} F^2(t)\, dt = \sigma^2$$

then clearly as t_0 becomes large

$$\int_0^{t_0} F^2(t)\, dt$$

becomes large and so the method of Fourier analysis does not apply. However, let us look at this method.

Suppose we have a stationary, continuous random function of time $F(t)$ and that we record a sample of length t_0. This may be analysed into a Fourier series to give:

$$F_1(t) = \sum_{n=1}^{\infty} \{a_n \sin n\omega_0 t + b_n \cos n\omega_0 t\} \qquad (4.1)$$

where $\omega_0 = 2\pi/t_0$ and $a_0 = 0$ because $\overline{F_1(t)} = 0$ and this function $F_1(t)$ will be equal to the random signal $F(t)$ over the interval $0 < t < t_0$. The interval is significant because $F_1(t)$ is a synthetic periodic function.

If this analysis is repeated over many samples of length t_0 on each occasion different values may be obtained for the a_n and b_n, which incidentally are called the Fourier spectrum coefficients. In fact, of course (assuming that the noise has a zero mean value) a_n and b_n will be independent random variables distributed about zero in some way. Measurement of the a_n and b_n would be a tedious process, but it so happens that there is an important statistical property which can be measured fairly easily.

If $F(t)$ is a random function which has a zero mean value, then the square of the standard deviation is:

$$\sigma^2 = \lim_{T \to \infty} \frac{1}{T} \int_0^T F^2(t)\, dt.$$

The mean square of the synthetic function is:

$$\overline{F_1^2(t)} = \sum_{n=1}^{\infty} \sum_{m=1}^{\infty} \overline{\{a_n \sin n\omega_0 t + b_n \cos n\omega_0 t\}\{a_m \sin m\omega_0 t + b_m \cos m\omega_0 t\}}$$

$$= \sum_{n=1}^{\infty} \sum_{m=1}^{\infty} \{\overline{a_n a_m \sin n\omega_0 t \sin m\omega_0 t} + \overline{b_n b_m \cos n\omega_0 t \cos m\omega_0 t}$$

$$+ \overline{a_m b_n \sin m\omega_0 t \cos n\omega_0 t} + \overline{a_n b_m \sin n\omega_0 t \cos m\omega_0 t}\}.$$

As the sine and cosine functions are orthogonal, all cross terms and terms where $n \neq m$ will become zero after the time integration. Thus the above equation reduces to

$$\overline{F_1^2(t)} = \sum_{n=1}^{\infty} \tfrac{1}{2}\{a_n^2 + b_n^2\}$$

$$= \sum_{n=1}^{\infty} \tfrac{1}{2} c_n^2 \qquad (4.2)$$

If $F(t)$ were a current, then the power developed by this current as it passed through a resistance of one ohm would be $F^2(t)$. Equation 4.2 defines a set of quantities known as the power spectrum coefficients of $F(t)$.

It is most important when considering the harmonic analysis of a random function to remember that there are no periodic components in the signal. The periodic components in equations 4.1 and 4.2 are fictitious. To make this clearer we go into the matter more deeply.

Suppose we have a sample of the noise signal and that its length is $2t_0$. We may arbitrarily choose an origin, so we say that this signal stretches from $-t_0$ to $+t_0$. We now construct a Fourier series in the form of equation 4.1 for a function the periodic time of which is $2T_0$ and one period of which stretches from $-T_0$ to $+T_0$. We make $T_0 > t_0$ and the function we synthesize is equal to $F(t)$ over the interval $-t_0$ to $+t_0$ and zero over the rest of the period. We now consider what happens if we allow T_0 to tend to infinity so that we have a function equal to $F(t)$ over the range $-t_0$ to $+t_0$ and zero elsewhere (see Fig. 4.3).

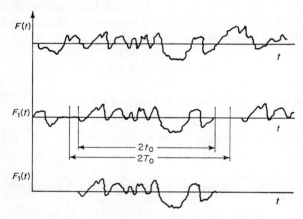

Fig. 4.3   Fourier analysis of random noise.

By substituting the complex form of $\sin n\omega_0 t$ and $\cos n\omega_0 t$ in equation 4.1 it can be transformed into

$$F_1(t) = \sum_{n=-\infty}^{+\infty} f_n\, e^{in\omega_0 t}$$

where

$$\omega_0 = \frac{2\pi}{2T_0} \quad \text{and} \quad f_n = \frac{1}{2T_0} \int_{-T_0}^{T_0} F_1(t)\, e^{-in\omega_0 t}. \tag{4.3}$$

The quantities f_n are known as the complex Fourier spectrum coefficients.

As $T_0 \to \infty$, the 'fundamental frequency' $\omega_0 = \pi/T_0$ tends to a small vanishing quantity $\Delta\omega$ and $n\omega_0$ becomes $n\Delta\omega$ which we may write as ω. Physically, this means that as T_0 becomes very large the frequencies of the 'components' become multiples of a very low frequency, i.e. the discrete frequencies become closer and closer together until a continuum is formed.

The expression for the complex spectrum coefficients now becomes

$$\lim_{T_0 \to \infty} f_n = \lim_{T_0 \to \infty} \frac{1}{2T_0} \int_{-T_0}^{T_0} F_1(t) \, e^{-in\omega_0 t} \, dt = \frac{\Delta\omega}{2\pi} \int_{-\infty}^{\infty} F_1(t) \, e^{-i\omega t} \, dt$$

and it is seen that these have become very small. Although the amplitude for any frequency tends to zero, the relative values for different frequencies remain the same, so we write this last equation as

$$\frac{f_n}{\Delta\omega} = f(i\omega) = \frac{1}{2\pi} \int_{-\infty}^{\infty} F_1(t) \, e^{-i\omega t} \, dt.$$

Thus we have the Fourier integral representation

$$F_1(t) = \int_{-\infty}^{\infty} f(i\omega) \, e^{i\omega t} \, d\omega$$

where

$$f(i\omega) = \frac{1}{2\pi} \int_{-\infty}^{\infty} F_1(t) \, e^{-i\omega t} \, dt.$$

If we now form an expression for the mean square value of the random function we get

$$\overline{F^2(t)} = \lim_{t_0 \to \infty} \frac{1}{2t_0} \int_{-\infty}^{\infty} F^2(t) \, dt = \lim_{t_0 \to \infty} \frac{1}{2t_0} \int_{-\infty}^{\infty} F(t) \int_{-\infty}^{\infty} f(i\omega) \, e^{i\omega t} \, d\omega \, dt$$

$$= \lim_{t_0 \to \infty} \frac{1}{2t_0} \int_{-\infty}^{\infty} f(i\omega) \int_{-\infty}^{\infty} F(t) \, e^{i\omega t} \, dt \, d\omega$$

$$= \lim_{t_0 \to \infty} \frac{1}{2t_0} \int_{-\infty}^{\infty} f(i\omega) 2\pi f^*(i\omega) \, d\omega = \int_{-\infty}^{\infty} \lim_{t_0 \to \infty} \frac{\pi |f(i\omega)|^2}{t_0} \, d\omega. \qquad (4.4)$$

The function under the integral sign is known as the power spectral density $S(\omega)$ and is a continuous function of ω.
Therefore

$$\overline{F^2(t)} = \int_{-\infty}^{\infty} S(\omega) \, d\omega. \qquad (4.5)$$

Once again it should be emphasized that there are no periodic components in a random noise signal. If periodic or zero frequency components are present the power spectral density becomes infinite at the appropriate frequency. The harmonic analysis of the signal or its power spectrum describe the variability of the signal.

Finally, we note above that there is an important statistical property which is easily measured. This is an approximation to the power spectral density and is measured by passing the unknown signal through a series of filters and integrating their output. This results in a set of quantities like

$$\int_{\omega_1}^{\omega_2} \frac{\pi f(i\omega)^2 \, d\omega}{t_0}$$

where ω_1, ω_2 define the pass-band of the filter (see also section 4.15).

4.10 Relationship between autocorrelation and power spectral density

One feels intuitively that, since both the autocorrelation function $R(\tau)$ and the power spectral density $S(\omega)$ describe the variability of a function, there should be a connexion between them. This idea is strengthened when one recalls that frequency and time are reciprocal. The relationship may be demonstrated as follows:

$$R(\tau) = \lim_{t_0 \to \infty} \frac{1}{2t_0} \int_{-t_0}^{t_0} F(t)F(t + \tau) \, dt$$

$$= \lim_{t_0 \to \infty} \frac{1}{2t_0} \int_{-t_0}^{t_0} F(t) \int_{-\infty}^{\infty} f(i\omega) \, e^{i(t+\tau)\omega} d\omega \, dt$$

$$= \lim_{t_0 \to \infty} \frac{1}{2t_0} \int_{-\infty}^{\infty} f(i\omega) \, e^{i\omega\tau} \int_{-t_0}^{t_0} F(t) \, e^{i\omega t} \, dt \, d\omega$$

$$= \lim_{t_0 \to \infty} \frac{1}{2t_0} \int_{-\infty}^{\infty} f(i\omega) 2\pi f^*(i\omega) \, e^{i\omega\tau} \, d\omega$$

$$= \int_{-\infty}^{\infty} S(\omega) \, e^{i\omega t} d\omega.$$

Hence

$$R(\tau) = \int_{-\infty}^{\infty} S(\omega) \, e^{i\omega\tau} \, d\omega$$

and   (4.6)

$$S(\omega) = \frac{1}{2\pi} \int_{-\infty}^{\infty} R(\tau) \, e^{-i\omega\tau} \, d\tau$$

In these expressions $S(\omega)$ is an even function having values for theoretical negative frequencies as well as positive and we know that

$$\int_{-\infty}^{\infty} S(\omega) \, d\omega = \overline{F^2(t)} \qquad \text{from equation 4.5.}$$

If we now restrict ourselves to the practical case of positive frequencies only we still have the area of the power spectrum equal to the mean square value of the signal. Introducing the symbol $G(\omega)$ for the spectral density for positive frequencies only we have:

$$\int_0^\infty G(\omega)\, d\omega = \overline{F^2(t)}$$

thus

$$G(\omega) = 2S(\omega).$$

$R(\tau)$ is also an even function and therefore we can write

$$R(\tau) = \int_0^\infty G(\omega) \cos \omega\tau\, d\omega \quad \text{from equation 4.6.}$$

Now-the power spectral density in terms of power per cycle per second, $G(f)$, for positive frequencies only is given by

$$G(f) = 2\pi G(\omega).$$

Hence

$$R(\tau) = \int_0^\infty G(f) \cos 2\pi f\tau\, df \tag{4.7}$$

and

$$G(f) = 4 \int_0^\infty R(\tau) \cos 2\pi f\tau\, d\tau. \tag{4.8}$$

Thus the autocorrelation function is the Fourier transform of the power spectrum. Functions which are Fourier transform pairs have an inverse spreading relationship. In the present case this means simply that if the power spectrum is wide the autocorrelation function drops rapidly to zero (a fact which we have already deduced physically, for if the power spectrum is wide this means that the signal is fluctuating rapidly and with little coherence). Conversely, if the power spectrum is very narrow then the autocorrelation function has non-zero values for very large values of τ. To be more precise, the autocorrelation function oscillates, its amplitude decreasing at a rate which is greater the greater the width of the power spectrum (see Fig. 4.4).

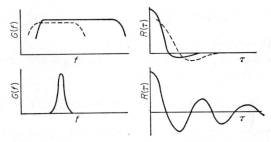

Fig. 4.4   Inverse spreading of $G(f)$ and $R(\tau)$.

If it is borne in mind that the autocorrelation function and the power spectrum both describe the variability of the signal this inverse spreading characteristic is not difficult to understand. A very narrow power spectrum centred on an angular frequency ω_0 means that the signal, although random and therefore unpredictable, fluctuates in time rather like a sine wave of frequency ω_0. Now a sine wave of frequency ω_0 has an auto-correlation coefficient given by $\cos \omega_0 \tau$. In fact we can calculate the autocorrelation function for a signal whose power spectral density $G(\omega_0)$ is constant from $(\omega_0 - \frac{1}{2}\Delta\omega)$ to $(\omega_0 + \frac{1}{2}\Delta\omega)$. It is:

$$R(\tau) = G(\omega_0) \int_{\omega_1}^{\omega_2} \cos \omega\tau \, d\omega = G(\omega_0) \left[\frac{\sin \omega\tau}{\tau}\right]_{\omega_1}^{\omega_2}$$

$$= G(\omega_0) \frac{2}{\tau} \sin \left(\frac{\omega_2 - \omega_1}{2}\right)\tau \cos \left(\frac{\omega_2 + \omega_1}{2}\right)\tau$$

$$= \left\{\frac{2G(\omega_0)}{\tau} \sin \frac{\Delta\omega\tau}{2}\right\} \cos \omega_0\tau. \qquad (4.9)$$

This is a decaying cosine wave as we suspected.

The duality of the power spectrum and the autocorrelation function does not mean that one of them is not needed. It is true that both present the same information, but it is in a different form. In theoretical work it is very often much easier, given information about the probability distribution of a process, to calculate the autocorrelation than to calculate the power spectrum; if it is the power spectrum which is required it is easier to calculate it through the autocorrelation than directly.

4.11 Cross-correlation and cross-power spectrum

It will be realized from the way in which the autocorrelation function has been introduced (namely as one of the properties of a two-dimensional probability distribution), that a similar function can be defined for two different variables. If $F_1(t)$ and $F_2(t)$ are two random functions of time then

$$R_{12}(\tau) = \lim_{T \to \infty} \frac{1}{2T} \int_{-T}^{T} F_1(t)F_2(t + \tau) \, dt$$

is known as the cross-correlation function.

As an example suppose $F_1(t)$ and $F_2(t)$ are the sound pressures at two points x_1, x_2 in an acoustic field, then

$$R_{12}(x_1, x_2; \tau) = \lim_{T \to \infty} \frac{1}{2T} \int_{-T}^{T} F_1(x_1, t)F_2(x_2, t + \tau) \, dt$$

is a cross-correlation or space–time correlation of the sound pressure.

The cross-power spectrum can now be defined as the Fourier transform of the cross-correlation function. This cross-power spectrum is now a complex function, because $R_{12}(\tau)$ is an odd function, and has a real part given by:

$$C_{12}(f) = 2 \int_{-\infty}^{\infty} R_{12}(\tau) \cos 2\pi f \tau \, d\tau. \qquad (4.10)$$

$C_{12}(f)$ is sometimes known as the co-spectrum and represents the narrow band correlation between the functions $F_1(t)$ and $F_2(t)$. In future work we shall call this the correlation spectrum of the quantities $F_1(t)$ and $F_2(t)$. We shall usually work on a normalized basis and use $C_{12}(f)/\sqrt{G_1(f)G_2(f)}$ to form the ordinate of the correlation spectrum.

The imaginary part of the cross-power spectrum is given by

$$Q_{12}(f) = 2 \int_{-\infty}^{\infty} R_{12}(\tau) \sin 2\pi f \tau \, d\tau$$

and is sometimes known as the quadrature spectrum.

4.12 Measurement of statistical quantities

4.12.1 *Power spectrum by filters*

As mentioned at the end of section 4.9, the power spectrum is measured fairly simply by passing the signal through each of a set of filters and integrating their outputs. It is important to remember that the result is integrated: even if the output is indicated on a meter the value is the output integrated by the time-constant of the meter.

The power spectral density is defined as

$$G(\omega) = \lim_{T \to \infty} \frac{2\pi}{T} |f(i\omega)|^2$$

and its dimensions will be power per radian per second, or if we define $G(f)$, power per cycle per second.

The quantity we measure is

$$\int_{\omega_1}^{\omega_2} \lim_{T \to \infty} \frac{2\pi}{T} |f(i\omega)|^2 \, d\omega$$

and we notice two things from this expression, viz. the result is a function of the bandwidth of the filter and depends upon the time for which the output of the filter has been integrated.

Taking the second of these two points, we recall that $f(i\omega)$ is a complex spectrum coefficient resulting from analysis of a single sample of the signal. As various samples are analysed the value of $f(i\omega)$ will fluctuate in some way about an average value. We have seen that if $F(t)$ is stationary then examining an ensemble is the same as looking at $F(t)$ for a certain time, and the longer we look at the signal, the larger an ensemble are we effectively covering. Thus if we look at $F(t)$ for a length of time T and estimate its power spectrum we shall have a certain error. If we look at it for $5T$ our probable error will be less. Hence we have to look at the signal for an interval long enough for the probable error to be negligible as far as we are concerned. This is the problem of integrating time. It can be shown that the time required to get a result of given accuracy is inversely proportional to the bandwidth of a filter. A meter connected to the output of a filter will therefore fluctuate much more for narrow-band than for wide-band filters.

Turning now to the first point, we notice that the quantity we measure is not the quantity we require, but the integral of that quantity over a certain interval of frequency. If we assume the power spectrum to be flat the correction is quite simple, for the reading is

$$X = \int_{f_1}^{f_2} G(f)\, df = G(f_0) \int_{f_1}^{f_2} df = G(f_0)(f_2 - f_1).$$

If we are using the decibel scale we read

$$\mathrm{dB} = 10 \log X = 10 \log G(f_0) + 10 \log (f_2 - f_1). \tag{4.11}$$

Hence the power spectrum is given in dB by

$$10 \log G(f_0) = 10 \log X - 10 \log (f_2 - f_1). \tag{4.12}$$

The result in dB is usually referred to as the spectrum level.

If the results of an analysis with an octave-band analyser are plotted against frequency and compared with those from a third-octave-band analyser, although the shape of the spectra will be similar, the levels will be different. Equation 4.12 shows how the results of one analyser can be compared with the results of another.

When the spectrum level is plotted against frequency its shape can be very different from a graph of results of a constant percentage bandwidth analyser.

The last two paragraphs emphasize the importance when presenting results of giving complete details of the analyser used and the way in which the results are plotted.

In order to save time during experiments signals are often recorded on magnetic tape. A given sample may then be spliced into a continuous loop and analysed later at leisure. It must be borne in mind here that splicing the sample into a loop does not change it from a sample of length T to one of infinite length; there is only a limited amount of information on the loop and the amount is not changed by looking at it many times. This technique can also introduce a further error in that one now generates a periodic signal as one plays back the loop. This can give rise to spurious components in the measured power spectrum. If one considers a sample of length T, of a single sine wave with arbitrary phase at the splice one finds that if the loop is analysed other frequencies appear. As the phase at the splice is changed to zero, the spurious components disappear, but it is unlikely that one would get an exact number of cycles in a sample. For any phase angle at the splice, as T is increased the spurious components are concentrated more and more around the time frequency, and so would be collected by a filter centred on that frequency. It is usually better to leave a blank piece of tape on the loop and then trigger the analyser each time it reaches the start of the signal. If tape loops are used then, there is another reason for a long sample.

Finally, it should be recalled that all the statistical theory above applies to stationary processes only. If there is any indication that, say, the r.m.s. value of a signal is varying, the problem should be carefully reviewed.

4.12.2 *Correlation*

For more accurate measurements of correlation coefficients an analogue computer is needed. The complete correlator will take the two signals, either directly or from a recording, delay one with respect to the other, multiply them together and integrate the resulting product. In other words it will, given $F_1(t)$ and $F_2(t)$, perform the calculation.

$$R(\tau) = \frac{1}{T} \int_0^T F_1(t)F_2(t + \tau) \, dt.$$

We note two elementary points. Firstly with regard to two electrical signals, the correlator effectively computes the phase difference between components common to both. Clearly then, there must be no phase difference between the two channels in the computer or this will appear as an additional delay time τ_0 which may be a function of frequency.

Secondly, the correlation function is correctly defined as a limit. Thus here again we have a problem of integration time, the calculation of small values of $R(\tau)$ becoming more unreliable as the integration time decreases.

4.12.3 *Power spectra by transformation of autocorrelation functions*

The limitations of the transformation process, represented by equations 4.8 and 4.10, for obtaining power and cross spectra are that the transformation involves an infinite integral which in turn requires a knowledge of $R(\tau)$ for all values of τ up to infinity. In practice the autocorrelation function must be truncated at some point τ_{max}. If the correlogram has reached steady zero values at this point, as in the case of broad-band noise, then the transformation can be carried out. However, for the type of spectrum which has narrow peaks (such as the strain in a panel excited by jet noise) the correlogram continues to oscillate for a considerable time and it may be necessary for practical reasons to truncate it before steady zero values have been reached. If this truncated autocorrelogram is transformed, a mathematical filter or 'spectral window' has been effectively introduced due to the fact that a weighting function has been applied to the correlogram.

In the simple case this function $D(\tau)$, say, has the form

$$D(\tau) = 1 \quad \text{for} \quad 0 < \tau \le \tau_{max}$$

$$D(\tau) = 0 \quad \text{for} \quad \tau > \tau_{max}.$$

The result of a transformation process is now an apparent spectral density given by:

$$P(f') = 4 \int_0^\infty D(\tau)R(\tau) \cos 2\pi f' \tau_d \tau$$

$$= \int_0^\infty Q(f - f')G(f) \, df \tag{4.13}$$

where the spectral window

$$Q(f - f') = \frac{2\tau_m \sin 2\tau_m(f - f')\pi}{2\tau_m(f - f')\pi}. \tag{4.14}$$

The main features of the 'filter' are that it has negative values extending to -0.2 for some frequencies, and a width which is inversely proportional to a maximum time delay. The negative portion can be removed, at the expense of bandwidth, by using Bartlett's weighting function which has the form:

$$D(\tau) = 1 - (\tau/\tau_{max}) \text{ for } 0 < \tau < \tau_{max}, \text{ and}$$

$$D(\tau) = 0 \text{ for } \tau > \tau_{max}.$$

This gives a spectral window

$$Q_B(f - f') = \tau_m \left\{ \frac{\sin \pi(f - f')\tau_{max}}{\pi(f - f')\tau_{max}} \right\}^2 \tag{4.15}$$

These functions are shown in Fig. 4.5.

Other weighting functions have been proposed which aim to reduce still further the value of the secondary peak. These functions are well described elsewhere.[2]

4.12.4 *Cross-power spectra*

The cross-power spectrum $G(x_1, x_2, f)$ of two signals can be measured directly by filtering the two signals by a narrow-band filter centred at frequency f and measuring the correlation of the filtered signals. Alternatively, it can be derived indirectly by measuring first the cross-correlation function $R(x_1, x_2, \tau)$ and then transforming this to $G(x_1, x_2, f)$ digitally.

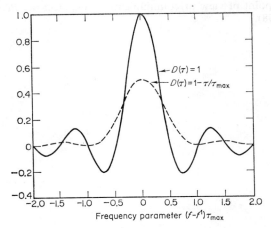

Fig. 4.5   Spectral windows $Q(f - f')$.

For broad-band noise the direct procedure can be used successfully but for structural response signals the filtering of the signals before correlating presents difficulties. The peaks in the spectra are narrow and some may be relatively close together. A filter bandwidth of the order of 1% is therefore required. Unfortunately, the two signals must be passed through identical filters to avoid spurious correlation due to relative phase shifts in the filters. Normal narrow-band filters will not be matched in this way. An alternative method is to use the same filter in both channels.

By using two tape recorders it is possible to play back the original signals, one of which passes through the filter before being re-recorded on the second recorder. The next step is to play back from the second recorder, this time filtering the second signal before re-recording on the first recorder. This system works successfully if the signals to be compared have peaks in the spectrum occurring at the same frequencies. If there is any frequency difference the differential phase shift in the side band of the filter will lead to spurious correlation readings.

The indirect method via the cross-correlation function eliminates the spurious phase shifts in the filter bands. There are still bandwidth problems associated with the spectral window and long time delays are required if high accuracy is to be achieved.

References

1. J. S. Bendat and A. G. Piersol. *Measurement and Analysis of Random Data*, Wiley, London (1966).
2. R. B. Blackman and J. W. Tukey. The measurement of power spectra from the point of view of communications engineering, *Bell System Tech. J.*, **37** (1958).

CHAPTER 5

General Theory of Aerodynamic Sound

Aerodynamic sound may be defined as sound which is generated as a direct result of an air-flow, without any essential part being played by vibrations of solid bodies. It is remarkable, but true, that in spite of the tremendous amount of literature on acoustics (as witnessed by the two volumes of Rayleigh's 'Theory of Sound', for example), it is only the effect of the considerable noise radiation from jet aircraft upon the general public which has led to any serious attempt to predict the intensity of the sound radiated from an air-flow.

The basic theory of aerodynamic sound generation, and its application to the noise radiated from a turbulent jet, was first given in two papers by Lighthill[1,2], who considered a fluctuating hydrodynamic flow, covering a limited region, surrounded by a large volume of fluid which is at rest apart from the infinitesimal amplitude sound waves radiated from the flow. We shall now attempt to see precisely why this problem cannot be approached on the basis of classical acoustic theory.

5.1 Classical acoustics: sources and dipoles

5.1.1 *The equations of acoustics*

The (perfectly general) Navier–Stokes equations of fluid flow may be written as

$$\frac{\partial \rho}{\partial t} + \frac{\partial}{\partial x_i} (\rho v_i) = 0 \tag{5.1}$$

$$\frac{\partial}{\partial t} (\rho v_i) + \frac{\partial}{\partial x_j} (\rho v_i v_j) = -\frac{\partial}{\partial x_j} (p_{ij}), \tag{5.2}$$

in the usual notation. Classical acoustics considers sound radiation in a medium which, apart from the radiation, is uniform and at rest. The following approximations are accordingly made:

(i) The particle velocities v_i are small, so all squares of velocities are neglected; for example, the term $\rho v_i v_j$ goes out.

111

(ii) The viscous terms in the stress tensor p_{ij} are neglected, so that only the pressure terms remain, and $\partial(p_{ij})/\partial x_j$ becomes $\partial(p\delta_{ij})/\partial x_j = \partial p/\partial x_i$.

(iii) It is then assumed that the small pressure and density fluctuations are proportional to one another, so that $\partial p/\partial x_i = a_0^2 \, \partial \rho/\partial x_i$, where the constant a_0^2 has the dimensions of velocity squared.

As a result of these approximations it follows that the approximate form of equation 5.2 is

$$\frac{\partial}{\partial t}(\rho v_i) + a_0^2 \frac{\partial \rho}{\partial x_i} = 0 \tag{5.3}$$

so that equations 5.1 and 5.3 are simultaneous equations for ρ and ρv_i. If we differentiate equation 5.1 with respect to t, equation 5.3 with respect to x_i, and subtract, then ρv_i is eliminated and we find that

$$\frac{\partial^2 \rho}{\partial t^2} - a_0^2 \frac{\partial^2 \rho}{\partial x_i^2} = \frac{\partial^2 \rho}{\partial t^2} - a_0^2 \nabla^2 \rho = 0. \tag{5.4}$$

5.1.2 *Spherical waves*

We now examine the physically interesting case of solutions of equation 5.4 which are spherically symmetrical. Thus ρ is a function only of $r = (x_1^2 + x_2^2 + x_3^2)^{1/2}$ and t. For functions which depend only on r (and t) it may be shown that

$$\nabla^2 \equiv \frac{1}{r^2} \frac{\partial}{\partial r}\left(r^2 \frac{\partial}{\partial r}\right).$$

Thus equation 5.4 becomes

$$\frac{1}{r^2} \frac{\partial}{\partial r}\left(r^2 \frac{\partial \rho}{\partial r}\right) = \frac{1}{a_0^2} \frac{\partial^2 \rho}{\partial t^2}. \tag{5.5}$$

This equation may be simplified by means of the transformation $\xi = r\rho$. If we write $\rho = \xi r^{-1}$, then equation 5.5 leads to the result

$$\frac{1}{r^2} \frac{\partial}{\partial r}\left(r^2 \frac{\partial \rho}{\partial r}\right) = \frac{1}{r^2} \frac{\partial}{\partial r}\left[r^2 \frac{\partial}{\partial r}\left(\frac{\xi}{r}\right)\right]$$

$$= \frac{1}{r^2} \frac{\partial}{\partial r}\left[r^2 \left(\frac{1}{r} \frac{\partial \xi}{\partial r} - \frac{1}{r^2} \xi\right)\right]$$

$$= \frac{1}{r^2} \frac{\partial}{\partial r}\left(r \frac{\partial \xi}{\partial r} - \xi\right)$$

$$= \frac{1}{r} \frac{\partial^2 \xi}{\partial r^2},$$

and since equation 5.5 shows this to be equal to

$$\frac{1}{a_0^2}\frac{\partial^2 \rho}{\partial t^2} = \frac{1}{a_0^2 r}\frac{\partial^2 \xi}{\partial t^2},$$

it follows that ξ satisfies

$$\frac{\partial^2 \xi}{\partial r^2} = \frac{1}{a_0^2}\frac{\partial^2 \xi}{\partial t^2}. \tag{5.6}$$

The most general solution of this equation is

$$\xi = \xi_0 + F_1\left(t - \frac{r}{a_0}\right) + F_2\left(t + \frac{r}{a_0}\right),$$

or

$$\rho - \rho_0 = \frac{1}{r}\left\{F_1\left(t - \frac{r}{a_0}\right) + F_2\left(t + \frac{r}{a_0}\right)\right\}. \tag{5.7}$$

5.1.3 *Simple sources*

We ask ourselves what the solution 5.7 represents physically. The answer is that the two terms represent respectively outgoing and incoming waves. We ignore the latter as being physically unrealistic, at least in the absence of reflecting surfaces, so that equation 5.7 becomes simply

$$\rho - \rho_0 = \frac{1}{r}F_1\left(t - \frac{r}{a_0}\right), \tag{5.8}$$

and turn our attention to the significance of the singularity at the origin. In fact this particular solution corresponds to a fluctuating source of mass at the origin, as we shall now show. Consider a sphere of radius r, centred at the origin. Then the flux of mass $q(t)$ per unit time through this sphere is

$$q(t) = \rho \cdot 4\pi r^2 \cdot v_r,$$

where v_r is the radial component of the velocity. Thus $q'(t)$ (primes denoting derivatives) is

$$q'(t) = 4\pi r^2 \frac{\partial}{\partial t}(\rho v_r)$$

$$= -4\pi r^2 a_0^2 \frac{\partial \rho}{\partial r} \quad \text{(by equation 5.3)}$$

$$= 4\pi r^2 a_0^2 \left\{\frac{1}{r^2}F_1\left(t - \frac{r}{a_0}\right) + \frac{1}{a_0 r}F_1'\left(t - \frac{r}{a_0}\right)\right\}$$

$$= 4\pi a_0^2 \left\{F_1\left(t - \frac{r}{a_0}\right) + \frac{r}{a_0}F_1'\left(t - \frac{r}{a_0}\right)\right\}$$

In the limit, as $r \to 0$, we see that the rate of change of mass flux $q'(t)$ through an infinitesimal sphere at the origin is

$$q'(t) = 4\pi a_0^2 F_1(t) \qquad (5.9)$$

With this value of F_1, we may re-write equation 5.8 as

$$\rho - \rho_0 = \frac{q'\left(t - \dfrac{r}{a_0}\right)}{4\pi a_0^2 r}. \qquad (5.10)$$

Such a solution is referred to as an acoustic source or pole.

Three things may be noted about this solution.

(i) In view of the presence of a singularity at the origin, this result only holds in a region which excludes the origin. In other words the region of the mass source is excluded.

(ii) It is not the rate of mass introduction, $q(t)$, but its time derivative, $q'(t)$, which really counts. In fact if there were a uniform rate of mass introduction then no sound would be radiated at all, since (mathematically) q' would then be zero in equation 5.10 and (physically) a uniform rate of mass introduction would merely lead to a new steady state being set up.

(iii) It will be noted that the density fluctuations $\rho - \rho_0$ reaching a distance r at time t are determined by the value of q' at an earlier time $t - (r/a_0)$. This is because of the time r/a_0 taken for a sound wave to travel this distance.

We can now generalize equation 5.10 to the case of distributed mass introduction. Suppose we postulate a rate of introduction of mass $Q(\mathbf{x}, t)$ per unit volume at position $\mathbf{x}$ and time t. Then the continuity equation 5.1 must be replaced by

$$\frac{\partial \rho}{\partial t} + \frac{\partial}{\partial x_i}\,(\rho v_i) = Q(\mathbf{x}, t). \qquad (5.11)$$

Elimination of ρv_i between equations 5.3 and 5.11 then yields

$$\frac{\partial^2 \rho}{\partial t^2} - a_0^2 \nabla_0^2 \rho = Q'(\mathbf{x}, t) \qquad (5.12)$$

In an unbounded medium the most general solution of this equation is

$$\rho(\mathbf{x}, t) - \rho_0 = \frac{1}{4\pi a_0^2} \int \frac{Q'\left(\mathbf{y}, t - \dfrac{|\mathbf{x} - \mathbf{y}|}{a_0}\right)}{|\mathbf{x} - \mathbf{y}|}\, d\tau(\mathbf{y}), \qquad (5.13)$$

which is, in effect, a summation of terms like equation 5.10 over the appropriate volume of mass introduction.

5.1.4 *Sound radiation by acoustic sources*

We may re-write equation 5.13 as

$$\rho(\mathbf{x}, t) - \rho_0 = \frac{1}{4\pi a_0^2} \frac{\partial}{\partial t} \int \frac{Q\left(\mathbf{y}, t - \dfrac{|\mathbf{x} - \mathbf{y}|}{a_0}\right)}{|\mathbf{x} - \mathbf{y}|} \, d\tau(\mathbf{y}). \qquad (5.14)$$

In dealing with acoustic radiation we are interested in what happens at large distances from the flow. Accordingly, since

$$|\mathbf{x} - \mathbf{y}|^{-1} = |\mathbf{x}|^{-1}\left\{1 + O\left(\frac{1}{|\mathbf{x}|}\right)\right\},$$

we may write equation 5.14 as

$$\rho(\mathbf{x}, t) - \rho_0 \simeq \frac{1}{4\pi a_0^2 |\mathbf{x}|} \frac{\partial}{\partial t} \int Q\left(\mathbf{y}, t - \frac{|\mathbf{x} - \mathbf{y}|}{a_0}\right) d\tau(\mathbf{y}). \qquad (5.15)$$

Now the maximum variations in the retarded time $|\mathbf{x} - \mathbf{y}|/a_0$ will be of order $\pm l/2a_0$ where l represents the scale of the region of mass introduction and provided $l < \lambda$, say, where λ is a typical acoustic wavelength (of the radiated sound) then the retarded time effect will in general be unimportant and equation 5.15 may be written as

$$\rho(\mathbf{x}, t) - \rho_0 = \frac{1}{4\pi a_0^2 |\mathbf{x}|} \frac{\partial}{\partial t} \int Q(\mathbf{y}, t) \, d\tau(\mathbf{y}). \qquad (5.16)$$

What this means, in effect, is that we have replaced the whole volume distribution of sources by a single point source, whose strength is equal to the total instantaneous source strength. It is clear, however, that this cannot be done if this total is zero (or close to zero) for in such cases the terms retained will be even smaller than the small neglected terms.

5.1.5 *Sound generation by distributed forces*

Suppose now that sources of mass are absent, but instead imagine a distributed externally applied force field of magnitude $F_i(\mathbf{x}, t)$ per unit volume. Then equation 5.1 holds, but a term F_i must be added to the right-hand side of equation 5.2 or 5.3, whence equation 5.4 becomes

$$\frac{\partial^2 \rho}{\partial t^2} - a_0^2 \nabla^2 \rho = -\frac{\partial F_i}{\partial x_i}. \qquad (5.17)$$

Suppose we treat this equation exactly as we treated equation 5.12. Then we write down the general solution as

$$\rho(x, t) - \rho_0 = -\frac{1}{4\pi a_0^2} \int \frac{\dfrac{\partial F_i}{\partial y_i}\left(y, t - \dfrac{|\mathbf{x} - \mathbf{y}|}{a_0}\right)}{|\mathbf{x} - \mathbf{y}|} \, d\tau(\mathbf{y}), \qquad (5.18)$$

and at large distances this becomes

$$\rho(\mathbf{x}, t) - \rho_0 = -\frac{1}{4\pi a_0^2 |\mathbf{x}|} \int \frac{\partial F_i}{\partial y_i} \left(\mathbf{y}, t - \frac{|\mathbf{x} - \mathbf{y}|}{a_0}\right) d\tau(\mathbf{y}).$$

If we now neglect retarded times, as we did above, then this yields

$$\rho(\mathbf{x}, t) - \rho_0 = -\frac{1}{4\pi a_0^2 |\mathbf{x}|} \int \frac{\partial F_i}{\partial y_i} (\mathbf{y}, t) \, d\tau(\mathbf{y}),$$

which is precisely zero. The reason why this is precisely zero is that, once the contributions in this last integral are instantaneous contributions, we may transform the volume integral into a surface integral, so that

$$\int \frac{\partial F_i}{\partial y_i} \, d\tau = \int \operatorname{div} F \, d\tau = \int F \cdot n \, ds,$$

where n denotes the outward normal from the volume. Since there are no internal solid boundaries, the appropriate surface is a large sphere surrounding the flow, where F vanishes. Accordingly a direct application of the concepts discussed above yields zero radiated sound. Why?

The answer is that this is a case where the total source strength is zero since, as we stipulated, there is zero rate of mass introduction. Accordingly the source distribution has degenerated into a dipole distribution, and unless account is taken of this fact in our mathematics, silly answers will be obtained. The appropriate way of taking such account is as follows. We transform equation 5.18 *before* neglecting retarded times.

We first transform the equation mathematically, and then interpret the result physically. We know that

$$\int \frac{\partial}{\partial y_i} \left(\frac{1}{r} F_i\right) d\tau = \int \frac{1}{r} \mathbf{F} \cdot n \, ds = 0,$$

since $\mathbf{F}$ is zero outside the flow. (Here $r = |\mathbf{x} - \mathbf{y}|$). Thus

$$0 = \int \frac{1}{r} \frac{\partial F_i}{\partial y_i} \, d\tau + \int F_i \frac{\partial}{\partial y_i} \left(\frac{1}{r}\right) d\tau$$

$$= \int \frac{1}{r} \frac{\partial F_i}{\partial y_i} \, d\tau - \int F_i \frac{\partial}{\partial x_i} \left(\frac{1}{r}\right) d\tau,$$

since $\partial(1/r)/\partial x_i = -\partial(1/r)/\partial y_i$, and noting that F_i depends on $\mathbf{y}$ but not on $\mathbf{x}$ this yields

$$\int \frac{1}{r} \frac{\partial F_i}{\partial y_i} \, d\tau = \frac{\partial}{\partial x_i} \int \frac{1}{r} F_i \, d\tau.$$

Upon using this relationship, equation 5.18 may be written as

$$\rho(\mathbf{x}, t) - \rho_0 = -\frac{1}{4\pi a_0^2} \frac{\partial}{\partial x_i} \int \frac{F_i\left(\mathbf{y}, t - \frac{r}{a_0}\right)}{r} d\tau(\mathbf{y}). \qquad (5.19)$$

This result may be simply interpreted as follows. On the right-hand side of equation 5.17 the expression $-\partial F_i/\partial x_i$ has zero integrated instantaneous strength and so may be replaced by two almost cancelling source distributions (as explained in detail by Lighthill[1]) which together form a dipole distribution. Since equation 5.17 is a linear equation we may thus add these solutions and conclude that the radiated density fluctuations are equivalent to the sum of two almost cancelling radiations, which in the limit equals the derivative of the radiation from the appropriate single simple source distribution. This both explains why the derivative in equation 5.18 may effectively be taken outside the integral, as in equation 5.19 and illustrates that most of the apparent radiation cancels, i.e. that equation 5.19 yields a non-zero but *small* amount of radiated density fluctuations.

Having transformed from equation 5.18 to equation 5.19, retarded times may now be neglected, and indeed a non-zero answer obtained, provided $\int F_i \, d\tau$ is non-zero, that is provided the total applied force is non-zero. If the total applied force is perchance zero, i.e. if the total dipole strength is zero, then positive and negative dipoles must be balancing, and the applied dipole field degenerates to a quadrupole field. We now summarize the results to date.

5.1.6 *Summary of results for sources and dipoles*

The most efficient means by which sound can be generated is by forcing the mass within a fixed region to fluctuate, and this corresponds to the acoustic pole or source. It may conveniently be visualized as a small balloon, successively inflated and deflated, so that the mass of *fluid* within a fixed region surrounding the balloon will fluctuate, i.e. $\int \rho \, d\tau$ fluctuates.

A second, less efficient, method is when the mass of fluid contained within a fixed region does not vary, but the momentum of the fluid within the fixed region (or the mass flux across fixed surfaces) varies, i.e. $\int \rho v_i \, d\tau$ varies. Since the rate at which the momentum within a region changes is equal to the overall force exerted on the fluid therein, we expect this mechanism to be relevant when solid bodies are present.

The total mass within the region does not vary, which may crudely be said to mean that mass is entering the fixed region across part of the boundary and is leaving at the same rate across another part. Accordingly this mechanism is equivalent to a positive source (where mass enters the

5

region) and a sink (where it leaves), and these together constitute an acoustic dipole.

5.1.7 *Acoustic quadrupoles*

It is fairly clear that if we consider a fluctuating fluid flow, away from solids, there is no physical mechanism by which either the mass or the momentum in a fixed region can vary. But the momentum flux, which is the rate at which ρv_i, the momentum in the x_i direction, is being convected with velocity v_j in the x_j direction, can vary. Expressed symbolically, $\int \rho v_i v_j \, d\tau$ will vary. Intuitively, since the momentum ρv_i which is entering the fixed region must be balanced by momentum leaving the region across a different part of the boundary, it may be expected that the sound generation will be that due to two almost cancelling acoustic dipoles, that is by an acoustic quadrupole.

5.2 General theory of aerodynamic noise in an unbounded medium

5.2.1 *Lighthill's theory*

In his now classical paper, Lighthill[1] considers a fluctuating fluid flow which occupies a limited part of an unbounded medium of which the remainder is at rest. The exact equations governing the density fluctuations in the real fluid are compared with the (approximate) acoustic equations 5.1, 5.3, and 5.4, appropriate to a uniform medium at rest. The difference between the two sets of equations is treated as if it were an *externally applied* field, known if the flow is known. This would seem to be a particularly appropriate method of analyses for at least two reasons.

(i) The fraction of the energy of a flow which is radiated as sound will be found to be extremely small, so it is extremely unlikely that (in the absence of solids particularly) the radiated sound can have any significant back-reaction on the flow itself, which may accordingly be replaced by a fluctuating external force field in the manner of a forced oscillation.

(ii) The sound radiated is calculated as in a uniform medium at rest so all such effects as convection with the turbulence, propagation at variable speed within it, etc. are incorporated into the hypothetical external field, and no further account need be taken of them.

To put these ideas on a precise mathematical basis, we begin with the equations of fluid dynamics 5.1 and 5.2.

$$\frac{\partial \rho}{\partial t} + \frac{\partial}{\partial x_i} (\rho v_i) = 0,$$

$$\frac{\partial}{\partial t} (\rho v_i) + \frac{\partial}{\partial x_j} (\rho v_i v_j + p_{ij}) = 0.$$

In order to compare these exact equations with the approximate acoustic forms 5.1 and 5.4, we write the second of the equations in the alternative but equivalent form

$$\frac{\partial}{\partial t}(\rho v_i) + a_0^2 \frac{\partial \rho}{\partial x_i} = -\frac{\partial}{\partial x_j}(T_{ij}) \tag{5.20}$$

where

$$T_{ij} = \rho v_i v_j + p_{ij} - a_0^2 \rho \, \delta_{ij} \tag{5.21}$$

Elimination of ρv_i between equations 5.1 and 5.20 then yields

$$\frac{\partial^2 \rho}{\partial t^2} - a_0^2 \nabla^2 \rho = \frac{\partial^2 T_{ij}}{\partial x_i \partial x_j} \tag{5.22}$$

If we compare this equation with equation 5.14 or 5.18 we can see that the sound is generated exactly as in a uniform medium at rest under the action of fluctuating applied simple sources of strength $\partial^2 T_{ij}/\partial x_i \partial x_j$ or dipoles of strength $-\partial T_{ij}/\partial x_j$ per unit volume. Thus the total instantaneous dipole strength may be seen to be zero since it is equal to

$$-\int \frac{\partial T_{ij}}{\partial x_j} d\tau = -\int T_{ij} n_j \, dS,$$

and thus zero, since T_{ij} is zero outside of the flow itself. Accordingly the sound generation is that due to *cancelling* dipoles, i.e. fluctuating applied quadrupoles of strength T_{ij}. The only reason why this situation produces any sound at all is that the signals emanating from two dipoles of opposite strengths (which together form a quadrupole) were not emitted simultaneously if they reach the listener simultaneously, and so the cancelling between the two dipoles is not complete. It may be noted that this mitigation of the cancellation increases with frequency.

The fact that the sound is generated as by quadrupoles was to be expected on physical grounds. In view of the assumption of an unbounded flow field there are no solids through which extraneous mass can be blown or across which external forces can act, so the sound generators must be of a higher order, i.e. they may be quadrupole or even a higher order still. In fact, such detailed numerical calculations as have been performed suggest that the total quadrupole strength is not instantaneously zero, so it does not seem likely that sources of higher order than quadrupole need be considered.

We now consider the question as to how we may approximate or estimate the strength T_{ij}. We shall concentrate our attention upon the appropriate procedure at relatively low Mach numbers when T_{ij} may be split into three parts.

(1) $\rho v_i v_j$. This part of T_{ij} represents a convection of momentum, momentum ρv_i being convected with velocity v_j. At low Mach numbers we may write $\rho \approx \rho_0$ so that this part of the tensor T_{ij} may be written as $\rho_0 v_i v_j$.

(2) $p_{ij} - p\delta_{ij}$. This contribution represents viscous stresses, which are known to be very small indeed in comparison with $\rho_0 v_i v_j$, for which reason they may usually be neglected.

(3) $(p - a_0^2 \rho)\delta_{ij}$. These terms represent the effects of heat conduction, which is responsible for departures of $p - p_0$ from $a_0^2(\rho - \rho_0)$. Again this is usually very small since $\rho - \rho_0$ is of order $\rho_0 M^2$ within the flow and $p - p_0 = a_0^2(\rho - \rho_0) \{1 + O(M^2)\}$, from which we deduce that

$$p - a_0^2 \rho = p_0 - a_0^2 \rho_0 + O(M^4).$$

Ignoring the uniform part, we see that $p - a_0^2 \rho$ is smaller at low Mach numbers than $\rho_0 v_i v_j$.

Thus we may usually approximate T_{ij} by the form

$$T_{ij} \simeq \rho_0 v_i v_j, \tag{5.23}$$

so that the external applied field depends only upon the velocity field within the hydrodynamic flow.

It is, of course, perfectly true that we could have obtained this result quite easily by making appropriate approximations right from the beginning. It should be pointed out, however, that had we done this it would have been very difficult to be quite sure we had not neglected small (but non-zero) source or dipole fields. The moral really is that approximations are absolutely essential, but should be delayed for as long as possible.

5.2.2 *The radiated sound field*

There is no essential difficulty in formally obtaining the solution of equation 5.22. It is simply

$$\rho - \rho_0 = \frac{1}{4\pi a_0^2} \frac{\partial^2}{\partial x_i \partial x_j} \int T_{ij} \left(\mathbf{y}, t - \frac{r}{a_0}\right) \frac{d\tau(\mathbf{y})}{r} \tag{5.24}$$

In writing down this result we have really jumped one or two steps as follows:

The formal solution is obtained from equation 5.13 upon replacing Q' by $\partial^2(T_{ij})/\partial y_i \partial y_j$ and the result so obtained may be shown to be exactly equal to equation 5.24 in one of two ways. Mathematically, we may twice apply the divergence theorem (Gauss' theorem). Alternatively, in a physically more enlightening manner, we may note that the applied quadrupole field may be replaced by four almost cancelling simple source fields (this is really the physical interpretation of a second derivative). Since the equation for the density is linear we may add together the almost cancelling solutions for these four simple sources, thus obtaining the second derivative of the solution for a simple source. By either of these methods the formal solution, with a second derivative *under* the integral sign is converted into the form 5.24 with the second derivative *outside* of the integral sign.

This solution is perfectly general and exact. If we knew T_{ij} with sufficient accuracy it would be possible to compute the radiated density fluctuations at any specified point. In practice this is not possible, and the following simplifications are both possible and necessary. We note that

$$\frac{\partial}{\partial x_i} \int T_{ij} \left(\mathbf{y}, t - \frac{r}{a_0} \right) \frac{d\tau(\mathbf{y})}{r}$$

$$= \int \frac{\partial r}{\partial x_i} \cdot \frac{\partial}{\partial r} \left[\frac{T_{ij} \left(\mathbf{y}, t - \frac{r}{a_0} \right)}{r} \right] d\tau(\mathbf{y})$$

$$= \int \frac{x_i - y_i}{r} \left[-\frac{1}{a_0 r} \frac{\partial T_{ij}}{\partial t} - \frac{1}{r^2} T_{ij} \right] d\tau(\mathbf{y}).$$

Thus

$$\frac{\partial^2}{\partial x_i \partial x_j} \int T_{ij} \left(\mathbf{y}, t - \frac{r}{a_0} \right) \frac{d\tau(\mathbf{y})}{r} = \int \frac{(x_i - y_i)(x_j - y_j)}{r^2}$$

$$\left[\frac{1}{a_0^2 r} \frac{\partial^2 T_{ij}}{\partial t^2} + \frac{2}{a_0 r^2} \frac{\partial T_{ij}}{\partial t} + \frac{2}{r^3} T_{ij} \right] d\tau(\mathbf{y}).$$

Now the relative magnitudes of the three terms in square brackets are

$$\frac{(2\pi n)^2 T_{ij}}{a_0^2 r} : \frac{4\pi n T_{ij}}{a_0 r^2} : \frac{2T_{ij}}{r^3},$$

where n is a typical frequency, and since $a_0 = \lambda n$, where λ = wavelength, this becomes $2\pi^2 : 2\pi\lambda/r : \lambda^2/r^2$, and if $r > \lambda$, the first term dominates. This condition will be satisfied at all points, far enough from the flow to be in the radiation field of each quadrupole, and in such a case equation 5.24 becomes

$$\rho - \rho_0 = \frac{1}{4\pi a_0^4} \int \frac{(x_i - y_i)(x_j - y_j)}{r^3} \frac{\partial^2 T_{ij}}{\partial t^2} \left[\mathbf{y}, t - \frac{r}{a_0} \right] d\tau(\mathbf{y}). \qquad (5.25)$$

The essential quadrupole nature of the sound generation is clearly seen in the fact that it is the second time derivative of T_{ij} which appears on the right-hand side of this equation.

Further, at points sufficiently far from the flow to be not only in the radiation field of each quadrupole, but also in the radiation field of the flow as a whole, that is at distances large compared with the dimensions of the flow, we may approximate equation 5.25 as follows. Provided the origin is within the flow, it follows that

$$\rho - \rho_0 \simeq \frac{1}{4\pi a_0^4} \frac{x_i x_j}{x^3} \int \frac{\partial^2 T_{ij}}{\partial t^2} \left[\mathbf{y}, t - \frac{r}{a_0} \right] d\tau(\mathbf{y}). \qquad (5.26)$$

In order to derive the local intensity of the radiated sound, we note that the intensity I is defined by

$$I = \frac{a_0^3}{\rho_0} \overline{(\rho - \rho_0)^2}, \qquad (5.27)$$

so we formally write down equation 5.26 twice (using different suffices) and take the mean value of the product. This yields

$$I = \frac{1}{16\pi^2 \rho_0 a_0^5} \frac{x_i x_j x_k x_l}{x^6} \int \int \overline{\frac{\partial^2 T_{ij}}{\partial t^2} \left[\mathbf{y}, t - \frac{|\mathbf{x} - \mathbf{y}|}{a_0} \right] \frac{\partial^2 T_{kl}}{\partial t^2} \left[\mathbf{z}, t - \frac{|\mathbf{x} - \mathbf{z}|}{a_0} \right]}$$
$$d\tau(\mathbf{y}) d\tau(\mathbf{z}). \qquad (5.28)$$

The total acoustic power output is obtained by integrating the intensity over the surface of a large sphere.

5.2.3 *Dimensional analysis of overall sound radiation*

We now consider geometrically similar mechanisms of flow generation, and examine the dependence of the sound field upon the parameters of the flow.

Within the flow itself, the value of T_{ij} will be proportional to $\rho_0 U^2$ where U is a typical velocity, together with a small dependence upon such parameters as Reynolds number and Mach number.

Accordingly the magnitude of the second time derivative of T_{ij} will be mainly proportional to $\rho_0 U^2 n^2$ where n is a typical frequency of the fluctuations in the flow. It follows from equation 5.26 that the density fluctuations are mainly given by

$$\rho - \rho_0 \propto \frac{1}{a_0^4} \frac{1}{x} \rho_0 U^2 n^2 l^3, \qquad (5.29)$$

where l^3 is a typical volume and l a typical length of the flow.

Now in a fluctuating fluid flow it is usually found that the non-dimensional frequency nl/U (usually referred to as the Strouhal number) varies with changing flow conditions far less than the frequency n itself. As an example of this fact, we may remark that for Reynolds numbers between 80 and 40,000 the frequency of eddy shedding behind a wire of diameter l in a stream of speed U is given to one significant figure by

$$n = 0.2U/l. \qquad (5.30)$$

We accordingly remark that in general n is mainly proportional to U/l so that equation 5.29 becomes

$$\rho - \rho_0 \propto \rho_0 \frac{l}{x} \left(\frac{U}{a_0}\right)^4. \tag{5.31}$$

We note that the radiated density fluctuations are proportional to the fourth power of the velocity, in contrast to the dependence on the second power of velocity within the flow itself. The additional factor of velocity-squared arises, of course, because of the quadrupole nature of the sound generation.

Since the intensity is equal to

$$I = \frac{a_0^3}{\rho_0} \overline{(\rho - \rho_0)^2},$$

we deduce that

$$I \propto \rho_0 U^8 a_0^{-5} \left(\frac{l}{x}\right)^2, \tag{5.32}$$

and the total acoustic power output, P, obtained by integration over a large sphere (of radius x) is

$$\rho \propto \rho_0 U^8 a_0^{-5} l^2. \tag{5.33}$$

Finally, the energy per unit volume is proportional to $\rho_0 U^2$, and the total rate at which energy must be supplied to maintain a steady flow is proportional to $(\rho_0 U^2)(Ul^2)$. Accordingly the acoustic efficiency defined as the ratio (acoustic power radiated)/(energy supplied) is

$$\eta \propto M^5. \tag{5.34}$$

For a turbulent jet it is found experimentally that the constant of proportionality is of order 10^{-4}, so turbulence at low Mach numbers is a singularly inefficient generator of sound.

5.2.4 *Analyses with respect to a moving frame of reference*

There are numerous situations in which the various terms which arise in the above theory can be calculated or estimated more readily if the analysis is carried out in a frame of reference which is not at rest. Without, at this stage, specifying the reasons for this we shall make the appropriate modification to the theory.

Suppose we take a fixed origin, with respect to which the listener has co-ordinates, $\mathbf{x}$, and the volume element has co-ordinates $\mathbf{y}$, and a moving origin with respect to which the volume element has co-ordinates $\boldsymbol{\eta}$, the origin moving with velocity $a_0\mathbf{M}$. Then in the new co-ordinates, T_{ij} represents momentum flux across *moving* surfaces, fixed with respect to the new origin. If the axes are so chosen that the origins coincide at time t (which is when the sound reaches the listener), then the appropriate value of T_{ij} in equations 5.24 must be

$$T_{ij}\left(\boldsymbol{\eta}, t - \frac{|\mathbf{x} - \mathbf{y}|}{a_0}\right) \quad \text{with} \quad \boldsymbol{\eta} = \mathbf{y} + M|\mathbf{x} - \mathbf{y}|. \qquad (5.35)$$

The reasoning is that if $\mathbf{y}$ is the position of the emitting quadrupole with respect to fixed axes, and the origin moves on a distance $M|\mathbf{x} - \mathbf{y}|$ during the time taken for the signal to reach the listener, then at the time of emission of the signal the position in the moving co-ordinates must be given by adding on this amount. Further, because of this kind of distribution, the volume element in equation 5.24 is altered, and we state without proof that

$$d\tau(\boldsymbol{\eta}) = d\tau(\mathbf{y})\left\{1 - \frac{M \cdot (\mathbf{x} - \mathbf{y})}{|\mathbf{x} - \mathbf{y}|}\right\}. \qquad (5.36)$$

Then equation 5.24 becomes

$$\rho - \rho_0 = \frac{1}{4\pi a_0^2} \frac{\partial^2}{\partial x_i \partial x_j} \int T_{ij}\left(\boldsymbol{\eta}, t - \frac{|\mathbf{x} - \mathbf{y}|}{a_0}\right) \frac{d\tau(\boldsymbol{\eta})}{|\mathbf{x} - \mathbf{y}| - M \cdot (\mathbf{x} - \mathbf{y})}. \qquad (5.37)$$

The crux of the matter is really that in the new frame of reference the variations of T_{ij} with time will be altered, since $\boldsymbol{\eta}$ changes with time for a fixed position in space.

As before we may make certain simplifications when we consider the sound at positions not too near the flow. Thus if $r > \lambda$, then we may apply the differentiation to T_{ij} alone. To do this requires a knowledge of $\partial(|\mathbf{x} - \mathbf{y}|)/\partial x_i$ at constant $\boldsymbol{\eta}$. Again without proof we will remark that this may be shown to be

$$\frac{\partial}{\partial x_i}|\mathbf{x} - \mathbf{y}| = \frac{x_i - y_i}{|\mathbf{x} - \mathbf{y}| - M \cdot (\mathbf{x} - \mathbf{y})},$$

so that equation 5.37 yields

$$\rho - \rho_0 = \frac{1}{4\pi a_0^4} \int \frac{(x_i - y_i)(x_j - y_j)}{\{|\mathbf{x} - \mathbf{y}| - M \cdot (\mathbf{x} - \mathbf{y})\}^3} \frac{\partial^2 T_{ij}}{\partial t^2}\left[\boldsymbol{\eta}, t - \frac{|\mathbf{x} - \mathbf{y}|}{a_0}\right] d\tau(\boldsymbol{\eta}). \tag{5.38}$$

We may compare this equation for radiated density fluctuations due to fluctuations in momentum flux across moving surfaces with that due to fluctuations in momentum flux across fixed surfaces. We note that the sound radiated at an angle θ is amplified by a factor $(1 - M \cos \theta)^{-3}$ so that more sound will be emitted forwards and less sound emitted backwards. There are two distinct physical factors working to produce this result. First of all, we recall that sound radiation from a distribution of quadrupoles arises only because the signals reaching the hearer from different parts of a quadrupole were not emitted simultaneously, so that the contributions do not exactly cancel even though the total source strength of a quadrupole is zero. Now for sound radiated in a forward direction, clearly the rear part of the quadrupole must emit first, since it is farthest from the listener. If the quadrupole were at rest the forward part would emit an instant later, but because of the forward movement of the quadrupole it has in fact moved slightly nearer to the listener during this time, and emission of its signal must be *further* delayed. It follows that the effective volume of the quadrupoles will be increased for forward emission of sound, and this produces one factor $(1 - M \cos \theta)^{-1}$. Equally, since the cancelling between signals is mitigated by this increase of the time lag, two further factors of $(1 - M \cos \theta)^{-1}$ are introduced, which correspond to the double differentiation.

We note that the additional sound radiated forwards is greater than the reduction in the sound radiated backwards, so the total acoustic power output will increase.

5.2.5 *Neglect of retarded times*

It will probably be fairly clear to the reader that the considerable complexities of the problem of turbulence, and of sound generated thereby, are such that there is little or no possibility of progress unless the differences in retarded times may be ignored. It has already been made abundantly clear that if such an assumption is made at too early a stage, completely spurious answers may be obtained. The question obviously arises therefore as to the circumstances in which retarded times may be neglected. The answer is as follows:

When dealing with a distribution of simple sources, provided the total source strength is non-zero, the distribution of sources may usually be replaced by a single point source, and retarded times neglected. However, if the total source strength is zero, then the simple source distribution in effect degenerates into a dipole distribution. In such circumstances, retarded times cannot be neglected whilst the equations are expressed in a form appropriate to simple sources, otherwise the nugatory answer of zero radiated sound will be obtained. However, if the equations are first expressed in a form appropriate to dipoles, then retarded times may formally be neglected.

If, again, the total dipole strength happens to be zero as well, then the various equations must be expressed in terms appropriate to quadrupoles before retarded times are neglected.

The above represents an essential condition for neglect of retarded times, but does not necessarily represent a sufficient condition for such neglect.

Clearly, even if we consider a distribution (of whatever type) having everywhere the same sign, so that there would be no possibility at all of the overall strength being zero, then radiated times cannot be neglected if the distribution covers too large a volume. To see how large a volume is permissible, we note that the maximum possible variations in retarded time will be of order $\pm l/2a_0$ where l is the maximum dimension of the flow. If $2\pi n(l/2a_0)$ is less than or equal to $\frac{1}{2}\pi$ say, i.e. $nl/a_0 \leqslant \frac{1}{2}$, then retarded times are not likely to be important. Since nl is typical of a velocity or velocity fluctuation in the flow, this condition usually holds at subsonic speeds.*

If the time scale of the turbulence is so small (and the frequency so large) that $l/2a_0$ is greater than the time scale then we seek to analyse the turbulence in a moving frame of reference in which time scales are maximized. This results in the addition of powers of $(1 - M\cos\theta)^{-1}$ in the formulae, but makes them more amenable to rough estimation. The reason why this can be done is that typically a pattern of turbulence is convected rather quickly whilst being distorted rather slowly. In an analysis in a fixed frame of reference the convection effect causes a spurious appearance of rapid changes in the turbulence, whereas in a frame of reference moving with the turbulence it is the distortion alone which appears. It is this latter effect which we wish to isolate anyway, since the convection part of the change is really octupole in nature, and does not generate sound with quadrupole efficiency.

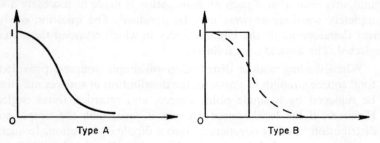

Type A   Type B

5.2.6 *Correlation volumes*

It is well known that acoustic sources with well-correlated strengths produce linearly-combining pressure fluctuations, whereas with uncorrelated sources it is only the mean square pressure fluctuations which add. Now it is an essential characteristic of turbulence that turbulent fluctuations are well correlated at near points but uncorrelated at sufficiently great separations. For this reason Lighthill suggested the division of a

*For a full discussion on the importance of retarded times when the flow is supersonic, reference should be made to the paper by Williams[5].

turbulent flow into a large number of eddy-volumes, such that quadrupole strengths at points within one volume are well-correlated, but strengths in different volumes are uncorrelated. Thus a correlation curve of the type A is approximated by one of the type B. One consequence of this idea is that in the condition $nl/a_0 \leqslant \frac{1}{2}$, for neglect of retarded times, the relevant value of l is a typical correlation radius, and not the (much larger) dimension of the flow.

Detailed calculations support this crude approximation, except when a region of positive correlation is surrounded by a region of negative correlation whose sum is zero—in effect the quadrupoles then become octupoles.

5.3 Summary of numerical calculations for special cases

As far as the problem of jet noise is concerned, the only quantitative results which have been derived have been for the two special cases of (a) isotropic turbulence with no mean flow, and (b) turbulence in the presence of a very large mean shear. These we shall now discuss briefly, both without quantitative detail other than quoting the answers obtained.

5.3.1 *Noise radiation from isotropic turbulence*

This problem was considered by Proudman[3], who investigated the noise radiated from a volume of isotropic turbulence (with zero mean flow) embedded in a medium otherwise uniform and at rest. The method used by Proudman starts from equation 5.28. When it is recalled that the stress tensor T_{ij} may be approximated by $\rho_0 v_i v_j$, it is seen that the integrand is a mean value involving fourth powers of velocity components. As nothing is known experimentally about this fourth-order velocity covariance, Proudman used an idea of Batchelor[4] in which the fourth-order covariance is expressed approximately in terms of the well-known two-point velocity-correlation function on the assumption that the velocities and their derivatives have a normal joint-probability distribution.

By using this idea several times, and making the appropriate simplifications because of the isotropy of the turbulence, Proudman shows that the total power radiated per unit volume of turbulence may be written as

$$P = \alpha \epsilon \frac{(\overline{u^2})^{5/2}}{a_0^5}, \tag{5.39}$$

where $\overline{u^2}$ is the mean square of a single component of the velocity fluctuation, ϵ is the rate of dissipation of energy per unit mass, and α is a constant, expressible as a rather complicated integral involving the two-point velocity-correlation function. Proudman estimated that α would be about 38 or so.

5.3.2 *The amplifying effect of a large shear*

It is reasonably clear that when there is a large mean velocity, say $\bar{v}_j$, there can be considerable variations in momentum flux, since there is a large mean momentum being shaken about by turbulent velocity fluctuations, and a large mean velocity to transport turbulent momentum fluctuations. In other words $\rho v_i v_j$ can fluctuate widely if fluctuations in ρv_i are amplified by a large mean value $\bar{v}_j$. *However*, this does not hold when $\bar{v}_j$ is uniform, for then

$$\frac{\partial}{\partial t} (\rho v_i \bar{v}_j) = \bar{v}_j \frac{\partial}{\partial t} (\rho v_i)$$

$$= - \bar{v}_j \frac{\partial}{\partial x_j} (\rho v_i v_j + p_{ij}),$$

which is an exact space derivative and so the quadrupoles degenerate into octupoles. Accordingly, subtracting out the uniform mean part, we conclude that maximum fluctuations in momentum flux can occur in the presence of a large mean shear, which makes $\bar{v}_j - \bar{v}_j(\mathbf{0})$ large.

More precisely, we examine $\partial(\rho v_i v_j)/\partial t$ in some detail. We know that

$$\frac{\partial}{\partial t} (\rho v_i v_j) \equiv v_i \frac{\partial}{\partial t} (\rho v_j) + v_j \frac{\partial}{\partial t} (\rho v_i) - v_i v_j \frac{\partial \rho}{\partial t},$$

and upon substituting for the derivatives from equations 5.1 and 5.2 it requires only a straightforward rearrangement of terms to show that

$$\frac{\partial}{\partial t} (\rho v_i v_j) = p_{ik} \frac{\partial v_j}{\partial x_k} + p_{jk} \frac{\partial v_i}{\partial x_k} - \frac{\partial}{\partial x_k} (\rho v_i v_j v_k + p_{ik} v_j + p_{jk} v_i).$$

$$(5.40)$$

The last term, being a space derivative, may safely be ignored, and the viscous contributions to the stresses p_{ik}, p_{jk}, are also neglected. Thus equation 5.40 becomes

$$\frac{\partial}{\partial t} (\rho v_i v_j) \simeq p \left(\frac{\partial v_i}{\partial x_j} + \frac{\partial v_j}{\partial x_i} \right) = p e_{ij}, \qquad (5.41)$$

which is the product of a pressure and a (large) rate of strain. Then, finally, we have

$$\frac{\partial^2}{\partial t^2} (\rho v_i v_j) \simeq \frac{\partial^2 T_{ij}}{\partial t^2} = p \frac{\partial}{\partial t} (e_{ij}) + \frac{\partial p}{\partial t} e_{ij}$$

$$\simeq e_{ij} \frac{\partial p}{\partial t} \quad \text{(if } e_{ij} \text{ is large enough)}$$

$$\simeq \bar{e}_{ij} \frac{\partial p}{\partial t}. \qquad (5.42)$$

The expression 5.42 for $\partial^2 T_{ij}/\partial t^2$ may be substituted into equation 5.28 and upon using the eddy-volume concept it is found that the intensity radiated per unit volume of fluid (in the presence of large mean shear) is

$$I \simeq \frac{x_i x_j x_k x_l}{16\pi^2 \rho_0 a_0^5 x^6} \int \overline{e_{ij}(\mathbf{o}) e_{kl}(\mathbf{z})} \; \overline{\frac{\partial p(\mathbf{o}, t)}{\partial t} \frac{\partial p(\mathbf{z}, t)}{\partial t}} \; d\tau(\mathbf{z}). \tag{5.43}$$

This expression, and the covariance of $\partial p/\partial t$ in particular, have been studied by Lilley[6]. Lilley sets up an equation relating pressure to velocity, and so obtains the approximate result that

$$\int \overline{\frac{\partial p(\mathbf{o}, t)}{\partial t} \frac{\partial p(\mathbf{z}, t)}{\partial t}} \; d\tau(\mathbf{z}) \simeq 0.6 \rho_0^2 \, \overline{l_{12}}^4 \, \overline{u^2} \int_0^\infty r^4 f(r) dr,$$

where $f(r)$ is the two-point velocity correlation function. Lilley's final result is that the intensity per unit volume is

$$I \simeq 0.02 \, \frac{\rho_0 \overline{l_{12}}^6 e_{11}^5 \overline{u^2} \sin^2 \theta \cos^2 \theta}{a_0^5 x^2} \tag{5.44}$$

where $l_{11} = \int_0^\infty f(r) \, dr$ is the longitudinal integral scale of the turbulence.

5.4 Comparisons between experiment and theory

5.4.1 *Principal experimental results on jet noise*

A considerable number of experiments have been done relating to various aspects of the noise radiated from turbulent jets. For simplicity, we shall here restrict our attention to the case of the cold subsonic jet, for which the following well corroborated results have been obtained:

(i) The acoustic power output of a turbulent jet varies as a high power of the jet velocity, usually near to the eighth power.

(ii) Almost all of the sound is radiated forwards, so much so that measurements behind the orifice are liable to error, because the relatively low intensity in that region can be swamped by reflections of the sound (of greater intensity) radiated forwards.

(iii) The higher frequency sound is emitted mainly from the region near to the orifice, where there is a mixing region of intense shear. The direction of maximum intensity is at an angle of 45° to the jet axis or slightly less.

(iv) The lower frequency sound has a directional maximum at a smaller angle to the jet axis; the angle decreases as either the frequency is reduced or the Mach number increases. This lower frequency sound emanates mainly from a distance 5–10 diameters downstream of the orifice.

It is necessary that all of these experimental findings should be satisfactorily explained by a theory before it can be regarded as acceptable.
This we shall now consider.

5.4.2 *Theoretical discussion of results*

We begin with result (i) relating to the acoustic power radiated by a
turbulent jet. As far as overall effects are concerned, we have already explained this result, for, by means of the dimensional analyses given earlier,
we have shown that the sound intensity is proportional to the eighth power
of a typical velocity of a flow. It is possible however to go further than this,
at least with regard to the mixing region of the jet.

We deduce from equation 5.28 using the concept of eddy-volumes, that
the sound radiated per unit volume of the flow will be of order $\overline{T^2}n^4V$
$\rho_0^{-1}a_0^{-5}$, where $\overline{T^2}$ is a mean square fluctuation of T_{ij}, so $\overline{T^2} \propto \rho_0^2 U^4$, n is a
typical frequency and V is an eddy volume. Now experiments by Laurence
indicate that over the first four diameters downstream of the orifice, before
the mixing region reaches the centre of the jet, the width of the mixing
region is about $0.24x$, where $x = $ distance downstream. We may deduce
that a typical frequency n is proportional to Ux^{-1} and an eddy volume to
x^3. It follows that the intensity per unit volume is proportional to
$\rho_0 U^8 a_0^{-5} x^{-1}$. But if we now re-write this result in terms of the intensity per
unit *slice* of the jet, then, remembering that the volume per unit slice will
be proportional to xd, it follows that the intensity per unit slice is proportional to $\rho_0 U^8 a_0^{-5} d$, which is constant.

It may likewise be shown that in the region far downstream of the orifice, where re-adjustment following the spread of the mixing region has
been completed, and similar profiles exist, the intensity is proportional to
$x^{-6}l_{11}^{-1}$. It is clear therefore that the intensity per unit slice of jet very
quickly falls to zero in this region.

We now turn to the second of the experimental findings, that almost all
the sound is radiated forwards. It must be remembered, in seeking to
explain this result, that it refers to experiments in which the Mach number
was far from negligible, since it was necessary to have a Mach number of
at least 0.3 or so in order to generate sufficient sound to be measured
accurately.

Because the Mach number is not small it is necessary to make allowance
for the effect of the convection of the quadrupoles. We recall that the
sound radiated from a single quadrupole is $(1 - M \cos \theta)^{-6}$ times what it
would have been in the absence of convection, but, since we have already
seen that the effective volume of a single quadrupole must be multiplied by
a factor $(1 - M \cos \theta)^{-1}$, it follows that the number of quadrupoles in a
given physical volume will be multiplied by $(1 - M \cos \theta)$. Another way of

expressing this fact is to say that contributions from only a limited volume can reach a given point simultaneously. It follows that the sound emitted per unit volume must be multiplied by a factor $(1 - \mathrm{M} \cos \theta)^{-5}$ and this factor leads to much more sound being emitted forwards and much less backwards.

We remark again at this point, that owing to the desirability of neglecting retarded times in all numerical calculations, it is essential to analyse the turbulence in a moving frame of reference if the Mach number of the jet is not small. The idea is that a varying pattern of turbulence is being convected downstream, and if we analyse in a frame moving with the turbulence, the time scale will be maximized, and the neglect of retarded times more readily justified. For a mixing region the convection speed is usually taken to be about midway between the speeds of the two streams.

As an example we note that the sound radiated forward is increased by a factor $(1 - M)^{-5}$ which is equal to 15 dB when the Mach number of convection is 0.5, i.e. the jet Mach number is about 1. At the same time, the sound radiated backwards is multiplied by a factor $(1 - M)^{-5}$ which is equal to a decrease of about 6 or 7 dB in the same circumstances.

Turning now to the higher frequency sound, which comes mainly from the mixing region, we have already noted that sound from the mixing region has a typical frequency of order Ux^{-1} (which is high when x is small), and that it has a maximum at 45° to the jet axis. The effect of convection at around half the speed of the jet reduces the angle of maximum emission a little, in agreement with the experimental results.

Turning now to the lower frequency sound, it is clear that this comes mainly from the adjustment region, where the turbulence will be (very roughly) isotropic. If in expression 5.28 we replace each velocity component by a mean value plus a turbulent fluctuation, then the product will involve terms representing interactions between the turbulence and the mean shear, and terms representing an interaction of the turbulence with itself. For approximately isotropic turbulence, and in the absence of convection effects, this 'self-noise' will be roughly non-directional. Equally, the sound generated because of the turbulence interacting with the mean shear will be rather less in this region, since the mean shear is much less than in the mixing region; nevertheless it will have a maximum at 45° to the axis in the absence of convection. However, the relative equality of the contributions from these two types of interaction will ensure that the overall peak will be at an angle much less than 45°. As the Mach number increases the noise radiated forwards will increase, and the direction of peak emission will be reduced further. Similarly, a reduction in frequency will lead to a further mitigation of cancelling in the case of forward emission, thus having the same basic effect (on directionality) as increases in the Mach number.

One final point must be noted, namely that there appears at first sight to be some inconsistency in the explanations given above. On the one hand we have apparently successfully explained the dependence of overall sound intensity upon velocity without introducing the idea of convection, whereas the convection effect was essential in explaining the directional distribution of the sound. In view of the fact that the effect of convection is to increase the overall sound generation, it would appear that when convection is accounted for, the theory would predict an overall sound intensity increasing with a higher power of the jet velocity than the eighth. In fact, it may readily be shown that given an otherwise uniform distribution of radiation, modified by convection of the quadrupoles at a Mach number M the acoustic power output would be increased by a factor $(1 + M^2)$ $(1 - M^2)^{-4}$. Now for jet Mach numbers between about 0.5 and 1, i.e. for values of M between 0.25 and 0.5, this factor varies roughly like M^2 or U^2. This then increases the eighth power to the tenth power, which does not agree with the experiment. The answer to this dilemma appears to be that there is in addition to this effect a second effect which is not adequately accounted for in the above analyses, namely that the intensity of the turbulence appears to decrease as the velocity increases. Indeed, experimental evidence suggests that $(u^2)^{1/2}U^{-1}$ decreases with increasing Mach number, associated with a decreased spreading of the mixing regions, and these effects are sufficient to account for a reduction of the basic law from the eighth to the sixth power of the velocity. The convection effect, velocity squared roughly, thus restores the dependence on the eighth power of velocity.

This explanation is supported by observations of sound radiated at right angles to the jet axis where no convection effects are present, and the sound is found to be proportional to the sixth power of the velocity. This steepens to the ninth power as the angle falls from 90° to about 20° and the integrated effect is roughly the eighth power.

5.5 Effects of solid boundaries

5.5.1 *General theory*

We have already given a general solution of equation 5.22 for the case of an unbounded flow. It is easy to see physically that there are two possible modifications when solid boundaries are present.

(i) The sound generated by the quadrupoles of Lighthill's theory will be reflected and diffracted at the solid boundaries.

(ii) The quadrupoles will be distributed, not over all space, but only over the region external to the solid boundaries. Thus it is possible that there could be a resultant distribution of dipoles (or sources) at the boundaries. All this is borne out in detail by the exact solution given by Curle[7].

The most general solution of equation 5.22 consists of a volume integral

$$\frac{1}{4\pi a_0^2} \int_V \frac{\partial^2 T_{ij}}{\partial y_i \partial y_j} \frac{d\tau(\mathbf{y})}{r}$$

as in Lighthill's theory, taken now over the volume occupied by the fluid, together with a surface integral. After appropriate (but exact) manipulations the solution may be put in the form

$$4\pi a_0^2(\rho - \rho_0) = \frac{\partial^2}{\partial x_i \partial x_j} \int_V \frac{T_{ij}}{r} d\tau(\mathbf{y}) - \frac{\partial}{\partial x_i} \int_S \frac{P_i - \rho v_i v_n}{r} d\tau S(\mathbf{y}) - \int_S \frac{\partial}{\partial t} (\rho v_n)$$

$$\frac{d\tau(\mathbf{y})}{r} \qquad (5.45)$$

where P_i represents the force per unit area exerted on the fluid in the x_i direction at the solid boundaries, and v_n is the normal velocity of the solid surface (measured towards the solid).

The four terms on the right-hand side of this equation are easily interpreted physically.

(i) The term involving T_{ij} represents the quadrupoles.

(ii) The term P_i appears in a form appropriate to dipoles, and represents the force which is (physically) exerted on the fluid.

(iii) The term $\rho v_i v_n$, which is zero when the solids are at rest, is again of dipole type. It represents the fact that when the boundary moves normal to itself momentum is imparted to the nearby fluid.

(iv) The term $\partial(\rho v_n)/\partial t$ is of a simple source type. It represents the fact that when a solid body moves, fluid is actually moved out of a region which it previously occupied, if the normal velocity of the solid is towards the fluid, or into a region previously occupied by the solid if the normal velocity has the opposite sign.

We note that the density fluctuations, as given by the theory, are radiated into a uniform medium at rest. All such effects as reflection and diffraction at solid boundaries are completely accounted for in the applied fields.

5.5.2 Boundary layer noise

We conclude this chapter by referring to the case of sound radiated from a turbulent boundary layer on a rigid flat plate. Since $v_n \equiv 0$ only the quadrupoles and the P_i dipole term remain. Now just as the intensity of radiation from a quadrupole distribution varies with velocity U like U^8, so may the radiation from a dipole distribution be shown to vary like U^6. Hence if the Mach number is small enough the dipoles will overwhelm the quadrupoles.

It has been found that the mean square pressure fluctuations in a turbulent boundary layer are much greater (by a factor of 30–40) than the mean square skin-friction fluctuation. Thus only the normal pressure fluctuations are retained in P_i and we write

$$P_i = p\delta_{i2}.$$

Equation 5.45 may then be shown to simplify to

$$p - p_0 = -\frac{1}{4\pi a_0^3}\frac{x_i}{x^2}\frac{\partial}{\partial t}\int_S \frac{\partial p}{\partial t}\left(\mathbf{y}, t - \frac{r}{a_0}\right) dS(\mathbf{y}) \qquad (5.46)$$

at large distances from the flow, and the intensity per unit area becomes

$$I = \frac{1}{16\pi^2 a_0^3 \rho_0}\frac{x_2^2}{x^4}\int \overline{\frac{\partial p}{\partial t}(\mathbf{o})\frac{\partial p}{\partial t}(\mathbf{z})}\, dS(\mathbf{z}). \qquad (5.47)$$

This equation will be discussed, in the light of the experimental results, in chapter 8.

References

1. M. J. Lighthill. *Proc. Roy. Soc.*, A. **211**, 564 (1952).
2. M. J. Lighthill. *Proc. Roy. Soc.*, A. **222**, 1 (1954).
3. I. Proudman. *Proc. Roy. Soc.*, A. **214**, 119 (1952).
4. G. K. Batchelor. *Proc. Camb. Phil. Soc.*, 47, 359 (1951).
5. J. E. F. Williams. *Phil. Trans. Roy. Soc.*, A. **255**, 469 (1963).
6. G. M. Lilley. *A.R.C. Rep.* **20376** (1958).
7. N. Curle. *Proc. Roy. Soc.*, A. **231**, 505 (1955).

CHAPTER 6

The Subjective Assessment of Aircraft Noise

6.0 Introduction

The derivation of a three decibel reduction in the noise output from a jet engine constitutes a significant achievement in reducing acoustic energy. Unfortunately, such a reduction is only just noticeable to the ears. It follows therefore not only that noise reduction of a much higher order is needed, but also that large subjective gains can be obtained from a real understanding of the reaction of the ear to noise and by concentrating on suppressing that part of the noise which is causing the adverse reaction. To understand the subjective laws of hearing it is first of all necessary to describe briefly the functioning of the ear.

6.1 The ear

The ear, drawn diagrammatically in Fig. 6.1, consists of the ear canal or

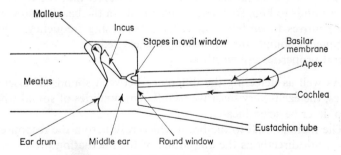

Fig. 6.1   Block diagram of the ear.

meatus which is terminated by the ear drum (tympanic membrane). In order to transfer as much energy as possible from the air pressure fluctuations striking the ear drum to the inner ear (cochlea), a mechanical transformer is necessary. The ossicular chain (malleus, incus and stapes) fulfils

135

this purpose and the footplate of the stapes transforms the pressure on the eardrum into a 22-fold greater pressure acting in the fluid of the inner ear. This causes the basilar membrane to be forced into vibration (Fig. 6.2) with the consequent excitation of the complex structure of the auditory nerve endings contained in the organ of the corti.

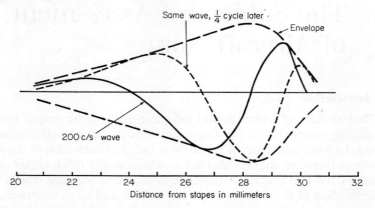

Fig. 6.2   Form of vibration of cochlear partition for 200 c/s tone (after Békèsy[1]).

The structure and damping of the basilar membrane are such that different frequency sounds excite different nerve cell areas along the membrane. These nerve cell excitations are complex and cause neural signals to transmit information to the brain, the details of which can be found elsewhere[1, 2, 3, 4]. It is sufficient to realize, however, that the basilar membrane responds to high frequencies maximally in the basal region, and this response moves towards the apex with decreasing frequency, where for low frequency sounds the whole length is excited.

Several properties are worth noting:

(1) As well as detecting the difference between sounds of discrete frequencies, the ear is very sensitive to different qualities of sound which are hard to describe scientifically.

(2) The main stimuli to the brain which result from the response of the ear vary, not directly as the amplitude of the fluctuating pressure, but as the fluctuating pressure raised to a relatively low order, i.e. almost as the logarithm of the pressure fluctuations. Thus, it has become the practice to quote sound pressures in terms of the logarithm of the root mean square pressure fluctuations, rather than to the r.m.s. pressure itself. Formally, Sound Pressure Level $= 20 \log_{10} \bar{p}/p_0$, where $\bar{p}$ is the r.m.s. pressure fluctuation and p_0 is usually 0.0002 dynes per square centimetre.

(3) Well away from a sound source, the pressure and particle velocity are in phase and proportional to each other. The Sound Intensity or Energy therefore differs only from Sound Pressure Level by virtue of using a slightly non-equivalent reference level. In the same way that Sound Energy can be the energy over the whole audible range, in an octave, third octave, fraction of an octave, or within a finite number of cycles, so the Sound Pressure Level can be quoted as that pressure level within the whole or just part of the audible frequency range.

(4) A sound at one frequency will mask a lower amplitude sound at a different frequency. Although a low frequency sound can be masked under certain conditions by a higher frequency, in general low frequency sounds are the more efficient maskers of high frequency sounds.

(5) The signal reaching the brain is, in some way, related to the duration as well as to the amplitude of the exciting signal. The process of assessing the subjective reaction to continuous noise therefore changes as the duration becomes smaller, and for short duration impulses such as sonic booms the ear is thought to integrate the energy over a short period of time and then form a judgement.

(6) Excessive exposure to sound can cause damage. For example, the normal ear drum will rupture at levels of about 185 dB, pain is felt deep in the ear at 130 dB, and above levels of 90 dB cumulative permanent damage to hearing can arise.

(7) Reaction to noise depends on the central nervous system coupled with the degree of interpretation needed. Among people judgements of loudness are reasonably uniform but judgements of annoyance are very varied.

Subjective units

It is obviously impossible to develop units of measurement which will indicate completely the true subjective response of the ear, since it would be necessary to have the correct weighting for different frequencies, and allowances made for masking and for duration. On the other hand, it would be very misleading to use sound intensity without some form of weighting of the noise energy in the different frequency ranges.

Several methods are available: (a) by using an electronically weighted overall sound level meter which accentuates the energy differently in the different frequency ranges; (b) by establishing a series of equal loudness curves or equal noisiness curves as indicated by the reaction of 'juries' of listeners, and computing the total loudness or noisiness of any composite noise from these curves; and (c) by a physiological study of the response of the ear and by establishing response curves to suit.

A, B, C and N Scale Meters

The first of the above methods is used extensively in industrial noise work, standard meters having been developed with appropriate filter characteristics for the various levels of sound. Thus in Fig. 6.3 the A scale is most suitable for low sound levels, the B scale for sound levels in the

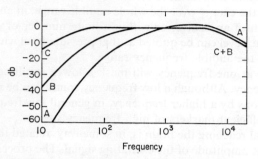

Fig. 6.3   Weighting filters on commercial meters.

70 decibel range and the C scale for loud noises. In practice, for motor car noise and the noise of machinery, experience suggests that the A scale gives the most consistent indication of subjective reactions and the report of the Wilson Committee on the Problems of Noise[5] has made recommendations in terms of these units. The filter characteristics are inter-

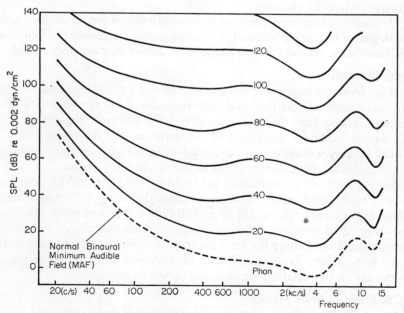

Fig. 6.4   Normal equal loudness contours for pure tones[6].

nationally agreed and are consistent from one make of meter to another. This is an unfortunate development in some ways since the weighting suitable for relatively low amplitude sound is not so representative for high intensity sound. Figure 6.4 shows a series of equal loudness contours[6] obtained by Robinson and Dudson using a jury of listeners. It may be seen that the slope of the equal loudness curves is not the same for all levels of loudness, the low frequency content contributing more to overall loudness for higher intensities. Thus, the C scale weighting is more representative for aeroplane noise work.

Experiments carried out in the U.S.A.[7] have indicated that people are more sensitive to high frequency noise than they are to high frequency pure tones, and this greater sensitivity must be allowed for in the filter characteristics of meters. Thus an N scale meter has been evolved which gives a better measure of loudness of aircraft noise. The differences are small, however, and the use of yet another scale may not be justified.

Loudness and noisiness computation techniques

While the above methods give a reasonably satisfactory method for comparing sounds and noises of similar general qualities, they cannot be used to differentiate between white noise and pure tones, nor to indicate the significance of composite sounds where some of the tones are masking others. Various methods of summation with varying degrees of complexity and difficulties of computation have been put forward to sum up such sounds and these are described in the literature. For example, the method of Stevens[8] based on octave band analysis is commonly used although in aircraft noise work preference is given to the equivalent and parallel method of 'perceived noise level'[7, 9]. These methods are described below together with that of Zwicker[10, 11], which is a procedure based on critical band analysis. The Stevens and Zwicker methods are stated formally in the British Standard procedures[12] for calculating loudness levels.

Perceived noise levels

When assessing the relative annoyance of piston engined aircraft and the new jet airliners coming into service, Kryter[7] and others found that to explain the significantly increased adverse public reaction to the jet, it was necessary to carry out a series of jury experiments similar to the loudness tests of Robinson and Dudson. They therefore subjected a fairly large number of people of various ages and backgrounds to a series of 'bands of noise' with different mid-frequencies and obtained a statistical series of equal octave-band noisiness curves. These have subsequently been modified[9] and the curves now used are shown in Fig. 6.5. The listening jury was also asked to fix, for each octave band of noise, values which were, in their opinions, twice as noisy as the previous one and hence to set up a series of similar curves each of which was recognized to be twice as noisy as the one below.

The total noisiness of a compound noise is calculated by the following procedure. The maximum value of the sound pressure level in each octave is converted into a level of perceived noisiness (NOYS) by reference to

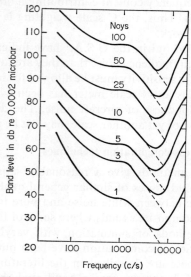

Fig. 6.5   Equal noisiness contours.

Fig. 6.5. The NOYS level corresponding to the measured octave band S.P.L. is obtained by considering the appropriate band centre frequency.

The NOYS values (N) in the various octaves are combined to give total noisiness N_T by using the relationship

$$N_T = N_M + 0.3(\sum N - N_M)$$

where N_M is the highest NOYS level obtained and $\sum N = N_1 + N_2$, i.e. $(\sum N - N_M)$ is the sum of the NOYS values for all n octave bands except that in which the maximum occurs, n depends on the measuring system being used; originally the audio range was covered by eight octave bands. Precision noise meters, and new equipment contain the new range of octaves which conform to the latest international recommendations. This system consists of 11 octave bands with the following centre frequencies (c/s)

31.5; 63; 125; 250, 500; 1,000; 2,000; 4,000; 8,000; 16,000; 32,000.

To establish perceived noise levels it is usual to consider only those octave-bands between 63 and 8,000 c/s in which case n is again eight.

In order to revert to a logarithmic measure of this 'perceived' noise, the conversion formula:

$$\text{Perceived Noise Level (PNdB)} = 40 + 33.3 \log_{10} N_T$$

is used.

As an example, assume the following octave band levels for a particular jet aircraft when flying over the first built-up area on take-off.

Octave Band   63   125   250   500   1,000   2,000   4,000   8,000

Sound pressure level (re 0.0002 dyne/sq.cm)

97   96   101   101   102   103   102   95

From Fig. 6.5 we find that

$$N = 24 + 32 + 59 + 75 + 80 + 105 + 159 + 75 = 609 \text{ NOYS}$$

$$N_M = 159 \text{ NOYS}$$

Hence

$$N_T = 159 + (0.3 \times 450) = 294 \text{ NOYS}$$

and

$$\text{Perceived Noise level} = 122 \text{ PNdB.}$$

This method may also be adapted for use with $\frac{1}{3}$ octave analyses.

Stevens phons

The Stevens Phons[8] are also based on octave measurements and are somewhat similar to the perceived noise levels. Although they are not often used in aviation acoustics, they are used in many of the other branches of acoustics for subjective assessment. The procedure for calculating the Stevens Phon loudness level is practically identical to that used for obtaining the PNdB values, except that the equal noisiness curves given in Fig. 6.5 are replaced by equal loudness curves shown in Fig. 6.6.

In this case the SONE and PHON are analogous to the NOY and PNdB used respectively in the previous section. The number of phons is obtained from the total number of sones either by using a relationship similar to that used to determine PNdB from the total noisiness or by using the conversion chart shown on the right hand side of Fig. 6.6. These sone curves shown are often referred to as 'loudness indices'.

Zwicker phons

A further method which has the advantage of being less empirical has become accepted recently, and it is recommended that, where perceived noise levels are found inadequate the method due to Zwicker be used.

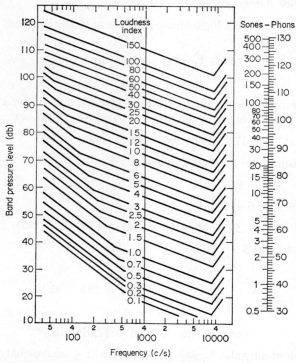

Fig. 6.6   Contours of equal loudness index (from ref. 12).

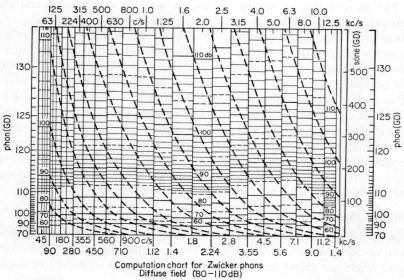

Fig. 6.7   Computation chart for Zwicker phons. Diffuse field, 80–
110 dB (from ref. 12).

Although the computational effort is much greater, the method has been programmed by the Applied Physics Division, N.P.L., and computation can be made by them on receipt of either octave band or third octave band data.

In Zwicker's method[10,11,12] the scale of sound pressure level is plotted separately for each band to allow for the different sensitivity of the ear to sounds of differing frequencies. In Fig. 6.7, for example, the fact that the 100 decibel ordinate for the one third octave centred at, say, 900 c/s is set below that centred at 4,000 c/s implies that this level of noise energy creates less subjective loudness than a similar level of sound energy at, say, 4,000 c/s. Thus if the one third octave band readings of any noise are plotted on the relevant scale the area under the curve is a measure of the total loudness.

Allowance for the complexities of the hearing mechanism is more satisfactory with this method. Zwicker found that a pure tone in one third octave gave a reduced but significant excitation in the neighbouring higher third octaves to an extent indicated by the dotted curves of Fig. 6.7. Thus it is necessary to add to the area of loudness these extra excitation areas, for example, in the way illustrated in Fig. 6.8. The Zwicker phon level is

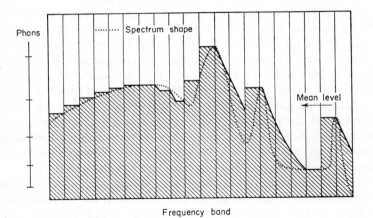

Fig. 6.8 Typical Zwicker phon calculation.

then found by establishing the rectangle with the same abscissa and area as that enclosed by the composite sound and the phon level being read on the ordinate scale.

This method is stated formally in the International Standards Organization procedures for calculating loudness levels,[12] and a loudness analyser incorporating the procedure has recently become commercially available.

Comparison of units

The Zwicker method is probably the best where discrete tones protrude above a general level of the noise. However, the method is laborious unless the N.P.L. facilities are available, and it is interesting to examine the differences of level obtained using these different units. Computations carried out by the Building Research Station with a variety of noises suggests that within a scatter of 2 or 3 decibels, Perceived Noise Levels and both Zwicker and Stevens phons can be equated and that overall decibels as measured on the A scale of a sound level meter give readings some 13 dBA less than the computed PNdB.

Impulsive noises

Impulsive noises such as explosions or sonic booms require a different assessment of loudness and of nuisance, since the ear tends to integrate the impulse over a short period of time. Depending upon the type of impulsive noise the impression of loudness becomes independent of duration after a certain time. However, the ear is very sensitive to rapid changes in the rate of build up of the impulse, and in particular to the rise of the initial pressure change.

A typical outdoor sonic boom N wave shape takes the form of a jump to the positive peak pressure in a time of 5–10 milliseconds, followed by a gradual fall to an equal negative pressure over a period of 50–250 milliseconds, with a similar return to ambient in the same time as the initial rise. Any subjective assessment of such a noise compared with a continuous noise having an rms level the same as the peak pressure of the initial jump, is subject to gross errors. Zepler and Harel[13] have put forward a method for assessing the loudness of an impulsive sound of any such arbitrary signature, and computations based on this method have been experimentally verified. The method also shows that low altitude turbulence in the atmosphere can affect the loudness of a sonic boom heard on the ground by over 10 decibels for distances apart as little as 200 ft.

When a boom occurs out of doors, a single reflection of the wave takes place, which tends to accentuate the pressure rise. If the ear is assumed to be 6 feet from the ground, this additional growth will occur some 10–20 milliseconds after the initial growth, and will provide in effect an additional rise in amplitude, but with an additional rise in time as well. The reflection effect has thus been found to increase the pressure jump by a factor of 1.7 or so, but the exact subjective effect must depend on the overall rise time. Experiments now going on at Southampton have already indicated that the slower the rise time, the greater the acceptability of the boom.

When a person is situated indoors, the boom deteriorates from being an N wave to a reverberant burst of noise lasting well over a tenth of a second.

The subjective loudness is a function of window and wall responses, the room reverberation time, and is also modified by any Helmholtz resonances which although too low in frequency to be heard themselves, may cause non-linear responses on crockery, structures and furniture which may in turn be heard as rattles[14].

Noise energy and noise number index

The noise emitted from an aeroplane is highly directional; when an aircraft flies overhead, the sound increases rapidly, continues for some 10 seconds, and then falls away over a period of some 30 seconds. Depending

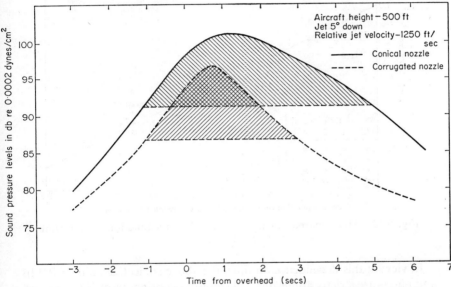

Fig. 6.9 Total noise versus time; typical curves for conical and corrugated nozzle (after Jenner[20]).

on the geometrical details of the aeroplane, its speed, and the kind of nozzle used, the duration of the sound can vary. Figure 6.9, for example, shows the time history of the sound from the same aircraft fitted with a normal and corrugated nozzle respectively.

Kryter[15] has shown that for sounds of relatively short duration (say 30 seconds), the apparent loudness of the sound depends very much on this duration, a doubling of the duration being subjectively equivalent to a 4.5 decibel increase in the peak level (Fig. 6.10). If this equivalence had been 3 decibels for doubling duration, it would be the same as saying that it is the total acoustical energy reaching the ear that matters, rather than the peak level; this is particularly true in view of the extremely uneven build up in sound heard by a listener on the ground. This variation with time arises from lower atmosphere turbulence and can cause fluctuations of 10 decibels or more over periods of a few seconds. Later work suggests that the effect of duration is not constant, but depends on the duration itself. Thus doubling very short duration is equivalent to 6 decibels increase in loudness, whilst doubling duration of the order of 30 seconds is equivalent to an increase of only one decibel in effective loudness.

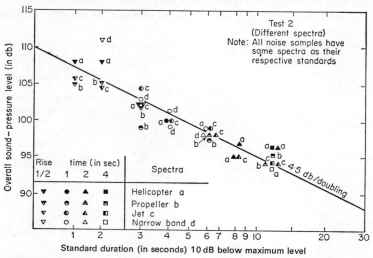

Fig. 6.10   The apparent change in loudness as a function of duration (after Kryter[15]).

In view of this dependence on time, the above subjective units of PNdB and phons are strictly applicable if the noise is continuous or if the pattern of noise growth is standard in its shape. For example, peak sound pressure levels (or average peak over a few seconds) are acceptable as a measure of aircraft loudness because aeroplane climbing speeds and approach speeds are fairly constant from one type to another so that the total duration is the same. On the other hand, body shielding, nozzle design, and aircraft attitude can make a difference to the noise signature of the aircraft and it is a pity that our units are unable to make allowances for this. There can be little doubt therefore that new units related to total noise energy will be introduced as our understanding of subjective acceptability progresses.

One such step in this direction has already been taken as a result of the findings of the Wilson Committee survey of noise nuisance at London Airport (Heathrow). Here it was found that people were sensitive to the number of times they were disturbed by aircraft *more* than they were aware of the peak noise level, and that a composite index of noise and number nuisance could be formulated which agreed closely with the duration experience described above. In fact, if N is the number of aircraft heard during a specified period, say a day, making a noise greater than 80 PNdB, then a new Noise and Number Index (NNI) can be established in the form:

NNI = Average Peak noise level of aircraft making more than 80 PNdB

$$+ 15 \log_{10} N - 80$$

where

$$\text{Average Peak noise level} = 10 \log_{10} \frac{1}{N} \sum_{1}^{N} 10^{L/10}$$

where L is the peak reading for each aircraft. The figure 80 is subtracted, because it was found that the level of noise annoyance had fallen to a level which was independent of noise level. This was also the reason for not bothering with aircraft making less noise than this number of PNdB.

In Germany[16] there is a move towards establishing limits in terms of what is called the Q factor. This is essentially a measure of weighted energy. In the U.S.A. new units are being considered which allow[17] for duration for the greater unacceptability of distinct tones within the Spectrum. Some suggested corrections for discrete tones are shown in Table 6.1. In view of the proposed noise certification of aircraft corrections for tonal content and duration in terms of Effective Perceived Noise Level (EPNL) are being devised.

Table 6.1

Correction to Peak Perceived Noise Level to Account for Discrete Frequency Components in a Random Noise Spectrum

Discrete tone frequency	Tone amplitude above random noise background (re: 1/3rd octave band) in decibels				
	1	6	11	16	21
20 c/s to 400 c/s	0.0	0.0	0.0	0.5	1.0
400 c/s to 2000 c/s	0.0	0.0	1.5	3.0	5.0
2000 c/s to 4000 c/s	0.0	1.5	3.0	5.0	7.5
4000 c/s to 9600 c/s	0.0	0.0	1.5	3.0	5.0

Note. The Correction for 2nd harmonic components is 1/5 the value shown on this table and is added to the correction due to the fundamental tone.

The need for a better method of indicating nuisance is exemplified by the fact that at London Airport less than 8 per cent of the people seriously annoyed were in a region where the noise level was greater than 103 PNdB. In this case this was also true to some extent in terms of NNI, the number of aircraft movements being greatest where the noise levels were also greatest. There is therefore a need to repeat this survey in a way in which noise and number are varied independently, before the old units are discarded. Even so it is as well to remember that for every doubling of traffic, there is to be expected a $15 \log_{10} 2$ increase in effective noise level, i.e. 4.5 decibels, at all points around the airport.

Index of community nuisance

More recently, and very tentatively, this concept of total daily noise energy has been examined[18,19] in its relationship to long term airport planning and an index of community nuisance (ICN) has been postulated which allows the airport and town planners to assess their contribution to the problem in parallel with the aircraft and engine designers. Thus the

Index of Community Nuisance at a point	Responsibility of:
$= $ Constant	
$+ \; 10 \log_{10}$ (Population Density)	Town Planner
$+ \; 10 \log_{10}$ (Total Daily Static Thrust)	Airport Planner
$+ \; 10 \log_{10}$ (Acoustic Conversion Factor)	Engine Designer
$+ \; 10 \log_{10}$ (Speed Factor)	Aircraft Designer
$+ \; 10 \log_{10}$ (Path Factor)	Airport Planner

Thus predictions of growth in nuisance can be made with the minimum of knowledge of the exact growth of aircraft size, speed and routes. It also indicates the need to separate our high density areas, from areas of high path factor.

Since the number of pounds of thrust per passenger flown is practically constant for jet airliners (except for supersonic aircraft where the figure is three times as large) and independent of their size, an extrapolation of the problems of nuisance growth can easily be made. The acoustic conversion factor of the engine on the ground is of course a function of the jet speed, the speed factor allows for the change in noise output with speed, and in the time over an area, and the path factor takes into account the lateral speed possible after the initial climb.

Acknowledgment

The figures from Refs. 5 and 6 are reproduced by permission of the British Standards Institution, 2 Park Street, London, W.1, from whom copies of the complete standard may be obtained.

References

1. G. von Békèsy. *Experiments in Hearing*, McGraw-Hill, New York (1960).
2. H. Spoendlin. *The organisation of the cochlear receptor*, Karger, Basel (1966).
3. E. G. Wever. Electrical potentials of the cochlea, *Physiological reviews*, 46/1, 102 (1966).
4. I. C. Whitfield. *The Auditory Pathway*, Arnold, London (1967).
5. *Noise*, Final Report of the Wilson Committee on the Problem of Noise, H.M.S.O. (July, 1963).
6. Normal equal loudness contours for pure tones and normal threshold of hearing, *Brit. Stds. Inst.*, **B.S. 3383**: 1961, I.S.O.—R. 226.
7. K. D. Kryter. Scaling human reactions to the sound from aircraft, *J. acoust. Soc. Amer.*, **31**, 1415 (1959) (See also *Noise Control* (Sept. 1960).)
8. S. S. Stevens. Procedure for calculating loudness: Mk. vi, *J. acoust. Soc. Amer.*, **33**, 1577 (1961).
9. K. D. Kryter and K. S. Pearsons. Modifications of NOY Table, *J. acoust. Soc. Amer.*, **36**, 394 (1964).
10. E. Zwicker. Ein verfahren zur berechnung der lautstärke, *Acoustica*, **10**, 304 (1960).
11. E. Zwicker. Subdivision of the audible frequency range into critical bands, *J. acoust. Soc. Amer.*, **33**, 248 (1961).
12. Method for calculating loudness, *Brit. Stds. Inst.*, **B.S. 4198**: 1967.
13. E. E. Zepler and J. R. P. Harel. The loudness of sonic booms and other impulsive sounds, *J. Sound Vib.*, **2**, 8249 (1965).
14. E. J. Richards. Sonic boom, *Science Journal* (May 1965).
15. K. D. Kryter and K. S. Pearsons. Some effects of spectral content and duration on perceived noise level, *J. acoust. Soc. Amer.*, **35**, 866 (1963).
16. W. Burck, M. Grutzmacher, F. J. Meister and E. A. Muller. Aircraft noise, its measurement and assessment, its bearing upon the planning of new housing estates, measures aimed at its reduction, Report drawn up for Federal Minister of Health, Germany (1965).
17. W. J. Galloway and H. E. von Gierke. Individual and community reaction to aircraft noise: present status and standardisation efforts, International Conference on reduction of noise and disturbance cause by by civil aircraft (1966).
18. E. J. Richards. The constraining order of airport noise, *ISAV* Report No. 148 (July 1966).
19. E. J. Richards. Aircraft noise—mitigating the nuisance, *Astronautics and Aeronautics*, **5**, 1, 35 (1967).
20. W. Jenner. Rig and flight noise tests on a corrugated convergent-divergent nozzle, *Rolls-Royce Report* (1958).

CHAPTER 7

Jet and Rocket Noise

7.1 Introduction

Chapters 1 and 5 contain the formal establishment of mathematical expressions for the noise associated with various source mechanisms in fluids. If there is any input of mass in the fluid (or the effective input of mass by a volume displacement of the existing fluid by the introduction of a solid body into the region), the condition is covered by the mass continuity equation

$$\frac{\partial \rho}{\partial t} + \frac{\partial}{\partial x_i} (\rho v_i) = Q(x, t)$$

where Q is rate of introduction of mass per unit time at time t and place x.

This can take the form of a periodic volume fluid input from, say, a piston engine exhaust, from the volume displacement of a propeller moving into the region, from the movement of the diaphragm of a loudspeaker, the vibration of the structure, or from the collapsing of cavitation bubbles. On jet engines, the jet efflux is continuous and uniform and therefore Q is a constant except where an unsteady efflux flow exists. This certainly occurs on a pulse jet, or on a reciprocity engine exhaust, but in general the deviation from constancy is small on jet engines.

Any fluctuating forces (F_i) exerted on the air are introduced into the momentum equations. For example, the momentum equation in the x_i direction ($i = 1, 2$ or 3) is:

$$\frac{\partial}{\partial t} (\rho v_i) + \frac{\partial}{\partial x_i} (\rho v_i v_j) = -\frac{\partial}{\partial x_j} (p_{ij}) + F_i(x, t).$$

(See Chapter 5, equation 5.2.)

Here $F_i(x, t)$ is the force exerted per unit volume in the 'i' direction at time t and place x. A typical example of such a force is that exacted as the blade of a very thin fan or propeller cuts the air at an angle of incidence. Both lift and drag (or torque) forces are imposed upon the air without any volume displacement occurring. Other mechanisms which can be considered as only exerting forces on the fluid rather than volume displacements are small vibrating objects (e.g. a violin string) in which the air can

150

flow from one side to the other of the source in spaces very small compared with a wavelength. The area near the trailing edge of an aerofoil also exerts a fluctuating force to react to the periodic shedding of eddies from the trailing edge and the radiated sound can be calculated from this 'dipole' type source.

A typical condition for this kind of source input is the existence in the fluid of a solid surface at that point. Similarly, electromagnetic forces can be applied in the fluid, if it is ionized, giving force type sources. A normal jet flow is devoid of solid surfaces and is generally non-ionized. 'Force' sources are therefore absent in jet flow.

If the above four equations are combined to eliminate ρv_i, the well-known wave equation is obtained in the form:

$$\frac{\partial^2 \rho}{\partial_t^2} - a_0^2 \nabla^2 \rho = \frac{\partial Q}{\partial t}(x, t) - \frac{\partial F_i(x, t)}{\partial x_i} + \frac{\partial^2 T_{ij}}{\partial x_i \cdot \partial x_j}$$

(see Chapter 5, equation 5.22) where

$$T_{ij} = \rho v_i v_j + p_{ij} - a_0^2 \rho \delta_{ij}$$

$$= \underset{\substack{\text{(convection} \\ \text{of momentum)}}}{\rho v_i v_j} + \underset{\substack{\text{(viscous} \\ \text{stress)}}}{(p_{ij} - p\delta_{ij})} + \underset{\text{(heat conduction)}}{(p - a_0^2 \rho)\delta_{ij}}$$

The formal derivation of the noise intensity at a point in terms of T_{ij} is given in chapter 5.

The viscous stresses are small, except possibly at the top of the frequency range, and the heat conduction terms vary only as M^4. Thus for moderate Mach numbers, T_{ij} can be equated to $\rho v_i v_j$, and again if the flow is near incompressible we can equate this to $\rho_0 v_i v_j$.

In the above, the flow momentum along one axis is $\rho_0 v_i$, v_j, being the velocity along the same or another axis, and $\rho_0 v_i v_j$ is the sum of nine terms. It is fairly easy to see that some terms are larger than others, however, and this leads to some simplification of the analysis. For example, if we replace x_i by (x, y, z) and v_i by $U + u, v, w$, where the general jet stream is in the x direction, then terms such as $\partial^2(\rho_0 v_i v_j)/\partial x_i \, \partial x_j$ will include terms $\partial U/\partial y \, \partial v/\partial x$ where U is the mean velocity in the x direction and v is the much smaller velocity in the transverse direction. Thus the term $\partial v/\partial x$ is magnified by the existence of a large velocity shear $\partial U/\partial y$ and the dominating sources of sound are to be found in this high shear region.

This is easy to see physically. If in Fig. 7.1 a fluid element moves downstream from A (where the velocity is high) to B (where the velocity is lower because of the gradient $\partial U/\partial y$), forces are set up owing to the deceleration. These forces in turn radiate sound; they are proportional to the gradient $\partial U/\partial y$ and to the random turbulent velocity v in the transverse direction.

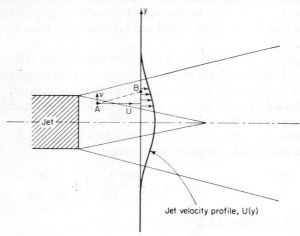

Jet velocity profile, U(y)

Fig. 7.1

It follows that for jet noise to be severe, it requires a region of high velocity shear to be associated with a region of high turbulence. Lilley[1] in a detailed treatment, has made computations of the typical strengths of the turbulence at various distances downstream and has estimated the relative strengths of the noise radiated from the shear layers at various numbers of chord lengths downstream of the nozzle. Fig. 7.2 shows this,

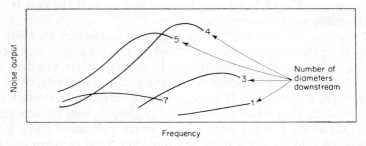

Fig. 7.2

and indicates that for a low-speed circular jet the noise output is small, close to the orifice (where the turbulence has not developed), and increases progressively for some five jet diameter lengths downstream and then subsides again.

An explanation of the change of noise characteristics with nozzle configuration can also be given in simple terms. As referred to in chapter 5, the turbulence, mean velocity shear and scale of turbulence have been measured for a circular and corrugated nozzle and the noise output computed (see Fig. 11.14, chapter 11). It may be seen that the noise from the corrugated nozzle is less than that from a circular jet simply because the reduction of velocity shear is not matched by the expected turbulence growth.

A further physical insight may be obtained, if the nature of the transverse velocity flow is considered. The outward transverse emission of x direction momentum discussed above takes place rather like a jet spurt which can occur sharply and in a short time. The inward flow into the high-speed area on the other hand must be taken from a greater volume of the outside air, and is more likely to take the form of a gradual and smooth entrainment; thus the inward v need not balance the outward component and the growth of turbulence can depend upon the exact degree of balance. Therefore, on a very highly supersonic jet the outward jet spurts can be such as to denude the mixing region of its turbulence and to contribute to the very slow spread rate generally observed as a characteristic of supersonic jet mixing.

If these jet spurts come from a region of the flow which has a supersonic convection speed, they will of course be convected supersonically downstream through the low-speed air surrounding the jet. Thus the acoustic radiation from them will be similar to those radiating from a supersonic aircraft or from a fan with a supersonic effective rotational source speed, and shocks will form. Thus, phase cancellation will be minimized and strong sound emission will occur in the direction of shock propagation.

The relative strength of this source of noise radiation as opposed to the general subsonic mixing sources further downstream is subject to some controversy. Williams[3] has shown theoretically that these sources can emit sound of an order equal to that measured. However, von Gierke[4] has obtained directional plots of sound transmission which suggests sound sources on rockets which are very many diameters downstream. The dominating source position is plotted against distance from the nozzle for some typical pressure ratios in Fig. 7.3. Other experiments in which a supersonic jet was directed into a water tank in such a way that differing lengths of the jet were uncovered also substantiates von Gierke's conclusion that the dominating source on rocket engines is in the region in which supersonic flow breaks down.

7.2 Shock-cell noise

Once the critical pressure ratio is exceeded in a nozzle, the physical nature of the jet flow depends on the degree of convergence or divergence

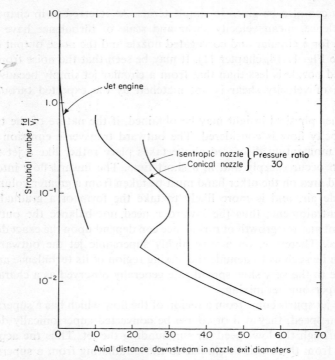

Fig. 7.3 Location of maximum sound pressure sources downstream of jet exit (after von Gierke[4]).

of the final nozzle. For any one pressure ratio, there is a single expansion ratio which allows a shock-free flow. Above and below this pressure ratio, the nozzle is either under- or over-expanded and the shock patterns of Fig. 7.4 occur[5]. These shock cells are sensitive to slight pressure or velocity changes and it requires only a slight change in the external pressure at the jet exit to involve a significant movement of the shocks.

These shocks involve very sharp discontinuities in the component of the velocity normal to themselves and this high $\partial U/\partial x$ value acts, as before, as an amplifier of any sound output arising from turbulence passing through it. Additionally, any longitudinal unsteadiness will cause a movement of the cell and the sharp change in the local pressure will give rise to a noise source which will radiate from the stream.

Both mechanisms probably constitute strong sources of noise in supersonic jets; these can, however, be aggravated by a 'back reaction' mechanism first studied by Powell[6] in 1952 and only now becoming important in practice. If the sound from such sources travels upstream to the nozzle

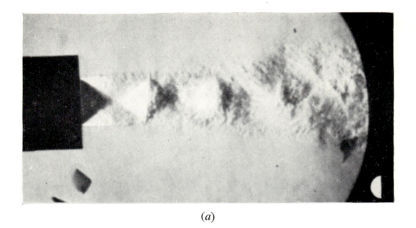

(a)

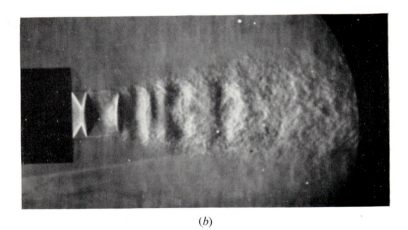

(b)

Fig. 7.4 Shock patterns in (a) under-expanded and (b) over-expanded
cold air jets (after Franklin[5]).

edge, the change of pressure there will be sufficient to cause a slight change in the cell pattern. Thus, depending on the dimensions and speed of the jet, a feedback system can occur which is tuned to certain frequencies and a noise generating system can be established which is far more efficient than the relatively inefficient system previously discussed.

These discrete tones have frequencies inversely proportional to the scale. Thus in early researches on models, the frequencies were very high. Fig. 11.16 of chapter 11 shows the variation of frequency with pressure ratio and shows how two separate cell patterns are compatible with some pressure ratios, and that a small disturbance can result in a modification to the frequency of such resonances. This phenomenon is very noticeable on the model, the frequency of tones being alterable by clapping the hands in the vicinity of the jet.

In 1952, no full-scale experience of this phenomenon was noted, possibly because of the relatively rough flow at the nozzle and the small excess pressure ratio. Recently, however, with smoother jet flows and the higher pressure ratio conditions, the mechanism has exhibited itself in high-altitude flights and has caused structural failures. The cell resonance frequencies are much lower than on models, and are now tending to coincide with natural frequencies of skin panels in the aeroplane tail structure. Since noise laws as high as V^{20} were measured in the model conditions, it is obvious that the energy transferred acoustically to the structure can be very intense, even though the normal noise radiation is small.

Whereas on the model it was relatively easy to suppress this cell noise, the rougher full-scale jet requires a much greater acoustical disturbance to set up the resonance, but requires more radical modifications to cause its elimination. No really satisfactory method of elimination has been evolved, though any system such as slant nozzles, or nozzles with internal vortex generators reduces the magnitude of the excitation. Since the mechanism is one of instability, methods of predicting frequency are satisfactory[7]. It is seldom possible, however, to predict magnitudes of instabilities, and this is unlikely to prove an exception.

7.3 Noise output formulae

Returning now to the problem of predicting jet noise, it is clear that if the relevant functions of the turbulence are known, the noise can be predicted by equation 28 of chapter 5, or by a similar one which makes allowance for convection velocity of the sources[8,9]. There is little likelihood, however, of our ever being able to calculate or measure all these functions, and we are forced to fall back on dimensional arguments and empiricism for both the magnitude and frequency spectral qualities of jet noise.

7.4 Noise laws

The V^8 law established in chapters 1 and 5 is a generalization which can be modified if the source is being convected downstream. The effect of convection is to increase the index, while the effect of reducing the turbulence level with speed or the effect of non-constancy of Strouhal number (frequency × distance ÷ velocity) is possibly to reduce the index. The actual index obtained experimentally is therefore of considerable interest.

The type of curve obtained of the total acoustic output against total jet power is shown[4] in Fig. 7.5. It is seen for unchoked jet engines and subsonic air jets, that the V^8 law is kept well and can be used for prediction.

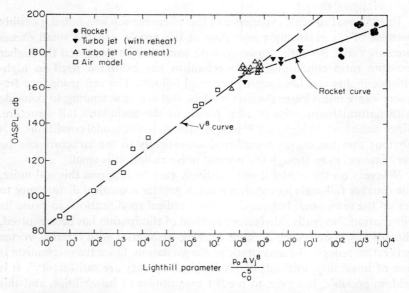

Fig. 7.5 Overall sound pressure level from various jets as a function of the Lighthill jet velocity parameter (after von Gierke[4]).

On the other hand, choked jets with supersonic speeds after expansion (see chapter 1) indicate a lower index near the 4th. The figure shows this index for a range of rockets, afterburning engines and normal jet engines at full military power.

For subsonic jet velocities the total acoustic power output has been expressed empirically in decibels relative to 10^{-13} watt as

$$60 \log_{10} V + 10 \log_{10} A - 48 \quad \text{(Ref. 10)}$$

and for supersonic velocities as

$$78 + 13.5 \log_{10} W_M$$

where $V = tg/w =$ effective exit velocity in feet per second calculated from the engine thrust t in lb

$w =$ total weight flow of air and fuel through the engine in lb/sec

$g =$ acceleration due to gravity

$W_M = 0.67w\, V^2/g =$ mechanical power in the jet

and $A =$ nozzle area in sq. in.

It may be seen that a rocket whose velocity is as high as 4000 ft/sec is quieter by some 20 dB than the Lighthill law would imply. This is not surprising since underlying the V^8 law is the essential assumption that acoustic power at each frequency can be plotted as a unique function of Strouhal number (fd/V) as indicated by the dotted line of Fig. 7.6. It has been found that in practice the spectra move to the left as the Mach number increases. This implies that frequency does not increase directly with velocity and that the quadrupole variation $V^2 f^6$ implies a law much lower than the V^8 law usually assumed.

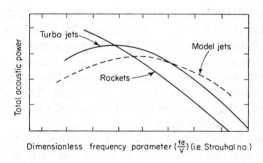

Fig. 7.6 Dimensionless frequency parameter (fd/V).

Von Gierke[11] has found that spectra covering all the above types of engine can be non-dimensionalized satisfactorily if a frequency parameter

$$\frac{fd}{V} \times \frac{c_j}{c_0}$$

is used and the power output is plotted

$$10 \log_{10} \left(\frac{1}{\rho_0 A} \times \frac{dW}{df} \times \frac{V}{d} \times \frac{c_0}{c_j} \right)$$

where c_j = velocity of sound in the jet

dW/df = spectrum of acoustic power output from the jet

and c_0 = the velocity of sound in the surrounding air.

The curve thus obtained is shown in Fig. 7.7.

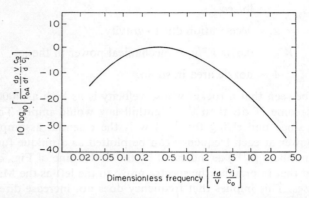

Fig. 7.7 Non-dimensional acoustic power spectrum plotted against non-dimensional frequency.

In the United Kingdom, Rolls-Royce have established an empirical method of estimating jet noise which agrees with practice both on static engines and in flight[12]. This method relies on two empirical relationships, one of which relates total overall sound pressure level to jet speed for engines on the ground, or relative speed in the air. The other is a relationship between spectral density and forward speed. This latter relationship does not introduce size, and limits the method to jets of sizes similar to those used in developing the method.

The maximum overall sound pressure level (corrected to a distance of 100 ft) has been analysed for a range of engines and static jet speeds. This is shown in Fig. 7.8 and illustrates the deviation of overall noise output from the V^8 law mentioned in chapter 1.

The equivalent curve for engines in forward flight is less well authenticated but is shown in Fig. 7.9. The relevant jet velocity is that relative to the air immediately adjacent to the jet flow, i.e. $V_{rel} (= V_j - V)$. The noise output is also related to this and will therefore decrease with some power of V_{rel}/V_j. Allowance is made for this in Fig. 7.9.

Once the maximum overall level is obtained, it is usually necessary to establish the distribution of the energy between the frequencies. This can be done by referring to von Gierke's generalized spectrum of Fig. 7.7 or by

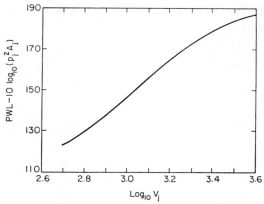

Fig. 7.8   Overall sound pressure level (dB) at 100 ft from a stationary jet as a function of jet velocity V_J (after Coles[12]).

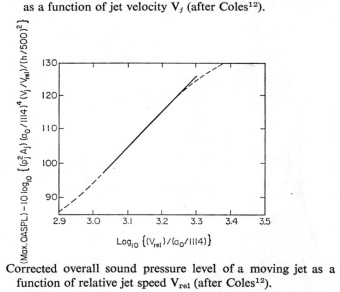

Fig. 7.9   Corrected overall sound pressure level of a moving jet as a function of relative jet speed V_{rel} (after Coles[12]).

using Rolls-Royce curves of Fig. 7.10 which indicate the distribution of third octave content in the Zth band for a range of jet (static case) or relative (flight case) speeds.

7.5 Directional patterns

Knowledge of the total acoustic power output does not, as such, tell us anything about the manner in which this power is radiated around the source, i.e. the directional characteristics of the noise from the jet. However, the assumption of lateral quadrupoles as the source of sound, as

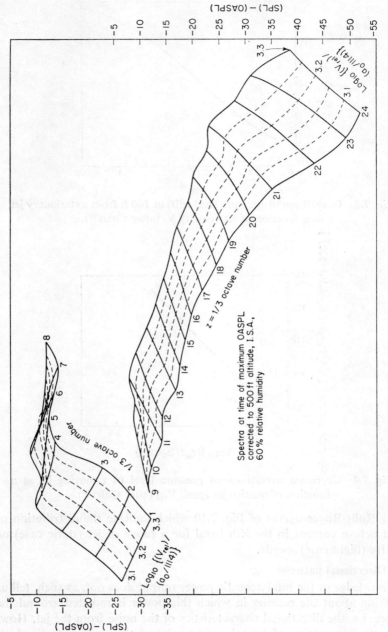

Fig. 7.10   Effect of jet velocity on spectrum (Avon flight tests)—after Coles[12].

occurs in the shear region of a jet, implies a four-lobe directional pattern of noise, with the downstream lobes amplified greatly by the convection of the eddies in that direction.

A typical directional distribution from a full-scale jet engine is given in Fig. 7.11. This indicates that the maximum noise occurs at about 35° to the rearward-pointing jet axis, and thereafter falls off steadily to a level some 15 dB lower at 90°.

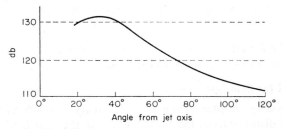

Fig. 7.11   Distribution in direction of noise from a jet engine.

The levels are shown relative to the space-average SPL (=SPL on a hemisphere enclosing a non-directive source of the same power level). The curve is actually for a jet velocity of 1,800 ft/sec, a typical value for the maximum rating of such an engine.

The sound pressure level for this particular engine at an angular position θ may be adequately determined for a velocity V in the range 1,250 ft/sec to 2,000 ft/sec by applying a correction to the dB level at the appropriate point on the curve of amount $10n \log_{10} V/1800$ where n is given by

$$5.2 + 3.4 \exp\left(-\frac{3}{5}\right)\left(\frac{\theta - 41}{7}\right)^2.$$

This empirical relationship indicates that the higher the jet velocity the wider the angle at which the peak noise is radiated. (Indeed some investigators estimate the noise level for a velocity V_2 from the polar curve of a basic velocity V_1 by applying a constant correction of $80 \log_{10} V_2/V_1$ and then applying a slight rotation to account for this change in directivity, the amount of rotation being dependent upon $(V_2 - V_1)$.)

For velocities less than 1,250 ft/sec sources other than jet noise begin to intrude and for velocities higher than 2,000 ft/sec mentioned, as with reheat, the change in polar shape is more radical than given above.

With a suppressor nozzle the spectrum shape as well as the overall noise level is modified. The directivity changes too, the greater the overall reduction of the suppressor nozzle, the greater is the angle of the peak noise from the jet axis. Fig. 7.12 shows the type of change which can be

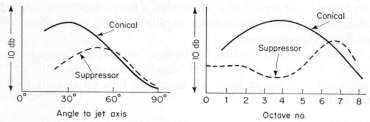

Fig. 7.12   Effect on noise directivity and spectrum of fitting a suppressor to a conical jet pipe.

expected in the directivity pattern for a typical suppressor nozzle and in the spectrum shape at, say, 30° angle to the axis.

7.6 Near and far field intensities

As explained in chapter 1, the noise intensity falls off as the inverse square of the distance from the source except in the near field. Here large reactive terms exist which do not radiate energy but which do contribute to the mean square pressure on the aircraft structure. If, for example, the noise is measured at varying distances from the aerodynamic source in the jet, the type of curve shown in Fig. 7.13 is obtained.

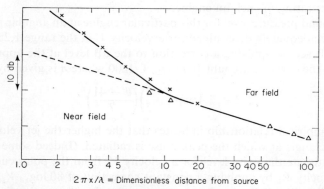

Fig. 7.13   Variation of narrow-band jet noise with distance from source (x = distance from source; λ = wavelength).

It is seen that when the measuring point is more than two wavelengths away from the source it may be considered as being in the far acoustic field[13], but that for points nearer than this no general scaling law can be used. Thus, for example, on a structure responding to say 150 c/s noise, we can estimate the noise by far field noise laws at distances greater than 10 ft from the source, and for noise at 750 c/s the near field extends only a distance of 2 ft or so. Thus, in most cases of structural fatigue the high-frequency noise can be treated as acoustically propagated, but this is not so for low frequencies less than 500 c/s. In such cases, and indeed in most practical cases, it is safer to take experimental surveys of the fluctuating pressure levels.

7.7 Pressure correlation

For estimates of structural response, it is necessary to know the pressure correlation in relatively narrow frequency bands. The only comprehensive series of measurements on a full-scale jet are those made at the N.A.C.A. Cleveland. This presents correlation measurements at several positions along the jet boundary.

The correlation patterns in the region around the jet efflux vary with position relative to the jet, but from the point of view of structural loading the region can be split up into two as follows:

Region 1 (close to the jet boundary). In this region the pressure fluctuations are strongly influenced by the convection of the hydrodynamic turbulent field. This region extends to about two diameters out from the jet boundary and downstream to a plane about twelve to fifteen diameters down from the nozzle exit plane.

Region 2 (near field). In this region the spatial distribution of the noise sources is on a scale comparable with the dimensions being considered in the correlation measurements. The region extends to about ten diameters out from the jet boundary, twenty diameters downstream and ten diameters upstream of the nozzle exit plane. The position of the outer surface of this region is not clearly defined and is probably frequency-dependent.

Unfortunately, from the scaling point of view, the pressure correlations in these two regions are likely to be strongly affected by the detailed structure of the turbulent mixing regions. It is conceivable, for instance, that varying amounts of initial or core turbulence in different jets will cause variations in the correlation patterns.

The general form of a narrow-band spatial pressure correlogram closely resembles a decaying cosine wave. In this case, the significant dimension of length is not the integral of the correlation function, but the distance to the first zero crossing or the wavelength of the cosine wave. As the decay rate may be large in some cases, we shall define the correlation length as the distance to the first zero crossing. The positions at which full-scale measurements have been taken are shown in Fig. 7.14.

7.7.1 *Pressure correlations close to the jet boundary*

Longitudinal pressure correlations along lines parallel to the jet centre-line are dependent primarily on the convection velocity of the noise sources

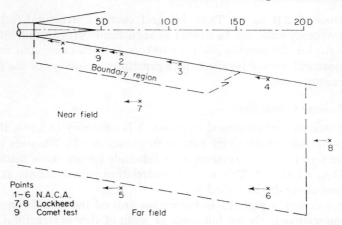

Fig. 7.14   Positions of full-scale pressure correlation measurements.

in the turbulent mixing region of the jet, whereas in lateral directions tangential to the surface of the jet, there is no convection effect. It is appropriate, therefore, to consider these two directions separately.

7.7.2 *Pressure correlations parallel to the jet centre-line (longitudinal)*

Overall space correlation. In this direction it is reasonable to assume that the distance to the first zero crossing point is given sufficient accuracy if we assume a 'frozen' convection pattern. That is to say, over the short distances involved the pressure pattern observed at one point is seen in exactly the same form at a point distance x downstream of the first observer at a time interval x/U_c later. The decay in the coherence of the pattern which does, in fact, take place in this distance is sufficiently small not to change the position of the first zero crossing point appreciably.

Close to the boundary of the jet, the power spectrum of the pressures has a peak at frequency f', say, and it has been found that the Strouhal number $f'D/U$ remains approximately constant. The peak frequency also varies with position downstream. At any one position the autocorrelation function will have a time scale such that the time to the first zero crossing, τ', will depend primarily on f', i.e. $f'\tau' =$ constant $A \approx 0.25$. If the pressure pattern is assumed to be convected rigidly, the space correlogram in the direction of convection can be obtained from the autocorrelogram by replacing τ by x/U_c. Now if x' is the distance to the first zero crossing, we have

$$x' = \tau' U_c$$

$$= \frac{A U_c}{f'}$$

Assume now that U_c is directly proportional to U. For a given position on one jet U_c/f' is constant and the distance x' is therefore independent of jet velocity. If the diameter of the jet is changed, the distance x' is changed in direct proportion to the change in diameter.

7.7.3 *Lateral correlations*

The lateral correlation scales for filtered bands of noise are likely to be influenced primarily by the circumferential length of coherence of the noise sources in the jet. There have been no systematic measurements made in this direction, but it seems likely that the correlation lengths will scale linearly with the scale factor.

7.7.4 *Pressure correlations in the near field*

There is little comprehensive data available for correlations in the near field. Full-scale measurements have been made at a limited number of positions, but there is no model data at identical scale positions available for comparison.

7.7.5 *Coplanar measurements*

In a plane passing through the jet centre-line the variation of the first zero crossing position with angle to the jet centre-line takes on an elliptical form. In the direction normal to the jet boundary the distance is given approximately by one-quarter wavelength. In the direction parallel to the jet boundary the distance is given from convection considerations. This simple model applies to reference stations that are located on a line which is approximately at right angles to the jet centre-line and passes through the source region for the particular frequency of interest.

For other positions the correlation scale depends on the angle θ which the correlation transverse line makes with the line joining the reference point to the source. The position of the apparent source region will also be affected by the Doppler shift. Longitudinal correlograms parallel to the jet centre-line for positions 8, 6 and 5 (Fig. 7.14) are shown in Fig. 7.15. Here the results have been plotted on the nondimensional base fx'/c. In Fig. 7.15a the direction of measurement traverse makes an angle θ of about 25° with the direction to the source region. In a travelling plane wave situation the distance to first zero crossing would be given by

$$\frac{fx'}{c} = \frac{0.25}{\cos \theta} = 0.28.$$

It can be seen from the figure that the value of fx'/c is about 0.25 for the higher frequencies, and 0.3 to 0.36 for the frequencies below 500 c/s. This difference is reasonable, as the higher frequencies are located farther up-

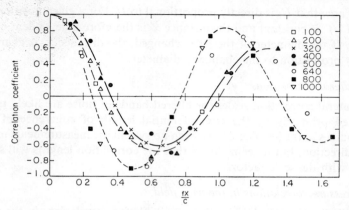

Fig. 7.15a   Correlograms for point 8 (Fig. 7.14).

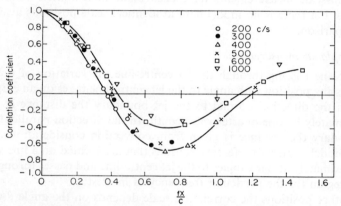

Fig. 7.15b   Correlograms for point 6 (Fig. 7.14).

stream and hence θ is smaller. For position 6 (Fig. 7.15b) the angle θ is approximately 44°, which gives a value of $fx'/c = 0.35$. The experimental value is approximately 0.4. For high values of θ approaching 90° the simple travelling wave approximation gives $fx'/c = \infty$ which is not true in the near field. The scale of the correlogram does become greater, as shown in Fig. 7.15c ($fx'/c = 1.5$; $\theta = 90°$).

The other trend shown by these three sets of results is that the decay of the correlogram appears to increase with increase in the angle θ. Also the scatter in results, particularly marked in Fig. 7.15a, is greater for positions closer to the jet, as would be expected because the effect of the angular variation in source position is greater. The measured values of fx'/c are compared with the theoretical (far field) case in Fig. 7.16. It can be seen

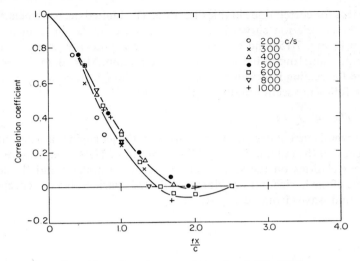

Fig. 7.15c   Correlograms for point 5 (Fig 7.14).

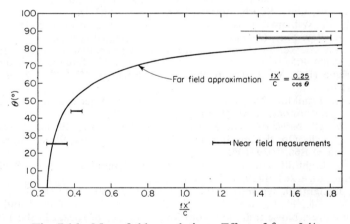

Fig. 7.16   Mean field correlations. Effect of θ on fx'/c.

that the limited measurements available show reasonable agreement with the simple (far field) approximations.

7.7.6 *Lateral correlations*

Pressure correlations in the direction normal to the plane passing through the jet centre-line are important in determining the excitation of interframe panel groups on stiffened fuselage structures. Very few data are available for correlations in this direction. The limited measurements show that the correlation curves in this direction are similar in form to those shown in Fig. 7.15c and that the correlation lengths are greater than the corresponding lengths in the longitudinal direction. More data are required before the scaling laws can be established, although some work described in the following section indicates the trends.

7.7.7 *The effect of reflection*

In several current designs of civil airliners, the jet engines are mounted at the rear and the tail unit is of the T configuration. The noise pressure levels and correlations on the surface of the vertical stabilizer and the under-surface of the horizontal stabilizer are affected considerably by reflection of the sound waves from nearby structure.

References

1. G. M. Lilley. On the noise from air jets, *A.R.C. Rep*, **20376-N** 40—FM 2724 (1958).
2. R. C. Potter. Measurements of the turbulence in interfering mixing regions from jets, Univ. Southampton M.Sc. Thesis (1963/4).
3. J. E. Ff. Williams. Noise from turbulence convected at high speed, *Phil. Trans. Roy. Soc.*, (*London*), A **255**, 469–503, April (1963).
4. H. von Gierke. Types of pressure fields of interest in acoustical fatigue problems, P.57 of *W.A.D.C.*—Univ. Minnesota Conference on Acoustical Fatigue (Eds. W. J. Trapp and D. M. Forney), *W.A.D.C. Tech. Rep.*, 59–676.
5. R. E. Franklin. Noise measurements on cold jets, using convergent divergent nozzles, *Aero. Quart.*, **8** (1957).
6. A. Powell. Noise of choked jets, *J. acoust. Soc. Amer.*, **25**, 385–389 (1953).
7. A. Powell. On the mechanism of choked jet noise, *Proc. Phys. Soc.*, (London) **B66**, 1039–1056 (1953).
8. M. J. Lighthill. On sound generated aerodynamically, *Proc. Roy. Soc.* (London) A **267**, 1329, 147–182 (1962).
9. M. J. Lighthill. Jet noise, *A.I.A.A. Jour.*, **1**, 7, 1507–1517 (1963).
10. J. N. Cole, H. E. von Gierke, D. T. Kyrazis, K. M. Eldred and A. J. Humphrey. Noise radiation from fourteen types of rocket in the 1,000 to 130,000 lb thrust range, *W.A.D.C. Rep. T.R.*, 57–354 (1957).
11. H. E. von Gierke, H. O. Parrack, W. J. Gannon and R. G. Hansen. The noise field of a turbo-jet engine, *J. acoust. Soc. Amer.*, **24**, 162 (1952).
12. G. M. Coles. Estimating jet noise, *Aero. Quart.*, **14**, 1 (1963).
13. K. Eldred. Prediction of sonic exposure histories. Proceedings of Symposium on Fatigue of Aircraft Structures, *W.A.D.C.-T.R.* 59–507, August (1959).

CHAPTER 8

Boundary Layer Pressure Fluctuations

8.1 Introduction

The flow of a turbulent boundary layer over the external surfaces of vehicles such as aircraft and submarines gives rise to several important engineering problems. These result mainly from the fact that the structure of such vehicles is elastic and not rigid and is caused to vibrate by the fluctuating pressure field associated with the turbulent boundary layer.

For example, if we are interested in the noise levels inside, or in the acoustic energy radiated from, a body moving in a fluid, we know that even if the surface could be made rigid the turbulent flow in the boundary layer would produce acoustic radiation. However, when the surface is flexible and vibrates as a result of boundary layer excitation the surface motion gives rise to an additional sound field in the fluid which may have a much greater intensity than that associated with flow over a rigid surface. Another result of boundary layer-induced vibration is that the material of the structure is subjected to fatigue stresses throughout the whole period of motion of the vehicle and this could be an important factor in designing high-speed aircraft to have an adequate fatigue life.

For purposes of calculating noise levels produced by a vibrating structure or assessing fatigue life it is necessary to calculate structural response, and before this can be done it is of course necessary to define the excitation. Therefore, in this chapter we will try to summarize knowledge of the character and statistical properties of the boundary layer wall pressure field.

8.2 R.m.s. values of fluctuating hydrodynamic stresses on the boundary surface

8.2.1 Pressure fluctuations

The earliest theoretical work on pressure fluctuations on the boundary surface was done by Kraichnan[1] who expressed the pressure fluctuations in terms of velocity fluctuations in the turbulence. He concluded that the

169

pressure was generated mainly by the interaction between turbulence and mean shear and estimated the root mean square value of the wall pressure fluctuation p' to be $6\tau_0$, where τ_0 is the mean shear stress at the boundary surface. The analysis was extended by Lilley[2] who obtained a value of $3\tau_0$.

At subsonic speeds experimental measurements of p' have been made by quite a large number of investigators under a variety of experimental conditions ranging from laboratory to flight tests. The r.m.s. pressure was usually expressed in terms of the dynamic pressure of the free stream as p'/q_0, and considerable variation in the value of this parameter was found. Reported values ranged from 1.5×10^{-3} to 12×10^{-3}, and the most generally accepted value was that obtained by Willmarth[3], namely 6×10^{-3}. Apart from the inevitable experimental scatter there are probably many reasons for these large discrepancies. For example, in many cases the microphone or pressure pick-up used was quite large in relation to the thickness of the boundary layer and the scales of the pressure field. In consequence they suffered from a lack of resolution at high frequencies; this probably accounts for many of the very low values which were obtained. Another important factor which has been omitted in merely comparing p'/q_0 values is the wall shear stress. More recent laboratory work has produced a more consistent set of results for p'/τ_0 (giving values between 2 and 3), but there is still some uncertainty about the effect of Reynolds number on this quantity. Corcos[4] has found that for pipe flow p'/τ_0 decreases with increasing Reynolds number whereas the University of Southampton 9 in $\times$ 6 in wind tunnel results show an increase with Reynolds number.

Willmarth's early work[3] and Von Gierke's flight test results[5] suggest that the variation of p'/τ_0 with Mach number over the subsonic speed range is not great, but at supersonic speeds the value increases. The measurements made by Williams[6], when corrected for transducer size, lie in the range $3 < p'/\tau_0 < 4.5$ for $1.2 < M_0 < 1.6$ and recent results obtained by Kistler and Chen[7] show p'/τ_0 increasing from 3.5 at $M_0 = 0.6$ to values of 5 to 6 at $M_0 = 5$. Kistler and Chen also found that the value of p'/τ_0 was fairly insensitive to Reynolds number variation over the Mach number range 1 to 5.

8.2.2 Shear stress fluctuations

The r.m.s. shear stress fluctuation on the boundary surface can be expressed as

$$\tau'_{0_f} = \mu \left[\left(\frac{\partial u'_1}{\partial x_2} \right)^2_0 + \left(\frac{\partial u'_3}{\partial x_2} \right)^2_0 \right]^{1/2} \tag{8.1}$$

(where u_1', u_3' are r.m.s. velocity fluctuations in the x_1 and x_3 directions respectively, μ is the fluid viscosity, and suffix 0 indicates a value at the boundary surface). This can therefore be obtained from measurements of the variation of the r.m.s. longitudinal and transverse velocity fluctuations with distance from the boundary surface (x_2). Laufer's experiments with flow in pipes[8] give $\tau_{0_f}' \approx 0.3\tau_0$ while similar experiments on two-dimensional channel flow[9] show slightly higher values.

The r.m.s. shear stress fluctuations are therefore considerably smaller than the pressure fluctuations and so attention will be focused on the pressure fluctuations in the remainder of this note. However, it should be noted that Kraichnan's analysis[1] indicates that the pressure fluctuations show strong cancellation effects over large areas and the possibility exists that if this does not also occur in the case of shear stress fluctuations, the shear forces over an area of the boundary surface could be as large as the pressure forces. Any importance attaching to shear stress fluctuations concerns the structural loading only, since noise generation depends essentially on transverse deflection of the structure.

8.3 Space–time correlations and spectra of the boundary layer pressure field

The statistical properties of the wall pressure field which are to be discussed are particular values of the pressure covariance on the boundary surface (considered as a flat plate in the $x_1 - x_3$ plane) given by

$$R_{pp}(\xi_1, \xi_3, \tau) = \langle p(x_1, x_3, t)p(x_1 + \xi_1, x_3 + \xi_3, t + \tau)\rangle \qquad (8.2)$$

or its normalized value, the correlation coefficient

$$\rho_{pp}(\xi_1, \xi_3, \tau) = R_{pp}(\xi_1, \xi_3, \tau)/\langle p^2\rangle. \qquad (8.3)$$

Here ξ_1 is the spatial separation in the x_1 direction, t represents time, τ time delay, and $\langle\ \rangle$ indicates a statistical mean value. The pressure field has been assumed statistically homogeneous in the $x_1 - x_3$ plane and stationary in time.

We have already considered one particular value of R_{pp}, namely the r.m.s. pressure given by

$$(p')^2 = \langle p^2\rangle = R_{pp}(0, 0, 0). \qquad (8.4)$$

The Fourier transform of R_{pp} with respect to time yields the spectral function

$$S_{pp}(\xi_1, \xi_3, \omega) = \frac{1}{2\pi}\int_{-\infty}^{\infty} R_{pp}(\xi_1, \xi_3, \tau)\,e^{-i\omega t}\,d\tau. \qquad (8.5)$$

The power spectral density of the pressure fluctuations as measured by a wave analyser is given by

$$G_p(\omega) = S_{pp}(0, 0, \omega) + S_{pp}(0, 0, -\omega)$$

$$= \frac{2}{\pi} \int_0^\infty R_{pp}(0, 0, \tau) \cos \omega\tau \, d\tau. \qquad (8.6)$$

8.3.1 *Power spectrum of the wall pressure fluctuations*

The results of power spectral density measurements made in the Southampton 9 in × 6 in boundary layer tunnel under a variety of flow conditions are shown in Fig. 8.1. They are plotted in non-dimensional form as $G_p(\omega)U_0/q_0^2\delta^*$ versus $\omega\delta^*/U_0$ where

U_0 = free stream velocity
q_0 = free stream dynamic pressure
δ^* = boundary layer displacement thickness
ω = circular frequency, radian/second.

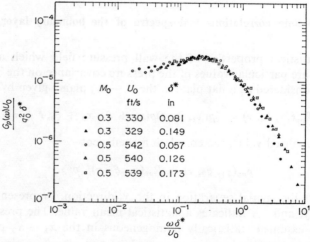

M_0	U_0 ft/s	δ^* in
• 0.3	330	0.081
▲ 0.3	329	0.149
○ 0.5	542	0.057
▲ 0.5	540	0.126
□ 0.5	539	0.173

Fig. 8.1 Frequency spectrum of wall-pressure fluctuations.

The pressure transducers used to obtain these data had pressure-sensitive elements 0.030 inches in diameter in all cases so that the value of the ratio of transducer diameter to displacement thickness varied with the thickness of boundary layer under investigation covering the range $0.17 < d/\delta^* < 0.52$. The transducer resolution was therefore best in the case of the thickest boundary layers, and this becomes evident at the higher frequencies, $\omega\delta^*/U_0 > 1$, when the measured values of spectral density in thin boundary layers fall below those for the thicker layers. It is believed that the results shown for the thickest boundary layers do not contain significant errors due to the finite size of the transducer, but even for transducers as small as these the work of Corcos[4] on transducer resolution effects indicates that very large relative errors in the spectral density measurements could be introduced for $\omega\delta^*/U_0 > 5$. (For further discussion of transducer resolution, see Willmarth and Roos[10].)

The results shown are fairly typical of recent subsonic measurements with small-pressure transducers and are in quite good agreement with those reported in reference 11. The experimental work therefore indicates a spectrum which is fairly flat (or falling slowly with decreasing frequency) at low frequencies ($\omega\delta^*/U_0 < 0.5$), and which falls off with increasing frequency at frequencies higher than this. It should be noted, however, that the high-frequency region where the spectrum is falling off still accounts for a large proportion of the total pressure energy.

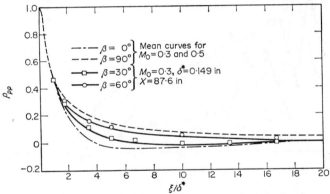

Fig. 8.2   Space correlations of the wall-pressure field,

$$\rho_{pp}(\xi \cos \beta, \xi \sin \beta, 0)$$

at various angles to the flow direction.

The only data available for supersonic speeds are those of Kistler and Chen[7], and, although they are not directly comparable with the subsonic spectra in the non-dimensional form used here, the supersonic spectra appear to be broadly similar in shape. The boundary layer momentum thickness and the free stream velocity were found to be the relevant scaling parameters.

8.3.2 *Space correlations*

The variation with spatial separation of the pressure correlation coefficient at zero time delay, $\rho_{pp}(\xi \cos \beta, \xi \sin \beta, 0)$, along lines at various angles, β, to the flow direction is shown in Fig. 8.2. These results were

obtained from the 9 in × 6 in boundary layer tunnel[12]. Mean curves are
shown for the streamwise correlation $\rho_{pp}(\xi_1, 0, 0)$ at $\beta = 0$ and the lateral
correlation $\rho_{pp}(0, \xi_3, 0)$ at $\beta = 90°$. They represent data covering a range
of boundary layer thicknesses at two free stream Mach numbers $M_0 = 0.3$
and 0.5.

The form of the curves indicates a lack of symmetry of the pressure
field—the longitudinal correlation falls to zero in a distance of about $4\delta^*$
and then becomes negative, while the lateral correlation remains positive
at the largest values of separation investigated, and the curves for the other
values of β have intermediate forms. The results show that the lateral
integral scale of the pressure field is about twice as great as the longitudinal
scale.

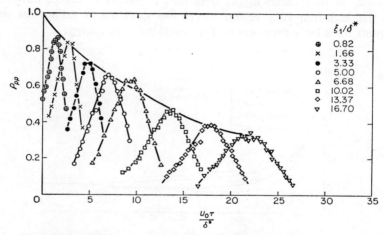

Fig. 8.3   Peaks of curves of longitudinal space–time correlation
$\rho_{pp}(\xi_1, 0, \tau)$. $M_0 = 0.3$, $\delta^* = 0.149$ in, $X = 87.6$ in.

8.3.3 *Space–time correlations and convection velocities*

Space–time correlations, $\rho_{pp}(\xi_1, 0, \tau)$, in the flow direction have the
form shown in Fig. 8.3. From the values of ξ_1 and time delay where the
correlation curves for constant values of ξ_1 are tangential to their envelope
curve the velocity of convection U_c of the pressure field can be found (or
U_c' from the slope of the ξ_1 vs τ curve). Most of the early measurements
showed values of U_c/U_0 in the vicinity of 0.8 but a more detailed examination at smaller spacings of the measuring points shows that this velocity
is not constant but varies with ξ_1. As can be seen from Fig. 8.4, when the
separation of the measuring points is very small the pressure field appears
to have a convection velocity in the region of 0.53 U_0, but the value in-

creases with increasing ξ_1 reaching 0.82–$0.83\ U_0$ at large separations. Similar results are given in reference 11.

This behaviour implies that the components of the pressure field with small longitudinal scales are convected relatively slowly at speeds of the order of $0.5\ U_0$ and quickly lose coherence so that at large separations the main contribution to the correlation comes from large-scale components which are convected quickly, at speeds typically of the order of $0.8\ U_0$.

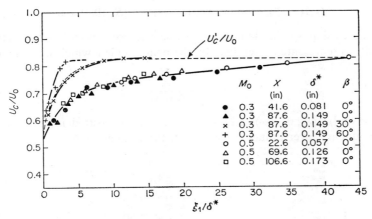

Fig. 8.4   Variation of convection velocity with spatial separation.

Figure 8.4 also shows convection velocities determined in a similar way from correlations for which $\xi_3 \neq 0$. In these cases the convection velocity rises more rapidly as ξ_1 increases, implying that in addition to losing coherence quickly the components of the field with small longitudinal scale are correlated only over small lateral distances. In consequence the greater ξ_3 the more predominant is the contribution of the large-scale components to the correlation and to the effective convection velocity even at $\xi_1 = 0$.

8.3.4 Narrow-band space–time correlations

We assume that a particular frequency at a fixed observation point arises essentially from the convection of a component of the pressure field with a particular scale having a characteristic convection velocity (an assumption which the measurements show to be broadly true). We can then obtain more detailed information about the convection velocities and the rate of loss of coherence of components of various scales from correlation measurements on the pressure signals after they have been passed through narrow-band filters.

It can be shown that the narrow-band correlation is given by

$$\rho_{pp}(\xi_1, \xi_3, \tau\|\omega) = \frac{|S_{pp}(\xi_1, \xi_3, \omega)|}{|S_{pp}(0, 0, \omega)|} \cos(\omega\tau + \alpha), \qquad (8.7)$$

S_{pp} being the complex spectral function given by equation 8.5. The way in which a particular component loses coherence is represented by the behaviour of the amplitude of the narrow-band correlation

$$|\rho_{pp}(\xi_1, \xi_3, \tau\|\omega)| = |S_{pp}(\xi_1, \xi_3, \omega)|/|S_{pp}(0, 0, \omega)|$$

with increasing spatial separation, while the corresponding convection velocity is represented essentially by the phase angle α which we can write as

$$\alpha = -\frac{\omega\xi_1}{U_c(\omega)}. \qquad (8.8)$$

Values of $|\rho_{pp}(\xi_1, 0, \tau\|\omega)|$ and $|\rho_{pp}(0, \xi_3, \tau\|\omega)|$ measured in the 9 in × 6 in boundary layer tunnel are shown in Figs. 8.5 and 8.6.

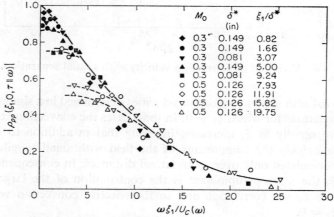

	M_0	δ^* (in)	ξ_1/δ^*
◆	0.3	0.149	0.82
●	0.3	0.149	1.66
▼	0.3	0.081	3.07
▲	0.3	0.149	5.00
■	0.3	0.081	9.24
○	0.5	0.126	7.93
◇	0.5	0.126	11.91
▽	0.5	0.126	15.82
△	0.5	0.126	19.75

Fig. 8.5 Amplitude of narrow-band longitudinal space–time correlations of the wall-pressure field.

Except at low frequencies the data tend to collapse when plotted in terms of the non-dimensional variables $\omega\xi_1/U_c(\omega)$ and $\omega\xi_3/U_c(\omega)$. This implies that in general a component of a particular scale ($= U_c(\omega)/\omega$) retains coherence in the flow direction for the time taken for it to be convected a distance proportional to its scale length, and similarly has coherence over a lateral distance which is proportional to its scale length. The value of $U_c(\omega)$ is found to decrease as the value of $\omega/U_c(\omega)$ increases, i.e.

as the scale decreases. It still shows some dependence on spatial separation of the measuring points but much less than that shown by the overall convection velocity. The results of narrow-band measurements are therefore in accord with the conclusions drawn from the overall correlation and convection velocity measurements.

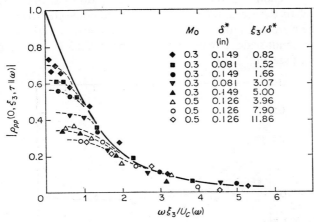

M_0	δ^* (in)	ξ_3/δ^*
◆ 0.3	0.149	0.82
■ 0.3	0.081	1.52
● 0.3	0.149	1.66
▼ 0.3	0.081	3.07
▲ 0.3	0.149	5.00
△ 0.5	0.126	3.96
○ 0.5	0.126	7.90
◇ 0.5	0.126	11.86

Fig. 8.6   Amplitude of narrow-band lateral space–time correlation of the wall-pressure field.

8.4 Probability distribution of the wall-pressure fluctuations

Measurement of the probability distribution of the pressure signal obtained from a turbulent boundary layer in a water flow[13] for a free stream velocity of 22.2 ft/s and boundary layer displacement thickness of 0.083 in given values of form factor of 1.27, skewness factor −0.082 and flatness factor 2.74. The corresponding values for a Gaussian distribution are 1.25, 0, and 3 respectively, from which it can be concluded that the signal is very nearly Gaussian.

8.5 Summary of the characteristics of the wall-pressure field of a turbulent boundary layer on a smooth wall

Summarizing, we can say that in subsonic flows the wall pressure field associated with a turbulent boundary layer developed on a smooth wall has the following characteristics.

The r.m.s. pressure fluctuation has a value of between 2 and 3 τ_0 depending on Reynolds number. It has a frequency spectrum very similar to that shown in Fig. 8.1, and the statistical distribution of pressure amplitudes at a given point is very nearly Gaussian. It has the character of a convected slowly varying random-pressure field which can be considered as being made up by the superposition of pressure components of various scales. With each component can be associated a reasonably well defined convection velocity or limited range of convection velocities. The various components, with the exception of those of the largest scale, are coherent over lateral distances proportional to their scale lengths. They retain coherence in the flow direction for times which are proportional to the times taken for them to be convected over distances equal to their scale lengths. The velocity of convection varies from about $0.53 U_0$ for small-scale components to about $0.83 U_0$ for large-scale components. (See Fig. 8.7.)

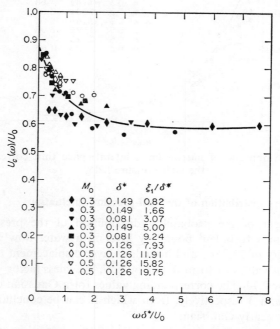

Fig. 8.7 Convection velocities derived from narrow-band longitudinal space–time correlations of the wall-pressure field.

The pressure field in supersonic flow has not been investigated in such great detail, but from available information appears to have a similar character to that found in subsonic flow. The r.m.s. pressure is greater, rising to 5 to 6 τ_0 at $M_0 = 5$, and the spectrum broadly similar. It still has the character of a convected pressure pattern but the correlation, as seen by an observer moving at the convection speed, tends to fall off much more rapidly with time than in the subsonic case.

8.6 Effects of surface irregularities

No really systematic investigation of the effects of surface roughness or discontinuities has been made but we do have some isolated pieces of data.

Willmarth and Wooldridge[11] report increases in p'/τ_0 of the order of 50% resulting from machine marks and small irregularities in the boundary surface in the vicinity of the transducer. A similar effect was observed in early pressure measurements made in the 9 in × 6 in tunnel at Southampton, in this case due to irregularities over the face of the pressure transducer itself, i.e. over a circle of radius $\frac{1}{8}$ in centred on the sensitive element. The value of p' was found to increase to as much as twice its smooth-wall value.

Measurements of r.m.s. pressure in the wall behind a step in a water flow have been reported[13]. Here, too, values up to twice the smooth-wall value were obtained.

In supersonic flows there are even less available data, although some measurements have been made by Kistler (reported at the AGARD Conference on Mechanism of Noise Generation in Turbulent Flow, April 1963) in the separated flow region in front of a forward-facing step. Quite dramatic increases in pressure level were observed—as much as 20 times the smooth-wall value. Thus while we have not many data on the effects of wall condition, we do have sufficient to emphasize the importance of achieving a good surface finish to minimize the intensity of surface pressure fluctuations.

References

1. R. H. Kraichnan. Pressure fluctuations in a turbulent flow over a flat plate, *J. acoust. Soc. Amer.*, **28**, 378 (1956).
2. G. M. Lilley. Pressure fluctuations in an incompressible turbulent boundary layer, *College of Aeronautics Rep.*, No. 133 (1960).
3. W. W. Willmarth. Space–time correlations and spectra of wall pressure in a turbulent boundary layer, *N.A.S.A. T.M.* 3–17–59 W.
4. G. M. Corcos. Pressure fluctuations in shear flows, *Univ. Calif. Inst. Eng. Res. Rep.* Series No. 183, Issue 2 (July 1962).
5. H. E. Von Gierke and K. Eldred. Proceedings of W.A.D.C.—University of Minnesota Conference on Acoustical Fatigue, 1959, *W.A.D.C. Tech. Rep.* 59–676, p. 57 (1961).
6. D. J. M. Williams. Measurements of the surface pressure fluctuations in a turbulent boundary layer in air at supersonic speeds, *Univ. Southampton Rep. A.A.S.U.* 162 (Dec. 1960).
7. A. L. Kistler and W. S. Chen. The fluctuating pressure field in a supersonic turbulent boundary layer, *Jet Propulsion Lab. Rep.* No. 32-277 (August 1962).

8. J. Laufer. The structure of turbulence in fully developed pipe flow, *N.A.C.A. Rep.* 1174 (1954).

9. J. Laufer. Investigation of turbulent flow in a two-dimensional channel, *N.A.C.A. T.N.* 2123 (July 1950).

10. W. W. Willmarth and F. W. Roos. Resolution and structure of the wall pressure field beneath a turbulent boundary layer, *J. Fluid Mech.*, **22**, 81 (1965).

11. W. W. Willmarth and C. E. Wooldridge. Measurements of the fluctuating pressure at the wall beneath a thick turbulent boundary layer, *Univ. Michigan Rep.* 02920–1–T (April 1962). (Also *J. Fluid Mech.*, **14**, 187 (1962).)

12. M. K. Bull. Properties of the fluctuating wall-pressure field of a turbulent boundary layer, *Univ. Southampton Rep. A.A.S.U.* 234 (March 1963). (Also *AGARD Rep.* No. 455 (1963).)

13. M. K. Bull and J. L. Willis. Some results of experimental investigations of the surface pressure field due to a turbulent boundary layer, *Univ. Southampton Rep. A.A.S.U.* 199 (Nov. 1961).

CHAPTER 9

Propeller, Helicopter and Hovercraft Noise

9.1 Introduction

The problem of propeller noise has long been recognized as one of considerable importance and has become a factor in propeller design. It was accentuated with the growth of engine power and the adoption of the high powered turbo-propeller engine. With the introduction into general service of the noisier pure jet engine the emphasis on propeller noise decreased and little work has been carried out on this topic for the last five to ten years. Recently, however, the problem of propeller noise has arisen again with the use of propellers in hovercraft. Its importance is even more acute than before because it is planned to use hovercraft for travel into highly populated areas. Researches carried out over a period of many years have established the physical nature of the various sources of noise on a propeller and reasonably accurate predictions, at least for high speed propellers, can now be made in the design stage.

The helicopter and the hovercraft both deserve special consideration because, although much of their noise can be explained in terms of propeller noise theory, there are a number of other sources which are exclusive to these machines and which can all make significant contributions to the overall levels.

The purpose of this chapter is to examine the physical characteristics and theoretical predictions of propeller, helicopter and hovercraft noise, and includes a discussion of some practical methods of noise reduction.

9.2 Sources of propeller noise

The noise spectra of two propellers, one with a subsonic and the other with a supersonic tip speed are shown in Fig. 9.1.

Although the shape of the two is different, the presence of a large number of peaks is common to both, suggesting that a strong source of propeller noise arises from periodic excitation at the blades. In fact these peaks are found to occur at the fundamental and successive harmonics of

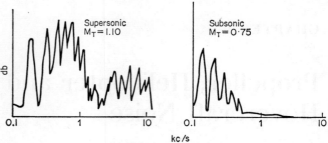

Fig. 9.1   Typical propeller spectra.

the blade passage frequency. The supersonic propeller is far richer in harmonics, suggesting that the surrounding air is subjected to a far sharper impulsive force than occurs on the slower subsonic blades.

In addition to the discrete frequency component of the spectra, there is found to be a region of broad band noise above one to two kilocycles; and at the lower range of tip speeds this can become comparable in magnitude to the discrete frequency noise. The presence of a broad band component indicates that random disturbances at the blades have also to be considered as possible sources.

It is convenient to discuss the origin of propeller noise under two headings

(1) Discrete frequency or 'rotational' noise arising from periodic disturbances of the air by the propeller.

(2) Broad band or 'vortex' noise arising from random disturbances at the propeller.

9.2.1 *Rotational noise*

(a) *Noise arising from the steady forces on the blades (thrust and torque)*

Each element of a propeller blade has a pressure distribution acting on it due to its motion through the air, which can be resolved into a thrust and a torque force. Conversely, the air in contact with the propeller has a force on it which can be resolved into the thrust and torque directions. This pressure field on the air is steady relative to the blade and rotates with it. At a fixed point on the propeller disc however the rotating field appears as an oscillating pressure—the frequency of the oscillation being the frequency with which blades pass the point, the waveform being determined by the chordwise distribution of pressure on the blades. A typical trace of the thrust or torque force on an element, $R.dR.d\theta$, of air in the propeller disc, due to the passage of a well loaded blade, might be represented by Fig. 9.2.

Such a distribution was considered by Gutin[1] in an analysis which has formed the basis of the present theoretical treatment of propeller noise. Gutin's approach was to replace the force at any point by the sum of the steady force and series of periodic forces (Fig. 9.2) which added up to the

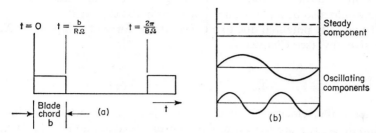

Fig. 9.2   (a) Pressure distribution on element; (b) Fourier analysis of (a).

actual torque and thrust at that point in space. The propeller was thus represented by an array of dipole sources from which the acoustic field could be calculated. The outline of the analysis is basically the same as that developed generally in Chapter 1 for force fields and can be outlined as follows:—

The thrust and torque loading density per unit radius for the rectangular distribution can be expressed as:—

$$\frac{1}{B}\frac{dT}{dR} \quad \text{and} \quad \frac{1}{BR}\frac{dQ}{dR}$$

where B is the number of blades, R is the radius of the element considered, and T and Q are the thrust and torque respectively.

Since the excitation is periodic it can be Fourier analysed and the harmonics considered separately, as indicated in Fig. 9.2(b). In the course of his analysis Gutin considers the blade width to be very small so that the excitation becomes effectively an impulse function (the effect of finite blade width is discussed later). Under these conditions the amplitude of the harmonics is constant and the expressions for the thrust and torque forces become

$$F_T = \frac{1}{\pi}\frac{dT}{dR}\, e^{i(\omega t - mB\theta)}$$

$$(9.1)$$

$$F_Q = \frac{1}{\pi R}\frac{dQ}{dR}\, e^{i(\omega t - mB\theta)}$$

where m is the order of the harmonic and $\omega = mB\Omega$, with Ω the angular velocity of rotation.

The phase angle $mB\theta$ arises from the fact that the forces on a general element in the disc at (R, θ) will be of the same amplitude as those at $\theta = 0$ but retarded by time θ/ω. In the general case of finite blade width there would also be a phase angle included in the exponent.

The velocity potential due to a concentrated force of components X, Y, Z is given by, (see Chapter 1)

$$\phi = \frac{-i}{4\pi k \rho c} \left\{ X\frac{\partial}{\partial x} + Y\frac{\partial}{\partial y} + Z\frac{\partial}{\partial z} \right\} \frac{e^{-ikr}}{r} \tag{9.2}$$

where c is the velocity of sound, $k = \omega/c$ is the wave number of the mth harmonic, x, y, z are the Cartesian co-ordinate axes and r is the distance of the observer from the source (see Fig. 9.3).

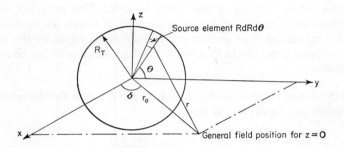

Fig. 9.3   Position of observation point relative to source.

It can be shown that if r is large compared with the propeller radius, and the observation point is in the x, y plane (this can always be attained by a suitable choice of origin when rotational symmetry exists) then

$$r = r_0 - R\cos\theta\sin\delta \tag{9.3}$$

where these quantities are defined in Fig. 9.3.

Using equations 9.1 and 9.3 the velocity potential can then be evaluated from equation 9.2. The analysis can be further simplified by considering only the far field radiation, i.e. where terms such as $1/r^2$ can be neglected in comparison with those involving $1/r$. It is then permissible to neglect the second term on the right hand side of equation 9.3 when r appears in the denominator. This term must be included in the exponent of equation 9.2, however, since it is comparable in magnitude to the wavelength. With these approximations, the expression for the velocity potential due to all the sources in the propeller is

$$\phi_{mB} = \frac{e^{ik(ct-r_0)}}{4\pi^2 \rho c r_0} \int_0^{R_T} \int_0^{2\pi} \left\{ \frac{dT}{dR} \cos \delta + \frac{1}{R} \frac{dQ}{dR} \sin \delta \sin \theta \right\} \times$$

$$\times \exp i(kr \sin \delta \cos \theta - mB\theta) \, dR \, d\theta \qquad (9.4)$$

The sound pressure is given by

$$p_{mB} = -\rho \frac{\partial \phi_{mB}}{\partial t} \qquad (9.5)$$

Using this and carrying out the θ integration in equation 9.4 yields

$$p_{mB} = \frac{mB\Omega}{2\pi c r_0} \int_0^{R_T} \left\{ -\frac{dT}{dR} \cos \delta + \frac{c}{\Omega R^2} \frac{dQ}{dR} \right\} J_{mB}(kR \sin \delta) \, dR \qquad (9.6)$$

where $J_n(x)$ is a Bessel function of the first kind of order n and argument x.

The integral in equation 9.6 can only be evaluated if the distribution of thrust and torque along the radius is known. Gutin, however, obtained an approximate solution in which the integral of each term is expressed in terms of a mean radius. It may be shown that for the first few harmonics these radii are approximately equal to an effective radius, R_e, at which the total torque and thrust forces may be assumed to act. The value of R_e will clearly vary with load conditions but as far as practical calculations are concerned a value of $0.8R_T$ appears to give good agreement with experiment.
Then

$$p'_{mB} = \frac{mB}{2\sqrt{2}} \frac{\Omega}{\pi c r_0} \left\{ -T \cos \delta + \frac{cQ}{\Omega R_e^2} \right\} J_{mB}(kR_e \sin \delta) \qquad (9.7)$$

where the prime denotes root mean square quantities.

This gives the r.m.s. pressure in the far field at a distance r_0 and azimuth angle δ for the mth harmonic of blade passage frequency from a single free running propeller. It is assumed that the forward speed is sufficiently low not to alter the directional pattern of the sound.

(b) *Noise due to the finite thickness of propeller blades*

In the previous section the disturbances considered were force fluctuations, set up by the pressure field rotating with the blade, and these were represented by an array of dipole sources. In addition to experiencing a fluctuating force, an element of air in the propeller will be physically moved aside by the finite thickness of the blade. This volume displacement of air is better represented by a distribution of simple sources (see Chapter

1) whose strength is determined by the component of displacement velocity normal to the propeller plane.

This approach was made by Deming[2] in an analysis in which a symmetrical blade section was considered to be operating at zero incidence. Because of the symmetry of the airflow it was assumed that only one half of the blade was operating and working next to a wall of infinite extent. The problem is then reduced to calculating the acoustic field from an array of simple sources distributed in an infinite baffle. This is obtained from an expression given by Rayleigh for the velocity potential due to one such source of velocity potential.

$$\phi = -\frac{1}{2\pi} \frac{\partial \phi_1}{\partial n} \cdot \frac{e^{-ikr}}{r} \, dS \qquad (9.8)$$

where dS is the area of the elementary source and $\partial \phi_1/\partial n$ is the component of the air displacement velocity v_n normal to the plane. This can be obtained from a knowledge of the pressure distribution over the blades, and may appear in the form of Fig. 9.4.

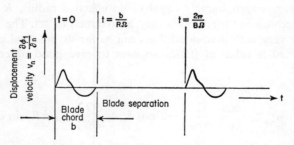

Fig. 9.4   Typical time history of the normal component of displacement
velocity on element of propeller disc.

Again since the excitation is periodic it is convenient to express the normal velocity v_n in the form of a Fourier series and to consider the separate harmonics.
Then

$$v_n = \frac{\partial \phi_1}{\partial n} = \dot{\xi}_0 \{ a_{1B} \sin (Bnt + \epsilon_{1B}) + \cdots a_{mB} \sin (mB\Omega t + \epsilon_{mB}) \} \quad (9.9)$$

where $\dot{\xi}_0$ is a parameter proportional to the section velocity ΩR, and dependent on section shape, and a_{mB} is the Fourier coefficient. Substituting the harmonic form of $\partial \phi_1/\partial n$ into equation 9.8 and integrating the effect of all the sources in the propeller disc,

$$\phi_{mB} = -\int_0^{R_T} \int_0^{2\pi} \frac{\dot{\xi}_0 a_{mB}}{2\pi r} e^{i(\omega t - mB\theta - kr + \epsilon_{mB})} \, d\theta \, dR \qquad (9.10)$$

If only the far field is considered, and the θ integration is carried out, the velocity potential becomes

$$\phi_{mB} = \frac{(-1)^{mB+1}}{r_0} e^{i(\omega t - kr_0 + \epsilon_{mB})} \int_0^{R_T} \dot{\xi}_0 a_{mB} J_{mB}(kR \sin \delta)R \, dR \qquad (9.11)$$

Now a_{mB} and $\dot{\xi}_0$ depend on the shape of the blade section and their variation with radius must be known before the integral in equation 9.11 can be evaluated. Deming, however, suggested that at least when considering lower harmonics, the typical normal velocity excitation could be approximated by triangular distributions so that the Fourier coefficients are known. In a later analysis of this case, Diprose[3] obtained an expression for the far field radiation which for the case of zero forward speed can be expressed in the form

$$p'_{mB} = \frac{\rho B \omega^2}{2\sqrt{2}\,\pi r_0} \int_0^{R_T} Ktb \, J_{mB}(kR \sin \delta) \, dR \qquad (9.12)$$

where t and b are the section thickness and chord at radius R, and K is a correction factor for finite solidity which is very nearly unity for small values of b/R.

Up to medium values of the resultant tip speed, thickness noise is generally found to be small compared with the noise arising from torque and thrust. At higher tip speeds however, it may assume equal importance. This is discussed further in section 9.3.2.

9.2.2 *Broad band noise* ('Vortex' noise)

The noise sources discussed so far have been periodic excitations resulting from a disturbance field which remains steady relative to the propeller blading. We now have to consider the effect of unsteady random disturbances which may be initiated at the propeller.

One such source for instance might be the turbulent region in the wake of the blades, but the reduced efficiency of this quadrupole radiation relative to the dipole radiation from unsteady forces on the blades themselves makes it unlikely to be significant. A discussion on how these dipole sources may be set up is included in the chapter on fan noise, and for a conventional propeller operating at normal incidence, away from the influence of solid bodies (wings, support struts and the like), one must draw the conclusion that the most likely source of such noise is the fluctuating forces exerted by the blades on the air during vortex shedding from the trailing edges.

Using the results of Chapter 10 it can be shown that the intensity can be expressed as

$$I = k(Re)^{-0.4} \cdot A\rho_0 \frac{U_T^6}{c^3} \cdot \frac{\cos^2\delta}{r_0^2} \tag{9.13}$$

where the constant k is of order 10^{-4}, and Reynolds number, Re, is based on tip speed, U_T, and on the mean chord. A is the blade area.

From the work of Yudin[4] on rotating rods in which a similar expression was obtained, Hubbard[5] has used an empirical relation to calculate the vortex component for a series of tip speeds (Fig. 9.6).

For the operating conditions of propellers considered in these charts, equation 9.13 gives values which are above but within 10 dB of Hubbard calculations. Allowing for (a) that the value of k in equation 9.13 was based on order of magnitude estimates, and (b) that the constant of proportionality used by Hubbard was obtained empirically from a single helicopter blade, equation 9.13 appears to provide a sufficiently accurate check on the available theoretical work on vortex noise.

The remarks so far apply only to propellers operating at normal incidence. If any part of the blades are stalled, local drag fluctuations set up by the separated flow may well appreciably increase the noise output. Tests on a model fan (Chapter 10) indicate that the increase could be of the order of 10 dB.

9.2.3 *Idealized directional patterns*

The theoretical distribution of the radiated sound around the propeller is shown in Fig. 9.5.

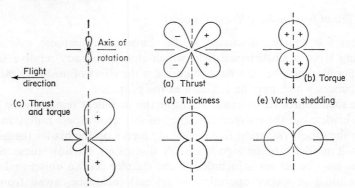

Fig. 9.5   Idealized directional patterns for the various noise sources.

The directionality of the torque component, (b), is determined by the function $J_{mB}(kR \sin \delta)$ and has the same phase all round. The thrust component, (a), is governed by $-\cos \delta \, J_{mB}(kR \sin \delta)$ and its phase is, therefore, reversed for angles between $\pm \pi/2$. The resulting distribution for lift noise is thus as shown in (c), with most of the radiation occurring to the rear of the propeller and a maximum between 100° and 120° depending on the harmonic considered.

The directionality of the thickness component, (d), is determined by the same function as for torque and hence is also a maximum in the plane of rotation. All the rotational components have zero radiation along the axis, which is to be expected under the ideal conditions where rotational symmetry exists in the propeller disc. The directionality of the noise from vortex shedding however, (e), varies as $\cos^2 \delta$ and this has its maximum on the axis and is zero in the plane of rotation. This may partly explain why many measurements of propeller noise show significant contributions on the propeller axis.

9.3 The effect of physical parameters

9.3.1 *Number of blades*

It has been shown during the discussion on rotational noise that the fundamental frequency of the source disturbances was that at which the blades passed any point in the disc. The effect of increasing the number of blades is to increase this fundamental frequency and to cancel all other harmonics except those which are integral numbers of the new blade number, the remaining harmonics being unaffected if the torque and thrust remains the same. This is shown in Hubbard's noise charts[5] which were calculated on the basis of equation 9.7. These represent the sum of the contribution of the first four harmonics at a value of δ of 105°, for various engine power ratings and numbers of blades. A typical chart is shown in Fig. 9.6 and shows clearly that, for a given tip speed, the noise is reduced by increasing the blade number.

The amount of relief obtained is seen to be a function of the tip Mach number M_T, being less at higher tip speeds. The reason for this can be found in the expression for thrust and torque noise (equation 9.7). Here the strength of the harmonics is governed by the dependence on mB, which for an effective radius of $0.8R_T$, can be expressed as $mBJ_{mB}(0.8M_T mB \sin \delta)$ where M_T is the tip Mach number. For values of $(0.8 \, M_T \sin \delta)$ less than about 0.7 the amplitude of the harmonics decreases with harmonic numbers, but above this the harmonics begin to assume greater importance than the fundamental. Under these conditions the cancellation of the fundamental and non-integral harmonics by adding blades will have less

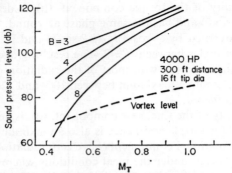

Fig. 9.6   Effect of number of blades and tip speed on noise from propellers at constant power.

effect. In an extreme case, if at high tip speeds, the net thrust and torque
is not left constant, an increase in these quantities due to extra blades
could result in an increase in noise, which is greater than the reduction
from cancelled harmonics.

An increase of blade number will tend to increase the intensity of vortex
noise. The unsteady force fluctuations on each blade is random in nature,
so that the total noise output will be proportional to the number of blades.
This is expressed in equation 9.13 by making A the total blade area.

9.3.2 *Tip speed*

The dependence of rotational noise on tip speed is not immediately
obvious from the equations. It is found, however, that with increasing tip
speed the harmonics of blade passage frequency increase at a greater rate
than the fundamental. This is clearly shown in the spectra of Fig. 9.1,
which are taken from the results of Hubbard and Lassiter,[6] where for the
supersonic propeller, the harmonics are greater than the fundamental. We
have seen in the preceding section that the harmonic amplitude function
is $mBJ_{mB}(0.8M_T mB \sin \delta)$. It was also noted that the question of whether
the amplitude of the harmonics was greater or less than the fundamental
(i.e. the shape of the spectrum), depended on the value of $(0.8M_T \sin \delta)$,
the critical value being about 0.7. (For directions near the plane of rotation
this would correspond to a tip Mach number of about 0.9.) Hubbard and
Lassiter compared these predictions with sound pressure measured in the
plane of rotation ($\sin \delta = 1$) for a two bladed propeller at subsonic and
supersonic tip speeds. The result is shown in Fig. 9.7.

For the subsonic case, the harmonic amplitudes calculated from equation 9.7 are confirmed by the measurements. For the supersonic case,
although agreement is good up to the third harmonic, the theory predicts
a continued rise in amplitude while the experimental curve indicates a

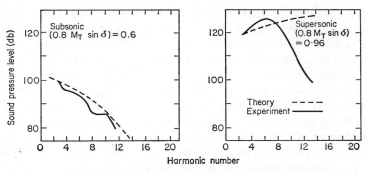

Fig. 9.7 Effect of tip speed on harmonic amplitudes.

fall-off. The reason for this is not known, but it has been suggested that the discrepancy arises from non-linear effects in the very high pressure fluctuations encountered at these speeds.

From the equation for thickness noise (equation 9.12) it is found that the harmonic amplitude parameter mB appears as m^2B^2 as opposed to the first power in equation 9.7 for thrust and torque. The effect of this is to lower the tip speed at which the harmonics become greater than the fundamental. It is difficult to measure separately the lift and thickness components of rotational noise from a given propeller, but Diprose has calculated the contributions to each that might be expected from a typical configuration.[3] The calculations include the effect of forward speed (see section 9.3.3), and the results are presented as the variation of the first three harmonics of lift and thickness noise with resultant tip speed (Fig. 9.8).

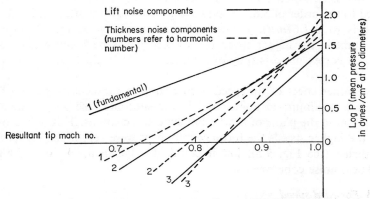

Fig. 9.8 Effect of resultant tip speed on lift and thickness noise. (*Crown copyright. Reproduced from Reference 3 by permission of the Controller, H.M.S.O.*)

Each harmonic of thickness noise has a greater slope than that of the corresponding harmonic of lift noise, and near a resultant tip Mach number of one, the harmonics of thickness noise dominate. Diprose suggests however that the theory of thickness noise overestimates the sound pressures at high tip Mach numbers because of non-linearities in the perturbation velocities over the blades. There is little evidence available on which an estimate of the discrepancy can be based and until this has been established it can only be assumed that thickness is responsible for some of the strengthening of harmonic content in the manner shown in Fig. 9.1.

As far as overall noise levels are concerned, Hubbard and Lassiter found that, for constant propulsive power, the measured overall noise (which includes all components) increased rapidly with increasing tip speed in the subsonic range, and in the supersonic range, at least up to a tip Mach number of 1.3 the noise was effectively independent of tip speed (Fig. 9.9).

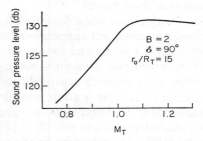

Fig. 9.9   Effect of tip speed on overall sound pressure level.

It has been noted that vortex noise intensity increases with tip speed raised to the order of 5.6, while Deming[7] has shown that the index for lift noise is $(6 + 1.67mB)$. Because of this, vortex noise is not normally important in the higher speed ranges, and it has been suggested that in practice vortex noise is only of significance below tip Mach numbers of 0.4.

One other effect of tip speed is to change the directivity pattern from having a maximum at about 120° in the subsonic case to one having a maximum in the plane of rotation at supersonic speeds. This would occur if at the higher speeds the torque component were very much higher than the thrust (see Fig. 9.5), but it could also be an indication of higher thickness noise components.

9.3.3 *Forward speed*

The effect of forward speed is to modify the length of the acoustic path between the source point and the observer. The necessary extensions to the theory of lift and thickness noise have been made by Garrick and Watkins[8] and Diprose.[3] The results for the far-field case can be expressed as

Lift noise

$$p'_{mB} = \frac{mB\Omega}{2\sqrt{2}\,\pi c S_0}\left\{-T\left(M_F + \frac{x}{S_0}\right)\frac{1}{\beta^2} + \frac{cQ}{\Omega R_e^2}\,J_{mB}\left(kR_e\,\frac{y}{S_0}\right)\right\} \qquad (9.14)$$

Thickness noise

$$p'_{mB} = \frac{\rho B\omega^2}{2\sqrt{2}\,\pi}\frac{1}{S_0\beta^4}\left\{1 + M_F\,\frac{x}{S_0}\right\}^2 \int_0^{R_0} K.tb\,J_{mB}\left(k.R\,\frac{y}{S_0}\right)dR \qquad (9.15)$$

where M_F is the forward Mach number,

$$\beta^2 = 1 - M_F^2$$

and

$$S_0^2 = x^2 + \beta^2 y^2.$$

The theory is derived for an observer moving with the same velocity as the source, but the results are applicable for the case of a stationary observer, provided that the correct instantaneous distances are used. The frequencies heard by the fixed observer will then be modified by the Doppler effect.

For real propellers operating at constant power, thrust decreases with increasing forward speed, so for low flight speeds where the value of β is nearly unity, the lift noise should decrease. This is indicated in the curves, calculated for a point near the tips, shown in Fig. 9.10, which are given in reference 9.

The calculations were based on the theory of Garrick and Watkins and show generally decreasing pressures up to a flight Mach number of about 0.5, and thereafter, as the terms in M_F become effective, a rapid increase with flight speed. General agreement with these trends has been obtained experimentally[10] for flight Mach numbers of up to 0.72. The increased slope of the higher harmonics shown in Fig. 9.10 suggests that as far as

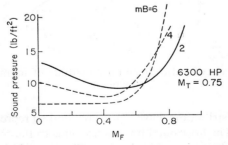

Fig. 9.10   Effect of forward speed on propeller noise. (*Courtesy of Elsevier Publishing Co. Inc., New York; see reference 9.*)

spectrum shape is concerned, an increase in forward speed has the same effect as an increase in tip speed.

9.3.4 *Blade width*

When considering the noise due to thrust and torque, it was noted that the assumption of impulsive excitation (zero blade width) led to the result that the amplitudes of the Fourier coefficients were independent of harmonic numbers (see equation 9.1). In fact, the coefficients for a rectangular excitation are

$$\frac{1}{\pi}\frac{dT}{dR}\left\{\frac{2R}{mbB}\sin\left(\frac{mbB}{2R}\right)\right\} dR\, d\theta$$

and

$$\frac{1}{\pi R}\frac{dQ}{dR}\left\{\frac{2R}{mbB}\sin\left(\frac{mbB}{2R}\right)\right\} dR\, d\theta.$$

The assumption that b is zero leads to the relationships of equation 9.1. This is still nearly true for finite blade width up to a value of $mbB/2R$ of approximately $\pi/4$, i.e. when the harmonic number m is of the order of $(4 \times \text{solidity})^{-1}$. Beyond this the actual function of b has to be used and the thrust and torque terms in the square brackets of equation 9.7 will each be a function of mB. Regier and Hubbard[11] have evaluated the effects of blade width and loading on the harmonic amplitudes and it is found that departures from the impulsive loading approximation increase with order of harmonic (see Fig. 9.11).

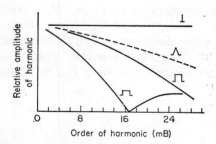

Fig. 9.11   Effects of pulse shape on amplitude of harmonics of blade loading.

All the load distributions have equal area and so represent equal forces on the element. The horizontal line corresponds to the constant harmonic amplitudes given by an impulsive loading. The base width of the triangular pulse is 3 per cent of the propeller circumference, and this is compared with rectangular pulses on 3 per cent and 6 per cent (lower curve) chords. All the curves are nearly equal to the impulsive loading curve for the first few harmonics, but above this the amplitudes decrease rapidly, the widest blade giving the greatest deviation from the impulse approximation.

For a given blade width, it is also possible to incur errors by representing the actual loading shape by a rectangular distribution giving the same force. This has been shown by Watkins and Durling[12] who examined two typical load distributions—a large positive loading and a small net positive loading with negative loads over part of the chord. Each case was first represented by a rectangular distribution and then considered in segments of chord length—each segment being represented by rectangular and triangular distributions. The results are shown in Fig. 9.12.

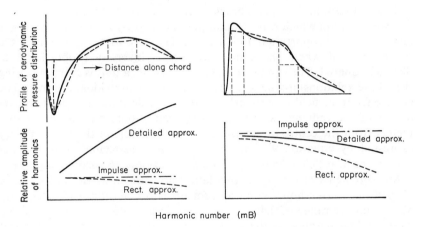

Fig. 9.12 Amplitude functions associated with typical force distributions.

For the well loaded blade a rectangular representation of the load is a good approximation to the actual distribution and both are close to the impulse approximation. For the small net loading, however, while the rectangular and impulsive approximations agree, they are found to seriously underestimate the true amplitude of all but the lowest harmonics. When considering noise from propellers operating under such conditions, it is therefore important to take into account the true shape of the load distribution.

9.4 Sound pressures in the near field

A knowledge of the oscillating pressure amplitude set up on the fuselage walls is of considerable importance in the design stage. The theoretical predictions discussed so far, however, have been based on the assumption of large distances between observer and source, and hence are not valid for the distances normally found between propeller tips and fuselage. To provide for this the theory for lift noise has been extended to give the free-space sound pressures at the general near field position.[13]

The instantaneous pressure for a given harmonic at any point is given by

$$p_{mB} = \frac{e^{ikct}}{4\pi^2} \int_0^{2\pi} \frac{1}{r_e}\left(Tx + \frac{Qy\sin\theta}{R_e}\right)\left[\frac{ik}{r_e} + \frac{1}{r_e^2}\right] e^{-i(mB\theta + kr_e)} d\theta \qquad (9.16)$$

where

$$r_e^2 = (x^2 + y^2 + R_e^2 - 2R_e y\cos\theta).$$

The first term in the square bracket represents the conventional acoustic radiation field in which r.m.s. sound pressure is proportional to frequency and inversely proportional to distance. The second term describes the induction field where the sound pressure is inversely proportional to distance squared. The exponential term represents a unit vector expressing the phase relationships between radiation from individual sources. The change from far field to near field conditions is dependent upon the wave-length of the radiated sound and is typically about two propeller dia-meters from the centre of rotation. Outside this region the Gutin theory expressed by equation 9.7 may be used with reasonable accuracy, but within, the near field effects expressed in equation 9.16 become important.

Measurements of free space oscillating pressures near the tips of static propellers[13] have shown good agreement with equation 9.16 over a tip Mach number range of 0.45 to 1.00, both in amplitude and distribution. Along a line parallel to the propeller axis, the important disturbances were found in the region near the plane of rotation, as shown in Fig. 9.13.

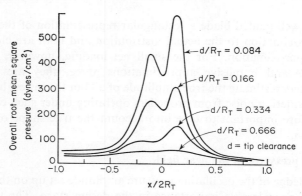

Fig. 9.13   Free-space pressures near propeller tips.

Maximum pressures were found to occur just ahead of the propeller plane and the amplitude of these peaks were roughly proportional to the inverse of the tip clearance. There was also found to be a phase difference of the order of 150° between corresponding points on either side of the plane of rotation. Another interesting result that emerged from the investigation was that the peak amplitudes of the lower harmonics of blade passage frequency decreased with increasing tip speeds up to $M_T = 1.0$, a result which contradicts the indications of far field investigations.

As mentioned earlier, the near field theory expressed in equation 9.16 relates to free space oscillating pressures, so that they do not represent the pressures on a solid surface in the acoustic field. Reference 13 includes a survey of the oscillating pressures on a flat vertical wall and a circular arc panel placed near the tips of a propeller to simulate a fuselage section. The results are shown in Fig. 9.14.

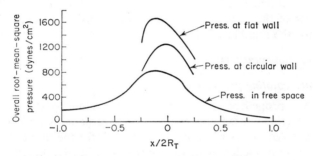

Fig. 9.14   Effect of reflecting surfaces.

The effect of the circular wall is to increase the overall sound pressure by a factor of about 1.5 while at the flat wall the pressures are approximately doubled.

In addition to the structural effects of pressures set up by the propeller, noise levels inside the aircraft are of interest. These could be calculated from known pressures outside the fuselage, but a general indication of the effect of the propeller parameters can be obtained from empirical results of a survey of cabin noise in a large number of military and civil aircraft of the period from 1941 to 1945.[15] The sound pressure level in the cabins was found

(a) to increase by about 2.7 dB for each increase of 100 ft/sec in propeller tip speed;
(b) to increase by approximately 5.5 dB for each doubling of horse power per blade;

(c) to increase rapidly as tip clearance between propeller and fuselage
was decreased below about 8″.

9.5 Empirical correlation of low tip speed noise

Although many of the ideas outlined in the previous sections allow
reasonably accurate prediction of propeller noise at high tip speeds, difficulty has been found in determining the noise for propellers running with
low tip speeds. For example, because of overall noise considerations it is
essential, if hovercraft are to operate successfully near built-up areas, that
propeller tip speeds are in the range Mach tip 0.4 to 0.6. The present
theories, however, fail to predict the higher harmonics of propeller noise
and, therefore, the overall noise levels. In an attempt to overcome these
limitations Trillo[14] surveyed practically all the available propeller noise
data and used this, together with measurements obtained from a hovercraft propeller test rig, to obtain an empirical method for determining the
noise. The relationship which is based on Mach Number at the tip and the
horse power of the system, gives the propeller noise in terms of dB after
allowing for the number of blades.

9.6 Special configurations

9.6.1 *Contra-rotating propellers*

When two propellers are mounted on the same axis and rotated in
opposite directions, it is found that the noise radiated is no longer axially
symmetric. A typical distribution about the axis for the case of two-bladed
contra-rotating propellers is of the form shown in Fig. 9.15.[16]

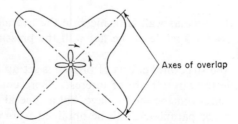

Fig. 9.15   Diagram of distribution of overlap noise in the vertical plane
from a dual-rotating propeller.

The noise is found to be a maximum in the directions of blade overlap
and minimum in directions mid-way between these. In the direction of
maximum radiation the noise has a magnitude and frequency spectra
which would be obtained from a single propeller having the same number

of blades as one stage of the dual (two blades in this case). In the direction of minimum radiation the levels and spectra are those which would be obtained from a single rotor propeller having the same number of blades as the total number on the dual (four). Similarly for two contra-rotating four-blade propellers, the noise would vary from that produced by a single four-blade propeller to that from a single eight-blade propeller.

If the speed of the two contra-rotating propellers is not the same, the overlap axes, shown as the dashed lines in the figure, will rotate. Then at a fixed observation point the noise will be amplitude modulated. If in addition the two propellers have different numbers of blades, then the only harmonics which are modulated in this manner are those which are integral numbers of both fundamentals, e.g. for a three-blade, two-blade configuration, the lowest modulated harmonic will correspond to $mB = 6$.

There was also found to be an additional component of noise arising from unsteady forces set up as the blades pass each other. A more detailed discussion on how such forces might be generated is included in Chapter 10; for the particular case of the free running propellers, however, this noise was found to be a maximum along the propeller axis and to be dependent on axial spacing of the propellers.

9.6.2 *Pusher propellers*

If a propeller is operating behind a solid obstacle, two effects can occur. Due to the flow over the obstacle the blades will pass through (a) a region of varying mean axial velocity associated with the potential flow around the obstacle, and (b) a region of turbulent air flow in the wake of the obstacle. Both these effects will cause local incidence charges and hence additional fluctuating forces on the blades which can seriously affect the noise output. In case (a) the additional noise will be periodic and for (b) additional broad band noise will result. The way in which these additional forces can be set up, and some experimental evidence obtained from an axial flow compressor rotor, is discussed in Chapter 10. The standard theory of propeller noise can still be applied for non-uniform loading around the disc providing one can express the variation of torque and thrust on a function of θ in equation 9.4. Again using the concept of equivalent radius and performing the radial integration first, yields for the far field radiation

$$p'_{mB} = \frac{mB\Omega}{4\sqrt{2}\,\pi^2 c r_0} \int_0^{2\pi} \left\{ T(\theta)\cos\delta + \frac{Q(\theta)}{R_e}\sin\delta\sin\theta \right\} \times$$

$$\times\, e^{-i(mB\theta - kR_e\sin\delta\cos\theta)}\, d\theta. \qquad (9.17)$$

A series of measurements made on an aircraft fitted with a pusher propeller[17] showed that the total noise from the aircraft was greater than that from a tractor aircraft operating at greater power and tip speed.

Most of this increase was found in the discrete noise with the harmonics being strengthened relative to the fundamental, but there was also some evidence of increased broad band noise. In the work on dual-rotating propellers referred to in the previous section, it was found that a strut placed upstream of the propeller produced more noise than one placed in the corresponding position downstream. Moreover, the strut upstream produced more noise when moved closer to the propeller, most of the increase again appearing in the discrete components of the noise.

9.6.3 *Shrouded propeller*

Shrouded propellers have been suggested as a means of silencing propellers and at the same time augmenting the thrust. An investigation has been carried out into the noise characteristics of a number of propeller-shroud configurations[18] and for unseparated flow over the shroud the overall levels around the propeller were reduced by about 50 per cent (see Fig. 9.16), which is equivalent to about 6 dB.

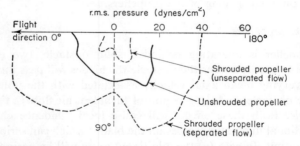

Fig. 9.16   Overall noise levels from shrouded propellers.

If the flow over the shroud separates ahead of the propeller, similar conditions exist to those discussed in the previous section of a pusher propeller operating in a disturbed flow. This is shown in Fig. 9.16 where the noise levels for separated flow are found to be greater than those for an unshrouded propeller.

9.7 Sources of helicopter noise

The noise spectrum from a helicopter is rather complex and indicates the large number of separate sources. Contributing to it (Fig. 9.17) the main sources can be identified and summarized as:

(i) *Main rotor noise*

As with the propeller this consists of (a) rotational noise, (b) vortex or

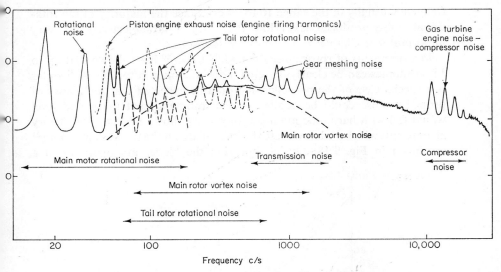

Fig. 9.17 Helicopter noise—narrow band spectrum.

broad band noise. Also associated with the main rotor, but not occurring in this particular case, is the discrete frequency noise associated with a certain operating condition known as 'blade slap'.

(ii) *Tail rotor noise*

The sources of noise from the tail rotor are basically the same as for the main rotor though the tail rotor is generally in such an unsteady flow area that discrete tones can occur due to unevenness of the general inflow through the disc.

(iii) *Power plant noise*

The characteristic of the power plant noise depends completely on the type of engine used, i.e. (a) Piston engine, (b) Gas turbine, and (c) Tip jets.

(iv) *Transmission noise*

This emanates from gearboxes and bearings and can be quite significant inside the cabin. Each of the above-mentioned sources are discussed in the following sections.

9.7.1 *Main rotor noise*

The mechanisms for the production of the rotational noise are similar to those outlined in the previous sections on propeller noise. The rotational noise is dominant below 200 c/s, with the fundamental blade passing frequency usually in the range 10 to 30 c/s. Since the majority of the rotational noise produced is of low frequency it is subjectively not very important (see Chapter 6 on hearing response).

Measurements of rotor noise on a whirl tower[20] have shown that up to 15 harmonics can be clearly detected. Since the main rotor of a helicopter is a special case of the propeller, it would be expected that the theory of propeller noise could be applied to the rotor and that the sound level would fall with harmonic numbers. In general, however, propeller theory gives results which underestimate the level of all but the fundamental note[21] as shown in Fig. 9.18. During hovering the blades are operating at a

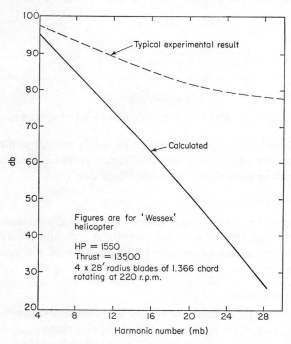

Fig. 9.18   Noise levels for a hovering helicopter at 50 ft, 15° behind rotor disc.

fixed collective pitch, so apart from possible near field interference from the fuselage, the blade loading on a single rotor helicopter would be expected to remain constant on the blades throughout each revolution. There should therefore be no difference in the accuracy of the noise calculations from that for the static propeller. Simons[22] and other investigators have shown that significant load fluctuations do exist even in the hover configuration, and this could account for the discrepancies already mentioned.

During forward flight, the difference in relative blade speed during forward and backward motion of the blade relative to the flight path requires a cyclic incidence variation to provide control and a reasonably uniform lift over the disc. To a first approximation, the forces on the air next to the disc would be constant under these conditions, the effects of incidence changes appearing only as variations of chordwise loading over the blade. The impulsive loading representation which leads to equation 9.7 would then give the levels of the first few harmonics, but because of the changing load distribution, this could, as pointed out in section 9.3.4, lead to significant errors in the calculation of the higher harmonics of the source waveform.

To sum up, therefore, the present difficulties of prediction of the high harmonic noise for a helicopter rotor almost certainly arises from inaccurate inputs regarding the force fluctuations on the blades. This can arise from poor aerodynamic flow, incidence changes, blade bending, self-excited blade oscillation and the like, all of which can contribute to the noise output. The existing theory can, however, be extended for these non-uniform effects by expressing the lift on the blade as a function of θ, as discussed in section 9.6.2.

In an attempt to overcome some of the limitations of calculating the blade loadings, Loewy and Sutton[23] have used measured blade loadings, but even so there is still a significant underestimation of the levels of the higher harmonics; the rapid fluctuations of loading were not measured in the tests, and a full representation of load fluctuations at the noise making frequencies is called for.

9.7.2 *Vortex Noise*

The frequency range for vortex noise is somewhat difficult to determine, but it is usual to consider that it is between 100 c/s and 1,000 c/s with the peak level occurring around 200 to 300 c/s. Since the directivity pattern for vortex noise rotates with the blades, the vortex noise will appear modulated to an observer situated away from the axis of rotation. The depth of this modulation is greatest at the frequency at which the peak frequency occurs.[24] Because of its frequency range and high level, this source is subjectively the most important on the main rotor. Although detailed evidence is sparse it appears also that the annoyance or loudness of such a noise is dependent on the depth of the modulation.

The theory developed for the propeller noise is again applicable, but it will be noted that equation 9.13 giving the vortex noise contains a constant k, which is not easily calculated. Davidson and Hargest[25] used actual helicopter measurements and data from reference 20 to develop a relationship for the sound pressure level of the vortex noise in terms of the blade parameters; i.e.

$$\text{Max. S.P.L.} = 60 \log_{10} V_t + 20 \log_{10} C_{Lt} + 10 \log_{10} S - 84.0 \text{ dB} \quad (9.18)$$

where V_t = blade tip speed—ft/sec

$\qquad C_{Lt}$ = lift coefficient referred to blade tip

$\qquad S$ = total blade plane area of the rotor—ft².

This S.P.L. refers to the noise 500 ft below the hub.

Goddard and Stuckey[20] obtained a more general relationship which gave the overall S.P.L. of the vortex noise at point (r, ϕ), viz.

$$\text{O.A.S.P.L.} = 16.6 \log T + 26.8 \log V_t$$
$$- 20 \log r - 20 \log \sec \phi + 2.8 \text{ dB} \quad (9.19)$$

where T = thrust per blade—lbs

$\qquad r$ = distance from source to observer—ft

$\qquad \phi$ = angle between tip located dipole axis and observer.

These formulae apply to hovering helicopters and appear to be reasonably accurate for the range of blade tip speeds in use on helicopters. In forward flight the noise levels would be expected to be higher, but little reliable evidence is yet available.

The vortex spectrum shape is fairly well defined, with the peak frequency t being given in terms of the Strouhal number, S_t.

$$S_t = \frac{fd}{V}$$

where V is the free stream velocity and d the projected blade thickness or more precisely the wake thickness. For the blade speeds under consideration S_t is about 0.2. Octave spectra presented in reference 26 show that the spectra for a tandem helicopter are considerably 'flatter' than those associated with single rotor helicopters and that the noise is most intense at positions to the side of the machine. This is quite understandable since on a tandem, particularly if the blade overlap is large, one rotor will be travelling through the turbulent downwash region associated with the other rotor. This will cause increases in the load fluctuations and hence noise. Measurements by Simons[22] and Leverton[27] of the loading and noise respectively has shown that they are affected by the slightest turbulence in the region of the blades (see also Chapter 10).

9.7.3 *Blade slap*

Blade slap or blade bang, as it is also called, is the sharp cracking or banging sound associated with the main rotor under certain operating conditions. When it occurs it is not only the predominant rotorcraft noise, but because of its impulsive nature (Fig. 9.19a) it is most objectionable

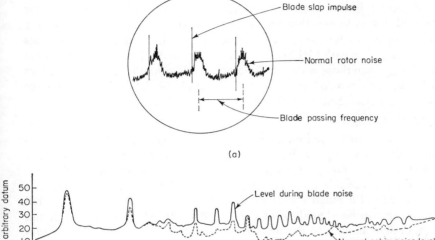

(a)

Blade passing orders

(b)

Fig. 9.19 (a) Blade slap—Belvedere helicopter (two 4-bladed rotors—250 r.p.m.); (b) narrow band analysis (1.5%).

and annoying. It occurs at the blade passing frequency and as illustrated in Fig. 9.19b, is very rich in high harmonics. This narrow band analysis is for a tandem helicopter and shows the normal cabin noise and that recorded when in a blade slap condition.

Bell Helicopter Co.[24] and Leverton and Taylor[28] have investigated blade slap in detail. There is general agreement that the noise is produced when one blade passes through a tip vortex shed by the previous or some other blade, but the actual mechanism of its generation is not fully understood.

A number of theories have been put forward. Bell Helicopter Co.[24] suggests that at low speeds it is caused by the rapid change in relative angle of incidence of the blade as it encounters the wake of a previous blade, while at high speed it is more likely to be due to local shock waves on the advancing blades. Schlegel,[29] however, associates blade slap with the retreating blade at the 270° position, this being the region where the angle of attack is greatest and the occurrence of stall is most likely. Other investigators have suggested that the mechanism is similar to the case put forward by Bell for high speed, namely that it is due to local supersonic flow that could occur as the outer portion of the blade intercepts a trailed vortex.

In the work at Southampton, blade slap was considered to arise from blade/vortex interaction. This condition was simulated on a model rotor rig and it was shown that the noise was proportional to V^6 and to the square of the vortex strength. The latter is most important, since it implies that helicopters with highly loaded blades are more likely to produce loud blade slap than those with light loading. In addition, the work showed that if the velocity profile 'seen' by the blade could be determined, the source power or noise could be estimated.

9.7.4 *Tail rotor*

The tail rotor is used on a single rotor machine to counteract the rotational couple set up by the main rotor. It consists of a sideways facing rotor usually running at constant speed, the yawing moment being obtained by a variable pitch control. Because of its smaller diameter it is usually run at a much higher r.p.m. than the main rotor. As illustrated on the narrow band analysis of Fig. 9.17 it produces a series of harmonics in a similar manner to the main rotor, but because of the higher fundamental frequency (40 to 120 c/s) it is subjectively more important. On many helicopters this is the most noticeable and disturbing source, especially from high speed rotors. The vortex noise associated with the tail rotor is generally of a lower level than the discrete or rotational tones.[24] Propeller theory, while adequate to calculate the noise from the tail rotor in free air conditions, is likely in practice to be in serious error. The tail rotor usually operates fairly close to its supporting structure, which means that the blades may be working in a disturbed airflow. Under these conditions the noise may well be appreciably increased.

9.7.5 *Transmission noise*

Gear and transmission noise is important internally, although on some of the smaller types of helicopter such as the Bell UH-1A gear noise can be heard externally. The range of noise is usually between 500 c/s and 1,500 c/s, the most serious speech interference frequencies. Internally, it is usually possible to identify the meshing (number of gear teeth × rotational frequency) and twice the meshing frequencies for all the major loading gears including the tail gearbox. Since in most helicopters the mechanical transmission from engine to the rotor head is in close proximity to the cabin, it is not surprising then that inside the aircraft noise arising from these sources can dominate.

Gearbox noise arises in several ways. The gear teeth, even when carefully designed, can give rise to impact loads which on poorly damped wheels and shafts, can lead to noise from both gears and casings. This is particularly so with the lightweight high duty steel gears which are frequently used in helicopters. Judder, air and oil pocketing can also be significant sources of noise. A fairly comprehensive outline of noise in gears is to be found in reference 30, and more recent work is reported in reference 31.

In addition to noise radiated directly from these sources, it is also possible for the supporting structure, say the cabin ceiling, to be excited by some unbalance in either the gearbox or the transmission. This in itself is a source of noise. If the forcing frequency coincides with one of the many natural frequencies associated with the structure it will be particularly severe.

On a tandem helicopter the drive system vibrations often become significant due to the relatively long inter-connecting shaft between the two main rotors. Since it is normally located above the cabin, this area is often subjected to considerable amounts of high frequency noise both air- and structure-borne. For this reason cabin levels in unsoundproofed helicopters are usually higher on a tandem than on single rotor helicopters.

9.7.6 *Power plant noise*

(1) *Piston engine*

Piston engines have been replaced in the majority of helicopter types by the less noisy gas turbine engine. However, there are still a large number of piston-engined helicopters in service and this type of power plant is still being used in many small helicopters. On large helicopters the engine is well isolated from the cabin and the noise is only a problem externally. On small helicopters it is, however, a major source both internally and externally and is particularly annoying since it occurs in the 200–1,000 c/s range. The primary source of the noise is the exhaust, the noise emanating

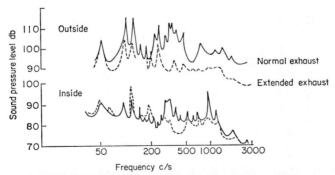

Fig. 9.20   Effect of long exhaust pipe on noise spectra.

from the periodic expulsion of the hot gases of combustion. As shown in Fig. 9.17, this produces harmonics of the engine firing frequency. Reference 32 reports the results of some exhaust silencing tests on the Whirlwind helicopter and these are shown in Fig. 9.20.

When the exhaust is silenced, the internal spectrum is virtually unchanged, but the external noise is reduced by as much as 10 dB, except below 200 c/s where the rotor noise dominates.

9.7.7 *Gas turbine*

The main source of noise on the turbine engine is the 'compressor whine' from the inlet. Additional gears are necessary on this type of engine to give a higher gear ratio; thus turbine engines indirectly bring with them increases of transmission noise.

The exhaust noise of a turbine engine is small because the exit velocity of the gases is relatively low and continuous. The origin of compressor whine is discussed in detail in Chapter 10. Usually only the noise from the first few stages of the compressor can be identified. As with fixed wing propeller turbine aircraft, these high frequency noises from the compressor can be extremely annoying. They are also troublesome on helicopters which are used for detecting submarines where the high frequency sound masks that of the search equipment.

To indicate the overall effect of the choice of engine on the noise, some results of a comparison[32] between two turbine engined (Westminster and Wessex) and two piston engined helicopters (Widgeon and Whirlwind), are shown in Fig. 9.21. The octave band spectra which were measured 500 ft below the aircraft show that although the absolute overall levels of

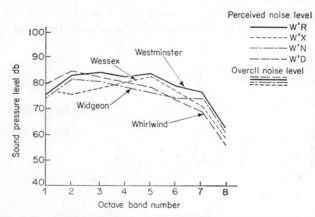

Fig. 9.21   Comparative spectra for turbine and reciprocating engined helicopters.

the turbine engines are not significantly different from the others (in spite of the much greater engine power developed), there is some difference in the subjective levels due to the greater high frequency content of the turbine engined machines.

9.7.8 *Tip jet noise*

There are several advantages to be gained from replacing the orthodox systems of mechanical rotor drive by jet propulsion units at the tips. However, one of the greatest objections to their use arises from the high noise levels generated. There are four main types of tip jet, pulse jet, ram jet, pressure jet and exhaust jet. All except the pulse jet, which, as its name implies, emits a high velocity jet for short periods of time, are continuous systems, the difference between them being primarily one of jet velocity and uniformity of flow.

Jet noise has been discussed to some extent in other chapters. It is sufficient here, therefore, to summarize the possible sources of noise on tip jets as follows:

(i) The pulse jet acts as a simple acoustic source. This will be of the same frequency as the pulse and has a mean noise intensity[33] given relative to 10^{-16} watts/cm^2 by

$$I = 10 \log_{10} \left\{ \frac{1}{32} \frac{\rho_0 c}{r^2 c^2} (V_{\max} - V_{\min})^2 f^2 s^2 \times 10^9 \right\} \text{ dB} \qquad (9.20)$$

where f is the frequency, and the velocities, V, refer to the velocities in a jet of exit area s.

(ii) On continuous jets, noise arises from the turbulence in the region of velocity shear and varies as some high power of the velocity of the jet relative to that of the surrounding air. On low speed jets this power is close to the eighth. On high speed jets, however, it is reduced until at the speeds found in rockets the index is about three. For an efflux velocity around 3,000 ft per second typical of pressure jets, the noise law has fallen to the sixth power or less.

(iii) When the nozzle is well choked, with a higher supersonic velocity downstream, a large increase of noise can arise from the interaction of the turbulence with the standing shock pattern. There is however little evidence available of this source recurring on tip jets.

(iv) Combustion noise takes two forms, noise from the increased turbulence in a rough burning jet, and that arising from a definite resonance in the jet pipe. The second effect is energized by periodic burning and is typified by a quite definite frequency with a wavelength of the order of the jet pipe diameter. While there is no evidence of this tone being present on small tip jets, there is a considerable noise increase resulting from burning generally. Comparative tests[34] between hot and cold jets have shown that there is a general increase from heating, and this appears at most frequencies in the spectrum and over a range of azimuth angles.

The effect of rotation on tip jet noise will be twofold; first there will be a change in the noise characteristics due to the varying position of the jets, and secondly there will be a reduction in the absolute level of jet noise because of the reduced relative velocity of the jet stream.

The first effect is more noticeable very near the jet, and results from the fluctuating distance of the noise sources and the highly directional character of the aerodynamic jet noise. The overall noise for instance at an angle of 45° to the thrust line is about 10 dB higher than at right angles to the jet axis. This, added to the effect of fluctuating distance, can modulate the noise at the observer by as much as 10 dB at frequencies of two, three or four times the fundamental frequency depending on the number of tip jets used.

The second effect, the reduction of the jet efflux velocity relative to the air around it, is wholly favourable and of quite a significant magnitude. Experimental results[34] have shown a reduction of some 6 dB at a tip speed of 400 ft/sec below that measured on a static rig (see Fig. 9.22).

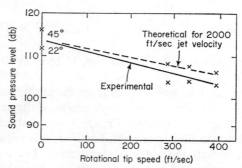

Fig. 9.22   Effect of rotational speed on the noise of tip jets.

The noise alleviation agrees well with the theoretical curve based on the eighth power law, but more recent tests in the static rig have implied a variation more like a sixth power law. Until more evidence becomes available then it would be wiser to assume a sixth power law in calculating the effect of rotor speed.

9.8 Hovercraft noise

Like the helicopter, the hovercraft has a number of important noise sources. As an illustration, Figs. 9.23(a) and 9.23(b) show narrow band

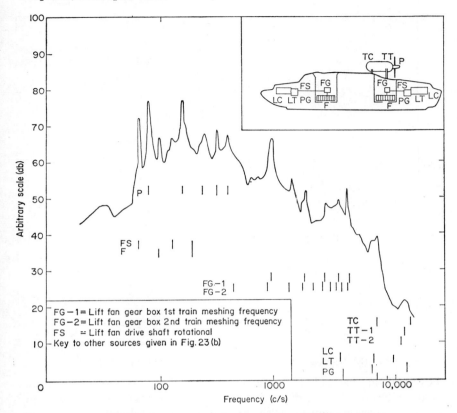

Fig. 9.23 (a) Narrow band analysis of cabin noise.

spectra recorded on the Vickers VA-3 hovercraft.[35] This shows clearly the predominant sources and the approximate frequency range over which they occur. Externally the major source is obviously the propeller rotational noise which occurs between 80–800 c/s. Wheeler and Donno[36] have shown that up to 14 harmonics of this rotational noise can be detected on the SRN5. The only other source which appears in the low frequency range is the first harmonic of the fan blade passing frequency. Both for the VA-3 and the SRN5 (given in reference 36) the level of the fan noise is 20 dB below that of the propeller noise.

In the mid- and high-frequency range there is transmission and engine noise as well as a fairly high level of broad band noise of uncertain origin.[35]

Although the transmission noise can be detected externally, it is most significant in the cabin, as shown in Fig. 9.23(a). Transmission systems in

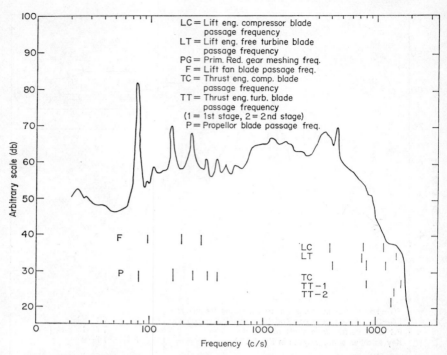

Fig. 9.23   (b) Narrow band analysis of external noise (100 ft, 90° port).

hovercraft are basically the same as that on helicopters and there is little that can be added to the discussion on the mechanisms of noise generation than was given previously. At the present stage of development, the lifting fan is the only noise source which is present on a hovercraft and not found on the helicopter. Again, a full discussion on the sources of noise in fans can be found in Chapter 10.

Control of many of the large hovercraft includes the use of variable pitch propellers often in a region of poor airflow. When a hovercraft is manoeuvring the pitch is being continuously varied, with the result that noise level is increased. When the hovercraft approaches the terminal it is 'braked' or slowed down by reversing the pitch of the blade and this causes an accompanying change in frequency of the propeller noise and an increase in noise.

References

1. L. Gutin. On the sound field of a rotating propeller, N.A.C.A. TM 1195 (1948).
2. A. F. Deming. Noise from propellers with symmetrical sections at zero blade angle II, N.A.C.A. TN 679 (1938).

3. K. V. Diprose. Some propeller noise calculations showing the effect of thickness and plan form, R.A.E. Tech. Note M.S. 19 (1955).
4. E. Y. Yudin. On the vortex sound from rotating rods, N.A.C.A. TM 1136 (1947).
5. H. H. Hubbard. Propeller-noise charts for transport airplanes, N.A.C.A. TN 2968 (1953).
6. H. H. Hubbard and L. N. Lassiter. Sound from a two blade propeller at supersonic tip speeds, N.A.C.A. *Rep.*, 1079 (1952).
7. A. F. Deming. Propeller rotation noise due to torque and thrust, *J. acoust. Soc. Amer.*, **12**, 173 (1940).
8. I. E. Garrick and C. E. Watkins. A theoretical study of the effect of forward speed on the free space sound-pressure field around propellers, N.A.C.A. *Rep.*, 1198 (1954).
9. E. G. Richardson (ed.). *Technical aspects of sound*, Vol. II, Chapter 8, Elsevier (1957).
10. M. C. Kurbjun and A. W. Vogeley. Measurements of free space oscillating pressures near propellers at flight Mach numbers to 0.72, N.A.C.A. *Rep.*, 1377 (1958).
11. A. A. Regier and H. H. Hubbard. Status of research on propeller noise and its reduction, *J. acoust. Soc. Amer.*, **25**, 363 (1952).
12. C. E. Watkins and B. J. Durling. A method for calculation of free-space sound pressures near a propeller in flight including considerations of the chordwise blade loading, N.A.C.A. TN 3809 (1956).
13. H. H. Hubbard and A. A. Regier. Free-space oscillating pressures near the tips of rotating propellers, N.A.C.A. *Rep.*, 996 (1950).
14. R. L. Trillo. An empirical study of Hovercraft propeller noise, *J. Sound Vib.*, **3**, 476 (1966).
15. H. W. Rudmose and L. L. Beranek. Noise reduction in aircraft, *J. Aeron. Soc.*, **14**, 79 (1947).
16. H. H. Hubbard. Sound from dual-rotating and multiple single-rotating propellers, N.A.C.A. TN 1654 (1948).
17. J. P. Roberts and L. L. Beranek. Experiments in external noise reduction of a small pusher-type amphibian airplane, N.A.C.A. TN 2727 (1952).
18. H. H. Hubbard. Sound measurements for five shrouded propellers at static conditions, N.A.C.A. TN 2024 (1950).
19. Vertol Division Boeing Aircraft Co. Study to establish realistic acoustic design criteria for future army aircraft, *U.S. Army Tech. Rep. T.R.E.C.*, **61–72** (1961).
20. J. O. Goddard and T. J. Stuckey. Investigation and prediction of helicopter rotor noise (part I—Wessex Whirl Tower Results), *J. Sound Vib.*, **5**, 50 (1967).
21. M. V. Lowson. Some observations on the noise from helicopters, *Univ. Southampton Intl. Rep.* I.S.V.R. (April 1964) (unpublished).
22. I. A. Simons. Oscillatory aerodynamic loads on helicopter rotor blades in the hover, *Univ. Southampton Intl. Rep.* I.S.V.R. (Feb. 1966) (unpublished).
23. R. G. Loewy and L. R. Sutton. A theory for predicting the rotational noise of lifting rotor in forward flight including a comparison with experiment, *J. Sound Vib.*, **4**, 305 (1966).
24. Bell Helicopter Co. A study of the origin and means of reducing helicopter noise, *United States Tech. Rep.* T.C.R.E.C., **62–73** (Nov. 1962).

25. I. M. Davidson and T. J. Hargest. Helicopter Noise, *J. Roy. Aeron. Soc.*, (May 1965).
26. D. C. Casaway and J. L. Haffield. Noise problems associated with the operation of U.S. Army Aircraft, *U.S.A.A.R.U. Rep.*, 63-1.
27. J. W. Leverton. *Univ. Southampton Rep.* I.S.V.R. (unpublished) (June 1966).
28. J. W. Leverton and F. W. Taylor. Helicopter blade slap, *J. Sound Vib.*, **4**, 345 (1966).
29. R. G. Schlegel. *Helicopter noise generation*, Sikorsky Aircraft Division, United Aircraft Corporation, U.S.A. (1965).
30. C. M. Harris (ed.). *Handbook of noise control* (Chapter 23), McGraw-Hill (1957).
31. G. Berry. N.P.L. Symposium No. 12, The control of noise, *H.M.S.O.* (London, 1961).
32. H. B. Irving. Helicopter noise suppression, *J. Helicopter Assoc. of Great Britain*, **13**, 187 (1959).
33. A. Powell. Noise of a pulse jet, *J. Helicopter Assoc. of Great Britain*, **7**, 32 (1953).
34. E. J. Richards. Problems of noise in helicopter design, *J. Helicopter Assoc. of Great Britain*, **9**, 225 (1955).
35. E. J. Richards and I. J. Sharland. Hovercraft noise and its suppression, *J. Roy. Aeron. Soc.*, **64**, 654 (June 1965).
36. R. L. Wheeler and G. F. Donno. The Hovercraft Noise Problem, *J. Sound Vib.*, **4**, 415 (1966).

CHAPTER 10

Fan Noise

10.1 Introduction

The turbojet engine introduced new noise problems into commercial aviation. For some time after its inception, the most important was that of the jet itself. It became apparent however that as much annoyance was being caused to people living near airports by aircraft on landing approach, as from aircraft taking off. Under approach conditions, engine power and hence jet noise is low. Attention was therefore directed to the engine intake, and subsequently the dominant source of noise was found to be the engine compressor. Since then a number of research programmes have been initiated on the sources of noise in axial flow compressors, and, more recently, with the advent of the by-pass engine and studies of VTOL aircraft, on the noise of ducted and lifting fans. The noise of ventilating fans has always been of interest to air-conditioning engineers. Besides the requirement of establishing a reliable test code, there is also the need to be able to specify the noise characteristics of particular fans so that suitable attenuating devices can be incorporated into the system.

The problem of fan noise may be summarized as one of identifying the various noise sources within the fan and relating the strength of these sources to the physical characteristics of the flow through the fan. With this knowledge it should be possible to estimate the noise characteristics of a particular fan or compressor at the design stage, along with its aerodynamic performance.

In this chapter the mechanisms of noise generation are discussed and some effects of the physical parameters are illustrated with experimental results. A secondary problem in the study of fan noise is the manner in which sound is propagated away from the fan, along the duct. Sound propagation in ducts is a subject in itself, but for completeness, a discussion on the applications to fan noise is also included.

10.2 Sources of noise in fans

Frequency spectra of ducted fans tend to have a characteristic shape. A typical spectrum, taken from a 24 in diameter single-stage ventilating fan at 1440 rev/min[1] is shown in Figure 10.1.

215

It can then be seen that there is a broad spectrum with a maximum occurring at frequencies in the order of 1 kc, and superimposed on this are a number of peaks or line spectra which are found to occur at the fundamental blade passage frequency and its harmonics. For other fans the relative contributions of the broad-band and discrete frequency components will depend on the type of fan considered; for example, the noise from low tip speed ventilating fans is largely broad-band, while the noise from high speed multi-stage aircraft engine compressors is characterized by a discrete frequency whine.

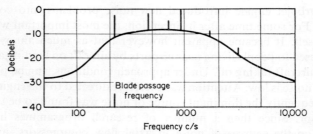

Fig. 10.1   Typical spectrum of noise from a ducted fan.

In general one must examine the origin of both components and, as with the case of propellers (chapter 9), it is convenient to consider them separately.

10.2.1 *Broad-band noise*

The shape of the spectrum indicates that broad-band noise arises from sources which are random in nature, and in fans these can be of two types—dipole sources arising from force fluctuations on the blades, and quadrupole sources associated with the turbulent flow in the wakes of the blades. In most practical cases the wake noise can be neglected in comparison with the dipole contribution,[2] so the problem is reduced to examining the mechanisms by which random force fluctuations can be set up on the blades.

It is known that there are two significant sources of broad-band noise in fans. One arises from the shedding of vorticity at the aerofoil trailing edge when the blade is operating in a smooth airflow, which induces local surface pressure fluctuations on the blade. The other is generated when the blade moves in a turbulent airflow. Under such conditions the turbulent velocity fluctuations ahead of the blade cause random changes of incidence of the relative flow, and hence a randomly fluctuating lift on the blade.

In order to be able to estimate the noise that would be produced by these sources in a given fan, we must first determine the parameters on which the output depends. The basis for theoretical analysis has been established by Curle,[3] in an extension of Lighthill's theory of aerodynamic sound, which includes the influence of solid boundaries. Curle showed that the total acoustic power of the dipole radiation from a solid surface in a flow, is governed by the distribution of fluctuating pressures on the surface. Specifically, for surfaces small compared with a wavelength,

$$W = \frac{1}{12\pi\rho a^3} \overline{\left[\int_S \frac{\partial p}{\partial t} \, dS \right]^2} \qquad (10.1)$$

where a is the velocity of sound, p is the local instantaneous surface pressure, the integration is over the surface S, and the bar denotes the time average of the integral.

The value of equation 10.1, for the sources mentioned above, may be obtained by writing the surface pressure distribution in terms of the local lift acting on a two-dimensional section of the blade. It can then be shown that equation 10.1 is equivalent to

$$W = \frac{\rho}{48\pi a^3} \int_{span} CU^4 . S_c \overline{\left[\frac{\partial C_L}{\partial t} \right]^2} \, dy \qquad (10.2)$$

where C is the blade chord at spanwise coordinate y, C_L is the fluctuating lift coefficient on the section, S_c is the correlation area for the local instantaneous pressure fluctuations, and U is the relative mean velocity of the blade section at y.

Quantitative estimates of the output from a fan blade can then be made by considering the mechanisms involved, and evaluating the integral in equation 10.2.

10.2.1.1 *Noise from a blade operating in a smooth flow*

If a bluff body is placed in a flow at low Reynolds number, vortices are shed from the trailing edge in a regular succession (the well-known Karman Street). This produces a periodic change of circulation which in turn results in a fluctuating lift on the body. At higher Reynolds numbers the vortex shedding becomes more diffuse and random in nature, although there may still be periodic components present, even for more streamlined sections.[4]

There is still some uncertainty about the precise mechanisms by which fluctuating forces are set up on the body, but it is possible to obtain an idea of the source strength by considering the orders of magnitude involved. These are discussed in some detail in reference 2 where it is concluded that the total acoustic power, from an aerofoil operating well away from the stall over all its span, is given by

$$W = \frac{\rho}{120\pi a^3} \int_{\text{span}} C \cdot U^6 (Re)^{-0.4} \, dy \qquad (10.3)$$

where Re is the Reynolds number based on the aerofoil chord.

10.2.1.2 *Noise due to initial turbulence.* When the force fluctuations on the blade are caused by random variations in the relative incidence due to velocity perturbations in the oncoming flow, the average mean square lift coefficient can be related to the tubulent velocity, by defining an average lift curve slope, ϕ, such that $C_L = \phi \cdot w/U$ with w the component of turbulent velocity normal to the chord. Then

$$\overline{\left[\frac{\partial C_L}{\partial t}\right]^2} = \frac{\phi^2}{U^2} \cdot \overline{\left[\frac{\partial w}{\partial t}\right]^2}. \qquad (10.4)$$

The physical significance of the lift curve slope, ϕ, has been discussed in some detail by Liepmann,[5] who showed that its value was determined by the scale of the oncoming turbulence. If this is large compared with the chord of the aerofoil, then ϕ approaches the quasi-steady value, i.e. about 2π for a thin section. If on the other hand the scale of turbulence is small compared with the chord then local lift fluctuations tend to cancel over the chord and the resulting average lift coefficient fluctuations are very much reduced. Again, order of magnitude estimates of the acoustic output from this mechanism are discussed by Sharland[2] where it is concluded that the total acoustic power is given by

$$W = \frac{\rho}{48\pi a^3} \int_{\text{span}} \phi^2 \cdot C U^4 \overline{(w)^2} \, dy. \qquad (10.5)$$

10.2.1.3 *Comparison with experimental results.* Equations 10.3 and 10.5 give theoretical estimates of the acoustic power which might be expected from the sources of broad-band noise examined. These estimates have been compared with a series of measurements of the noise from a small flat plate (1 in chord × 2½ in span) held at zero incidence in an air jet.[2] The plate was held first in the smooth flow of the potential core of the jet and then in the highly turbulent mixing region. The noise at right angles to the plate and 3 feet from it, was measured for a range of jet velocities. Using the relationship between total power and maximum sound pressure level for dipole directionality, the theoretical sound pressure level for each source was then calculated, using the above equations. The results of the comparison are shown in Fig. 10.2.

For the turbulent flow case the levels predicted by equation 10.5 are in very good agreement with the measured levels. In the case of smooth flow, allowing for the rather rough estimates used in the theoretical analysis (equation 10.3), a reasonable degree of accuracy is also obtained. For this case, however, both theoretical and experimental levels are well below those of the turbulent flow case, which shows clearly the dominant effect that incident turbulence can have on the noise produced.

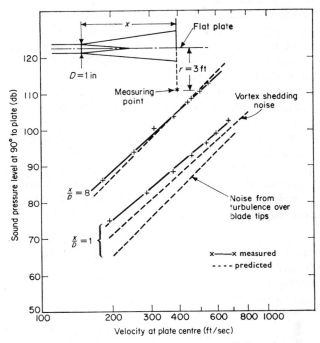

Fig. 10.2   Noise radiated from a small flat plate.

The experimental curves in Figure 10.2 are in qualitative agreement with the results of an investigation into the noise of rotating rods,[6] where it was found that the noise from rods over which the flow had separated was significantly higher than the noise from rods with no separation. This was explained as being due to an increased drag coefficient with separation, but it is possible that the separated wakes of each rod were passing over the following rod, so that all the rods were moving in a turbulent flow.

The results summarized in Figure 10.2 show that the equations derived in the preceding sections can be used to predict the noise for a fairly simple arrangement like the flat plate. It can also be demonstrated that they can be used to predict the output from a fan rotor. The situation in the fan rotor is more complicated in that the sources are moving, but as a first approximation, it can be assumed that the total acoustic power is not affected by the blade motion.

Much of the published data on fan noise are not in a form from which the broad-band noise can be calculated by the methods described here, but the results of tests on a small axial compressor carried out at the University of Southampton[2] show that at least for the case of initial turbulence (equation 10.5) the noise can be predicted reasonably well. This was demonstrated by placing, just inside the fan intake, a circular rod bent to form a ring of diameter approximately equal to the mean height of the blades. The increment of acoustic power arising from the passage of the blades through the turbulent wake of the ring was calculated using equation 10.5 for a range of tip speeds. This estimate was then compared with the levels obtained from free field measurements around the fan and agreement was within 3 dB.

Under normal conditions, of course, the flow should be everywhere smooth over the blades, at least for the first row. It is possible, however, that the tips may be moving through the turbulent flow contained in the boundary layer on the duct wall. On the basis of this possibility, the broad-band noise that might be expected from the Southampton compressor, arranged as a single rotor with no stators, was estimated in two parts:

(a) noise from vortex shedding (equation 10.3) assuming the flow was smooth over all the blade area;
(b) noise arising from initial turbulence (equation 10.5) assuming this to be caused by the annular duct boundary layer turbulence acting over only the tip region of the blade.

A comparison of these estimates with observed levels of broad-band noise is reproduced in Fig. 10.3.

If the estimates are of the right order, then it appears that most of the broad-band noise from such a configuration was due to vortex shedding, but allowing for the limits of accuracy which can be claimed, it is possible that the duct boundary layers make a measurable contribution.

So far, only small blade incidences have been considered, but it is well known from work on fan noise that when the blade is stalled the noise output rises significantly. This is illustrated in Fig. 10.4, where the static pressure rise through the fan and the overall acoustic power are shown as functions of the mass flow.

As the mass flow is decreased (i.e. blade incidence increased) the sound power tends to fall slightly, but as the stall is approached the noise increases rapidly—in this case by about 10 dB. On the basis of the previous discussion, the increase could be attributed to the increase of turbulence

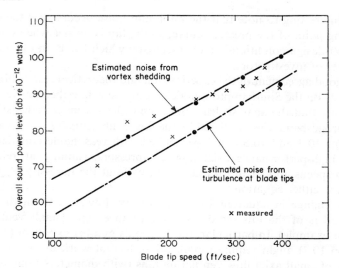

Fig 10.3 Broad-band noise from a single rotor row.

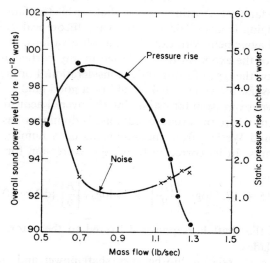

Fig. 10.4 Static pressure and noise characteristics from a single-stage fan (after Sharland[2]).

over each blade, the effect of which is to set up large amplitude pressure fluctuations on the blade itself behind the separation point, and possibly on adjacent blades.

It is interesting to note that the minimum noise occurs near the design operating point of the pressure curve, so the fan is noisier when operated under off-design conditions—a characteristic which has been observed by a number of investigators.

The good agreement of the experimental and theoretical results obtained from tests on the Southampton model compressor described above, seems to indicate that the analysis leading to the orders-of-magnitude estimates of the broad-band noise radiated by the fan is substantially valid, and that equations 10.3 and 10.5 could be applied with reasonable confidence to similar high-performance axial flow compressors, running at relatively high tip speeds and small tip clearances, without the necessity of grossly modifying either equation.

Some change in equation 10.3 is necessary, however, when the rotor concerned is of the low-speed, low-performance type (small ventilating fans, for example). In particular, the Reynolds number exponent (-0.4 in equation 10.3) might require modification. Recent work by Austin[7] on a number of small axial flow ventilating fans (with diameters ranging from 14 in to 22 in, and operating speeds from 600 to 1400 rev/min) has shown that the value of the exponent does not remain constant, but depends rather on the fan type. Austin observed that for his test fans, values of the exponent ranging from -0.05 to -0.2 gave theoretical estimates which were in good agreement with experimental observations.

The value of the exponent to be selected in any particular case, would thus depend on the type of rotor under consideration. For high-speed axial flow compressors, the value -0.4 should be a good guide.

An empirical expression for estimating the overall acoustic power from ventilating fans where the broad-band noise is dominant has been derived by Peistrup and Wesler[8] from measurements on a number of axial and centrifugal units. In its corrected form[9] the expression for sound power level is

$$PWL = 115 + 17.7 \log_{10}\left[\frac{HP}{B}\right] + 15 \log_{10}\left[\frac{B}{6}\right] \text{ [dB } re \text{ } 10^{-13}\overline{w}] \quad (10.6)$$

where HP is the shaft horsepower delivered to the rotor, and B is the number of blades.

Assuming a V_T^3 relationship between shaft power and tip speed, this expression predicts a sound power output proportional to $V_T^{5.9}$, which is of the same order as the $V_T^{5.6}$ to V_T^6 dependence obtained in the preceding discussion. The expression has been used to estimate broad-band noise from turbo-jet intakes[10] with reasonable accuracy.

It is somewhat limited, however, because it was derived from measurements of the noise radiated by only two centrifugal fans with a low horsepower range. For this reason, it has been superseded by a number of other empirical relationships put forward by various authors.

These empirical equations, in general, attempt to relate in as simple a manner as possible, the sound power output of a given rotor to its physical and operating characteristics, but as yet no really accurate or reliable general expression has been developed. Those that have been suggested vary in their range of applicability and accuracy, and apply in most cases to fans of specific type only.

The equation which superseded 10.6 was that of Beranek, Kamperman and Allen[9] who derived empirically the equation

$$PWL = 100 + 10 \log_{10} HP \tag{10.7}$$

from measurements of the inlet and outlet noise of a number of centrifugal fans of different sizes and covering a wide horsepower range. Except for a possible scatter of approximately ± 5 dB, it has been shown[11] that equation 10.7 does in fact give quite a reasonable estimate of the actual sound power output of centrifugal fans.

Equation 10.7 by itself includes only the rotor nameplate horsepower as the relevant parameter to use when estimating the corresponding sound power levels radiated by the fan. Allen,[11] improving equation 10.7, suggested that better accuracy could be obtained if the static pressure P developed by the rotor was included in the computation, viz.

$$PWL = 100 + 10 \log_{10} HP + 10 \log_{10} P. \tag{10.8}$$

The inherent error in Allen's equation 10.8 was not very much better than that associated with equation 10.7. Nevertheless, better accuracy in the estimate is possible with equation 10.8 when the static pressure developed is considerable.

For small centrifugal fans of similar design, empirical equations of slightly better accuracy than either equation 10.7 or equation 10.8 have been proposed. Maling,[12] for instance, suggested that the acoustic power $W_{75-10,000}$ radiated by a given rotor in the frequency band 75 to 10,000 c/s could be related to its diameter D, its 'aspect ratio', α (ratio of fan width to fan diameter), the static pressure developed, P, and the flow Q through the rotor, by the equation

$$W_{75-10,000} = C_1 \frac{P^3}{\alpha} + C_2 \frac{Q^5}{D^{10}\alpha^5} \tag{10.9}$$

where C_1 and C_2 are constants which have to be chosen for the best fit with measurements.

Numerous other examples of fan noise equations could be cited[13, 14, 15] including recent proposals by Van Niekerk,[15] in which an empirical expression is developed, to predict not only the overall sound power levels radiated by a given fan, but also the corresponding spectral distribution of the sound energy. In applying these empirical relationships, great care should be taken to ensure that the equation selected to give estimates of the radiated sound power is the valid one to use in that particular case.

10.2.2 *Discrete frequency noise*

In many types of axial fans and compressors the noise is dominated by a discrete frequency tone. There may be a number of harmonics present as shown in Fig. 10.1, but the fundamental frequency is generally the blade passage frequency. There appear to be two mechanisms by which periodic disturbances can be set up in the fan, these are classified as propeller noise, and noise generated by aerodynamic interaction effects between rotor and stator blades.

10.2.2.1 *Propeller noise.* Since the rotor of an axial flow fan is essentially a propeller acting in a duct, it might reasonably be expected that the rotating pressure field on the blades will act as a distribution of dipole sources in the manner discussed in chapter 9. This may well be true, except in so far as the higher values of blade solidity in fan rotors will have some effect on the source strength as pointed out in section 9.3.4, but the radiation from these sources will be considerably modified by the presence of the duct, so that the theory of lift noise from propellers is not valid. It might be argued that, for a given blade loading and neglecting duct attenuation, a fan rotor would generate the same amount of acoustic power when operated inside the duct as it would if operated as a free-running propeller; in which case the total noise energy could be calculated from the standard propeller noise theory given in chapter 9.

The validity of this argument, however, is still open to question. While good agreement between propeller noise theory and observed compressor noise has been obtained on some installations,[10] others have shown a considerable discrepancy.

Filleul,[16] for example, found from his experiments with axial flow fans having detachable blades (rotating in the absence of struts or other obstacles in an anechoic chamber), that estimates of the radiated noise based on the classical propeller noise theory of Gutin[17] (which Filleul generalized to cover many-bladed rotors) were in poor agreement with actual measurements for cases when the rotor under test had many blades (or high solidity)—being several orders of magnitude out. On gradually removing the blades in a symmetric fashion from the test rotor, however, Filleul observed that his measurements agreed reasonably well with estimates from Gutin's theory when the rotor had two blades. In other words, Gutin's theory appeared to hold only in the case where the solidity of the rotor was low and the corresponding blade loading high. In fact, Filleul noted that the discrete frequency whine, whose levels the classical theory predicts, became pronounced only when the number of blades on the test rotor was down to two. It would appear therefore, that Gutin's propeller noise theory is accurate only in very specific cases where the blade solidity is low. Its accuracy in cases such as the multi-bladed compressor rotor, having very high solidity, must remain somewhat doubtful.

10.2.2.2 *Interaction noise.* The pressure field rotating with the blades will be steady relative to the blades only if the flow near the rotor is uniform over the area of the annulus. If the rotor is operating in the vicinity of some solid obstacle, such as stator blades or bearing support struts, the blades will pass through regions of varying velocity associated with the potential field, and wake, of the stationary obstacle. This is shown diagrammatically in Fig. 10.5 (a) for the case of a rotor blade moving behind a stator cascade.

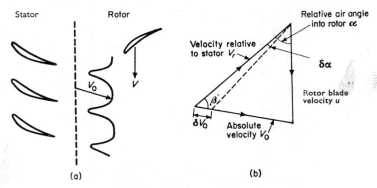

Fig. 10.5   Effect of stator on flow into rotor.

The effect on the rotor of this variation of velocity is shown in the velocity triangle (Fig. 10.5 (b)). The solid lines show a typical set of velocities which determine the incidence of the rotor blade. If there is a change of absolute velocity, the direction of the relative velocity into the rotor, shown by the dashed line in Fig. 10.5 (b), is also changed, with a resulting change of incidence and hence force on the rotor blade. This is the same situation as in the case of a blade moving in turbulent flow, where turbulent velocity fluctuations normal to the blade produce corresponding lift fluctuations, but in the case considered here the lift fluctuations are periodic.

If the stator cascade is close to the rotor, it is also possible for fluctuating pressures to be set up on the stator blades due to the passage of the rotor pressure field.

The amplitude of the force fluctuations on rotors and stators arising from aerodynamic interaction has been discussed by Kemp and Sears,[18] and an estimate of the noise produced by this mechanism has been made by Hetherington.[19] There is, however, little experimental evidence available, although some idea of the significance of interaction noise can be obtained from the results of measurements on a fan of differing geometry, reported by Sharland[2] and reproduced in Fig. 10.6.

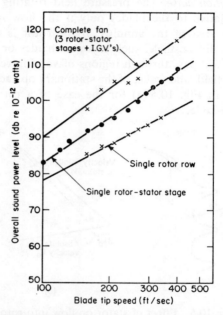

Fig. 10.6   Increase in fan interaction noise as stages are added.

Taking the noise from the single rotor row as a datum for the particular fan, the increase due to the addition of a downstream stator row was from 4 to 8 dB—the increase appearing mainly at the fundamental blade passage frequency. The addition of two more rotor-stator pairs downstream, and a row of inlet guide vanes upstream of the first rotor, produced a further increase of some 6 dB, again in the discrete frequency components.

The indications are then that interaction noise can be of greater significance than propeller noise, and so it is worth considering in a little more detail how it will be affected by blade design. The distance between rotor and stator rows will clearly have an effect on the interaction forces set up. One would expect that the force fluctuations are greater when the rows are close together, since then the disturbance fields are stronger. It has, in fact, been known for some time that fans are noisier when stator-rotor clearance is reduced (see, for example, Marks and Weske[20]). The shape of the stator may also be relevant; the flow over a thick profile for instance would be less uniform than that over a thin one. Moreover, if the stator itself radiates noise from the fluctuating pressure set up on it by the rotor pressure field, the area of the stator will be important.

Some of the effects of these parameters have been shown by the results of tests[2, 21] in which a single upstream stator blade was moved towards the rotor blades of an axial flow fan. The effect of rotor-stator separation on the noise at 30° to the fan intake is shown in Fig. 10.7, for some of the stator shapes examined.[21]

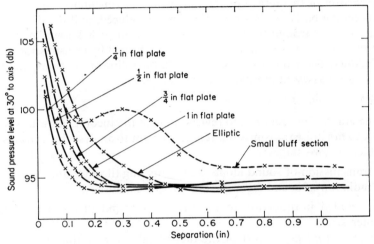

Fig. 10.7   Effect of rotor-stator separation.

For all section shapes the noise increased sharply as separation was decreased, most of this appearing at the fundamental blade passage frequency. The critical separation was larger for the elliptical section than for the flat plate. In addition, for the flat plates, Fincher[21] found that the individual curves for different stator areas could be represented quite well by a single curve, if the actual stator-rotor separation was normalized by expressing it as a proportion of rotor chord plus twice stator chord. The hypothesis here was that the rotor pressure field which appeared to be oscillating relative to the fixed stator, could be represented by an acoustic model consisting of a piston, of diameter equal to the rotor blade chord and oscillating at the rotor blade passage frequency, in a baffle of diameter equal to the rotor chord plus twice the stator chord. Fincher found further that the theoretical change with frequency and baffle size, of the acoustic output from a simple piston-baffle arrangement, correlated well with the observed changes on the fan with variation of rotor speed and stator size.

It was also found during these tests that the interaction noise was reduced steadily as the stator was rotated relative to the radial direction—a decrease of about 16 dB being observed at an angle of 60°. This is presumably because the effective velocity gradient sensed by the rotor is reduced with stator rotation.

10.3 Sound propagation

So far we have discussed only the strength of the sources in fans. Comparison of theoretical predictions with experimental results has been made on the basis of total acoustic power produced, and the distribution of this energy around the fan has been neglected. It is known, moreover, that the noise generated at the rotor can be affected by transmission effects in the duct. These factors are discussed in the following sections, and their significance assessed in relation to fan noise.

10.3.1 *Theoretical background*

The aim of this section is to review some of the theoretical results which apply to fan noise transmission. It is convenient to analyse the transmission problem in two stages:

(a) transmission of sound along the fan ducting (if any); and
(b) radiation of sound from the fan inlet or exhaust.

A detailed introduction to the theory of sound in ducts is given in chapter 2.

Duct transmission. Most of the theoretical work on duct transmission of fan noise[22-25] is based on a simplified model which nevertheless illustrates some of the important features. The model consists of a uniform straight duct, with rigid impervious walls, containing a uniform fluid at rest. Diffusion effects in the fluid (viscosity, thermal conductivity) are neglected, and the pressure fluctuations p' caused by sound transmission are assumed to be small compared with the isentropic bulk modulus ρa^2 of the fluid.

Two important consequences of these simplifying assumptions are:

(1) the fluctuating pressure is governed by a linear equation, so that the theory can be restricted without loss of generality to sinusoidal fluctuations at a single frequency. In complex (phasor) notation,

$$p' = \mathscr{R}(P); \qquad P \propto e^{-i\omega t}, \tag{10.10}$$

where $\omega = 2\pi f$ is the circular frequency.

(2) the boundary conditions permit solutions for P in which the spatial variation is given by two separate factors, one involving distance along the duct (x) and the other involving position across the duct section (y).

A typical solution of this form may be written

$$P = A \, e^{ik(\alpha x - at)} E(y), \tag{10.11}$$

which represents the propagation of a particular pressure pattern along the duct with phase speed a/α; a is the speed of sound in the fluid, and $k = \omega/a$.

Any solution for the fluctuating pressure must obey the wave equation; in addition, the rigid walls impose the boundary condition that the normal gradient should vanish at the boundaries of the duct cross-section. Functions $E(y)$ which meet both these requirements exist only for certain characteristic values ($k_0^2, k_1^2, \ldots k_N^2, \ldots$) of $k^2(1 - \alpha^2)$: the corresponding characteristic functions E_N are called the 'mode shape functions' for the particular duct section.

Cut-off phenomenon. From equation 10.11 the variation of the Nth-mode pressure pattern along the duct in the $+x$ direction is proportional to $e^{i\alpha kx}$, with α now determined by $k^2(1 - \alpha^2) = k_N^2$. It follows that α can be either real or imaginary, depending on the ratio $k^2/k_N^2 = \nu^2$:

$$\alpha = \pm(1 - 1/\nu^2)^{1/2}. \tag{10.12}$$

Imaginary values of α imply that the pressure pattern decays along the duct: thus $P_N \propto \exp -(k_N^2 - k^2)^{1/2}x$, ($\nu < 1$). Real values of α on the other hand (i.e. $\nu > 1$) imply propagation at constant amplitude; for this reason, the ratio $\nu = (k/k_N) = (2\pi f/ak_N)$ is called the cut-off ratio of the Nth mode of transmission.

Axisymmetric ducts. The transmission properties of axisymmetric ducts are of particular importance in the study of axial fan noise. In general, an axisymmetric cross-section is bounded by two concentric cylinders: in the axial fan case, the inner cylinder (radius r_1) would represent the fan hub and the outer cylinder (radius r_0) the casing.

The corresponding mode shape functions are best expressed in polar coordinates: for a given hub-tip ratio μ ($= r_1/r_0$),

$$E_{mn}(r, \theta) \propto e^{im\theta} Z_m(k_{mn}r),$$

where Z_m represents a cylinder function of order m.

The two mode numbers m and n replace the single index N of the preceding sections: they have a simple interpretation in terms of the modal pressure pattern over the duct section, in that $|m|$ is the number of zero-pressure diameters (i.e. the number of cycles circumferentially around the annulus) and n is the number of zero-pressure circles within the annulus (see Fig. 10.8).

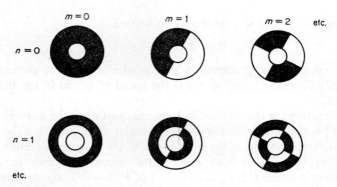

Fig. 10.8 The (m, n) series of modes for an axisymmetric duct section.

In the series of modes ($n = 0, 1, 2,$ etc.) corresponding to a given value of m, the one with $n = 0$ has the lowest cut-off frequency (i.e. the lowest k_{mn}). These $(m, 0)$ modes are of particular importance in fan noise transmission, not only because they propagate at frequencies where the others decay, but also because in axial fans they generally correspond best to the spatial distribution of the sound sources: the pattern does not change sign across the annulus. Their transmission can be visualized in terms of spiral waves (Fig. 10.9) which travel between the walls of the annulus in a direction normal to the wavefronts. The wavefront separation (being analogous to the wavelength of plane waves) will increase with reduction of frequency, until the limit shown in Fig. 10.9 (b) is reached: the wavefronts

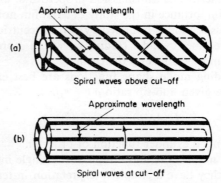

Fig. 10.9 The effect of frequency on spiral-wave transmission.

are then moving round and round the duct rather than along it, and cannot move any further apart. Any further frequency reduction leads to the breakdown of the wave picture, and the onset of decay in the axial direction.

Table 10.1

n	m = 0	m = 1	m = 2	m = 3	m = 4	m = 5	m = 6	m = 7	m = 8	
0	0	1.8412	3.0542	4,2012	5.3175	6.4156	7.5013	8.5778	9.6474	
1	3.8317	5.3314	6.7061	8.0152	9.2824	10.5199	11.7349	12.9324	14.1155	$\mu = 0$
2	7.0156	8.5363	9.9695	11.3459	12.6819	13.9872	15.2682	16.5294	17.7740	
0	0	1.5821	2.9685	4.1801	5.3130	6.4147	7.5011	8.5778	9.6474	$\mu = 0.3$
1	4.7058	5.1374	6.2738	7.7213	9.1526	10.4750	11.7214	12.9286	14.1145	
0	0	1.3547	2.6812	3.9577	5.1752	6.3389	7.4622	8.5586	9.6382	$\mu = 0.5$
1	6.3932	6.5649	7.0626	7.8401	8.8364	9.9858	11.2270	12.5095	13.7964	

Values of $k_{mn}r_0$ for axisymmetric ducts are given in Table 10.1, for hub-tip ratios of 0, 0.3, and 0.5. Reference 25 gives the following approximation for larger hub-tip ratios:

$$k_{mn}^2 \doteq \left(\frac{m}{R}\right)^2 + \left(\frac{n\pi}{r_0 - r_1}\right)^2, \qquad (\mu > 0.5) \qquad (10.13)$$

where R is the mean radius of the annulus.

Radiation from duct opening. The sound power which leaves a ducted fan installation depends as much on the radiation process from the duct terminations (inlet or exhaust) as on the duct transmission process described above. An important case which has been investigated theoretically is that of radiation from the open end of an axisymmetric duct.[24, 25, 26]

The starting point for these investigations is the following expression for the sound pressure radiated by an assumed normal-velocity fluctuation at the plane of the opening (Fig. 10.10). For an area element $dS = rdrd\theta'$ at

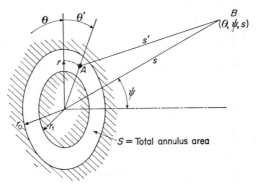

Fig. 10.10   Coordinate system for radiation from the open end of an axisymmetric duct.

the point A, moving with complex velocity U, the corresponding complex pressure at B is

$$dP = -i\omega\rho U \, dS \, \frac{e^{iks'}}{2\pi s'}. \tag{10.14}$$

Equation 10.14 assumes: (a) that the opening can be treated as if mounted in a rigid baffle of dimensions large compared with the sound wavelength; (b) that the mean flow through the opening has no effect on the sound radiation. As a result, the analysis cannot be expected to apply at angles from the axis greater than $\psi = 90°$, or at flow Mach numbers approaching 1.

An estimate of the far-field sound pressure radiated from the opening can be made, using this approach, for a velocity distribution $U(r, \theta)$ which corresponds to any of the annulus modes. In this way it is possible to calculate the radiated power, as well as its directional distribution.

Radiation resistance. It is well known (see for example reference 27) that the sound power radiated from a piston vibrating in a baffle is not simply $\rho a S$ times the mean square piston velocity, but differs from this by a frequency-dependent factor τ (the radiation resistance ratio). If the piston is circular, with radius r_0, then τ is small for values of kr_0 less than 1, and rapidly approaches 1 for larger values.

In the present context, the piston-like motion corresponds to just one mode of sound transmission in a uniform duct: the ratio τ is the open-end

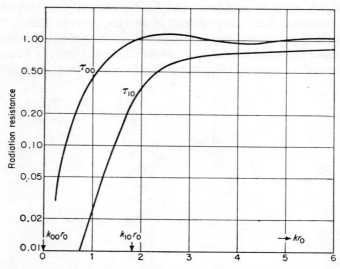

Fig. 10.11   Radiation resistance of (0, 0) and (1, 0) modes at a flanged circular opening.

resistance ratio for plane waves in the duct. The fact that τ is small for a circular duct when $kr_0 < 1$ implies a mis-match of impedances and hence a large reflection coefficient at the opening. The idea applies equally to other modes; thus the modal resistance ratio τ_{mn} at an annular opening is a measure of the effectiveness with which the opening radiates sound from the (m, n) mode.

Figure 10.11 shows the resistance ratios τ_{00} and τ_{10} for the $(0, 0)$ (i.e. plane wave) and $(0, 1)$ modes at a flanged circular opening, calculated by the method of the previous section. As the frequency parameter kr_0 increases from 0 to $(k_{mn}r_0 + 1)$, the resistance ratio rises steeply; at higher frequencies, the ratio fluctuates around 1.0.

This conclusion is supported by further calculations in references 25 and 28. In other words, the cut-off process marks a transition from ineffective to effective radiation of a given mode from the opening, just as it marks the change from decay to a propagation of the same mode within the duct.

Effect of non-rigid walls on duct transmission. The foregoing theory represents the duct walls as being rigid and impervious. A practical method of reducing fan noise, however, is to line the duct installation with sound-absorbent material. The specific acoustic admittance of the walls can then no longer be regarded as zero; but provided the admittance is still small compared with $1/\rho a$, the modes of transmission do not differ much from those of the rigid duct.

The main effect on the transmission of fan noise, where the wavelengths of interest are generally small compared with the transverse dimensions of the duct (cf. chapter 2), is to introduce an attenuation coefficient along the duct which is related to the real part $(\kappa/\rho a)$ of the local wall admittance. If the conductance ratio κ averages to K over the total perimeter L of the duct section, then the attenuation coefficient is approximately KL/S; the transmitted sound power falls off at $8.686 \, KL/S$ dB per unit length of duct.

Effect of mean flow. At high subsonic Mach numbers, a mean flow along a uniform duct has a significant effect on sound transmission. Three important departures from the zero-flow case discussed hitherto are:

(a) the transition from decay to propagation along the duct occurs at

$$\nu = (1 - M^2)^{1/2} \tag{10.15}$$

(cf. equation 10.12), where M is the flow Mach number along the duct;

(b) distances in the axial direction are effectively lengthened by a factor $1/(1 - M^2)$;

(c) the sound power travelling along the duct is increased (for the same pressure amplitude) for waves propagating in the flow direction. For any given mode, the power is multiplied at frequencies well above cut-off by $(1 + M)^2$ for downstream waves, and by $(1 - M)^2$ for upstream waves.

10.3.2 *Axial-fan applications*

The practical usefulness of the foregoing modal theory depends on a knowledge of which modes are excited by a particular noise source when it operates inside a duct. For example, when noise is radiated from the end of the duct, it is helpful to know which modes transmit most of the sound power: attention can then be concentrated on these modes alone, and their transmission and radiation studied in detail.

An important factor is the geometrical coupling between source and mode patterns over a duct cross-section. The sound power generated by a piston, fitted inside a uniform duct and moved up and down, would be confined exclusively to the plane-wave mode. Although rotating-machinery noise sources are more complicated than this, the combination of geometrical and cut-off considerations narrows the range to a finite number of modes, and even to one or two in some cases.

One way to identify predominant modes is to measure the far-field radiation pattern outside the duct opening. Then, if the duct is axisymmetric, theoretical predictions based on section 10.3.1 can be compared with the measured directional pattern to infer which modes are present. Examples of this approach appear in the following paragraphs.

Relative weighting of modes: broad-band noise. The broad-band noise sources in an axial-fan do not form a coherent spatial pattern over the blade disc; it is realistic, in view of their origin (section 10.2.1), to regard them as a large number of uncorrelated sources. Two further assumptions —that each correlated region is small compared with the sound wavelength, and that the sources are distributed uniformly over a cross-section of the surrounding duct—lead to the following conclusions:

(a) geometrically, there is no discrimination between modes;
(b) cut-off considerations preclude any significant power transmission in modes whose cut-off frequency is higher than the excitation frequency.

In axisymmetric ducts, in view of the results of section 10.3.1, the relative weighting of the sound power in different modes can be approximately represented by (cf. reference 22):

$$\left. \begin{array}{l} k_{mn} r_0 < (k r_0 - 1): \text{ equal weighting;} \\[2mm] k_{mn} r_0 \geq (k r_0 - 1): \text{ zero weighting.} \end{array} \right\} \qquad (10.16)$$

Such a weighting implies, for any given frequency, a definite radiation pattern in the far field outside the duct. Figure 10.12, taken from reference

26, shows how this prediction compares with measurements on high-speed axial compressors; at the lower non-dimensional frequency, the directivity follows the prediction closely up to an angle of 60 degrees from the duct axis, beyond which the predicted fall-off is rather sensitive to any inaccuracy in the modal weighting (equation 10.16). At the higher frequency, two of the compressors show much sharper directivity patterns than

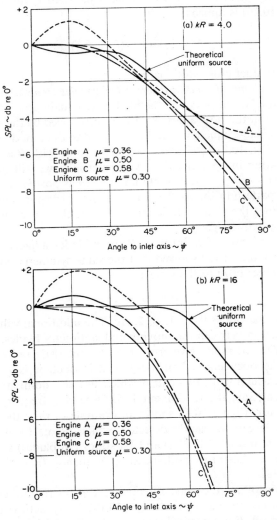

Fig. 10.12 Broad-band compressor noise: radiation patterns from axisymmetric inlets.

predicted; the inference is that the sound radiation is deficient in high-order modes. Such a deficiency could be due to correlated source regions which were not (as assumed earlier) small compared with the sound wavelength.

The total sound power radiated, by a source distribution of the type assumed, into a duct is proportional to the number of modes transmitting power. In an axisymmetric duct, the number N of such modes at any given frequency (i.e. the number of modes which satisfy the first condition of equation 10.16) is approximately

$$N \doteq \frac{k^2 S}{4\pi}, \quad (k > k_{11}), \tag{10.17}$$

where S is the duct cross-sectional area.

Relative weighting of modes: discrete frequencies (tones). In an axisymmetric fan installation, provided the blades are equally spaced and there is no asymmetry in the flow apart from that introduced by the blades themselves, the discrete-frequency sound is confined to duct modes with certain values of m (Fig. 10.8).[23-25] For example, a rotor with B equally-spaced blades, operating in an otherwise axisymmetric flow, will generate only modes with $m = rB$ ($r = 1, 2, 3$, etc.), taking m as positive for rotation of the pressure pattern in the shaft direction. The corresponding frequencies will be rB times the shaft rotational speed, i.e. the harmonics (r) of the blade-passing frequency.

In the general case of a fan with B rotor blades adjacent to V stator blades, Tyler and Sofrin[24] showed that the same symmetry considerations restrict the values of m to the series:

$$m = rB + sV, \tag{10.18}$$

where s takes the values 0, ± 1, ± 2, etc. The additional values of m are due to interaction between the moving and stationary blades.

Since each (m, n) mode gives rise to a characteristic radiation pattern on leaving the open end of an axisymmetric duct, it is possible from directivity measurements to deduce which are the dominant modes of transmission at, say, the fundamental blade-passing frequency of a particular fan rotor. Figure 10.13 shows an example of this: the presence of the $(1, 0)$ mode can be clearly detected, which points to an interaction between the 1st rotor (29 blades) and 1st stator (28 blades) as responsible for the tone at 1st-rotor blade-passing frequency ($r = 1$, $s = -1$ in equation 10.18).

Besides controlling the tone directivity, the restriction of the mode number m to certain values (equation 10.18) has practical implications because of cut-off. For a particular tone, only a few of the available modes will have cut-off frequencies below the excitation frequency: the others may be disregarded. As the fan speed is reduced, further modes will be successively cut off until only one remains: this will be the mode with $n = 0$ and the lowest value of m permitted by equation 10.18. Once this last mode is cut off, the tone will disappear almost completely.

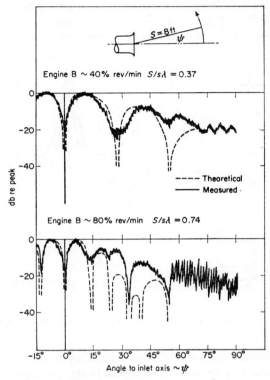

Fig. 10.13   Compressor-tone radiation patterns from an axisymmetric inlet.

It is possible to set down a design criterion to ensure that at a certain fan speed, the rth blade-passing harmonic will be cut off in this way. The basic requirement is that

$$g^* > kr_0$$

where g^* is the value of $k_{mn}r_0$ for the last mode to disappear; in terms of fan parameters, $k = \omega/a$ and $\omega = r(Bu/r_0)$ (u = rotor tip speed), so the required criterion is

$$\frac{kr_0}{g^*} = \left(\frac{rB}{g^*}\right) \cdot \left(\frac{u}{a}\right) < 1. \tag{10.19}$$

The ratio (rB/g^*) is a geometrical factor determined by the three quantities (rB, V, μ). Its value depends almost entirely on (rB/V), as Fig. 10.14 indicates; any additional variation with V or hub-tip ratio μ is contained in the shaded band on the graph. If the cut-off criterion equation 10.19 is to be satisfied over a wide speed range, the blade numbers must be chosen so as to minimize (rB/g^*); Fig. 10.14 implies that for this purpose, the number of stator blades should ideally exceed $2rB$.

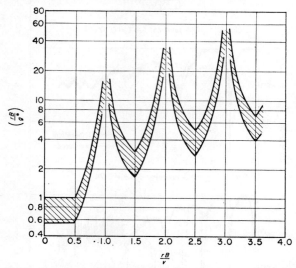

Fig. 10.14   Cut-off boundary for rotor-stator interaction modes.

These geometrical arguments do not depend on the detailed mechanisms of tone generation. They have been tested in the course of aircraft compressor research at Bristol Siddeley Engines Ltd.[29]; reductions of 10 dB in sound output at the blade-passing frequency were obtained, for the same aerodynamic performance, by altering the blade numbers to provide suppression by the cut-off phenomenon.

10.4 Noise reduction

Methods of reducing fan noise have so far been confined mainly to the acoustic treatment of the fan ducting–acoustic lining, splitters and the like. There is no objection to this (except perhaps in the additional cost involved), in such applications as air conditioning units or static compressors where such techniques can be employed without appreciable effect on performance. In aircraft engines, however, where weight and compressor efficiency are important operational parameters, the amount of acoustic treatment required often makes a really effective solution along these lines impracticable.

One form of silencing of engine intakes that is undoubtedly effective is the 'choked' intake. This incorporates a throat in the intake, at which the air reaches sonic velocity, leaving only the annular area of the duct boundary layer through which sound can be propagated upstream. Some results of investigations into this are reported in reference 10. As much as 28 dB reduction was achieved in a direction at 10° to the axis, but at 90°, there was no significant reduction in broad-band noise, although 10 dB was lost from the discrete frequency noise. There is, however, an inherent mechanical problem with this device. Since the high losses associated with the relatively small intake area would be unacceptable under all but approach conditions, the area would have to be adjustable in flight, and the complications of a reliable mechanism for this may not be justified.

It would clearly be more satisfactory if fan noise could be reduced to a minimum at its source, and from the preceding discussion on the physical nature of fan noise a number of points emerge which suggest that this is a feasible proposition.

For example, it is important that the airflow into the fan should be smooth. This is invariably the case in conventional aircraft engines, but in many industrial systems and possibly VTOL lifting fan configurations the flow into the fan is disturbed by nearby obstructions. It may be possible to reduce the turbulence at the rotor blade tips by control of the duct boundary layer, but there are no reports available at present of work on this. There is also the need for knowledge of the effect of blade shape on the amplitudes of lift fluctuations caused by random vortex shedding. There may well be an optimum shape for which these fluctuations are a minimum. The possibility of reducing interaction noise appears to be a combination of stator shape and clearance, and the indications at present are that slender profiles and separations of the order of one chord length are required. Some relief would also be obtained from tilting the stators, but here, significant reductions may only be achieved at the cost of aerodynamic performance.

As a second possibility, duct propagation effects can be used for noise reduction. Here, pending further research, one can say only that ducts should be designed to have cut-off frequencies as high as possible for those modes that are likely to be excited. In principle, very large reductions in noise can be achieved in this way.

References

1. B. B. Daly. *J. Inst. Heating Ventilating Eng.*, **26**, 29 (1958).
2. I. J. Sharland. *J. Sound Vib.*, **1**, 302 (1964).

240   *Noise and Acoustic Fatigue in Aeronautics*

3. S. N. Curle. *Proc. Roy. Soc.*, **A.211**, 505 (1955).
4. R. A. Davis. Univ. Southampton, Dept. of Aeronautics and Astronautics Rep. (1966).
5. H. W. Liepmann. *J. Roy. Aeron. Soc.*, **19**, 793 (1952).
6. E. Y. Yudin. Nat. Advisory Comm. Aeron. *Tech. Memo.*, 1136 (1944).
7. A. C. Austin. Univ. Southampton Inst. Sound and Vib. Res., M.Sc. dissertation (1965).
8. C. F. Peistrup and J. E. Wesler. *J. acoust. Soc. Amer.*, **25**, 322 (1953).
9. L. L. Beranek, G. W. Kamperman and C. H. Allen. *J. acoust. Soc. Amer.*, **27**, 217 (1955).
10. F. B. Greatrex. *S.A.E. Transactions*, **69**, 321 (1961).
11. C. H. Allen. *Noise Control*, **3**, 3 (1957).
12. G. C. Maling. *Noise Control*, **5**, 4 (1959).
13. R. B. Goldman and G. C. Maling. *Noise Control*, **1**, 6 (1955).
14. G. C. Maling. *J. acoust. Soc. Amer.*, **35**, 10 (1963).
15. C. G. Van Niekerk. *J. Sound Vib.*, **3**, 46 (1966).
16. N. le S. Filleul. *J. Sound Vib.*, **3**, 147 (1966).
17. L. Gutin. Nat. Advisory Comm. Aeron. *Tech. Memo.* (*Translation*), 1195 (1948).
18. N. H. Kemp and W. R. Sears. *J. Roy. Aeron. Soc.*, **20**, 585 (1953).
19. R. Hetherington. *AIAA J.*, **1**, 473 (1963).
20. L. S. Marks and J. Weske. *Trans. Am. Soc. mech. Eng.*, **56**, 807 (1934).
21. H. M. Fincher. *J. Sound Vib.*, **3**, 100 (1966).
22. I. Dyer. *J. acoust. Soc. Amer.*, **30**, 833 (1958).
23. J. H. Prindle. Boeing Airplane Co., Transport Division, *Document No. D6-7538* (issued as *A.R.C.* 24882) (1961).
24. J. M. Tyler and T. G. Sofrin. *Trans. S.A.E.*, **70**, 309 (1962).
25. C. L. Morfey. *J. Sound Vib.*, **1**, 60 (1964).
26. C. L. Morfey. American Society of Mechanical Engineers, 11th Annual Gas Turbine Conference, Zürich (1966).
27. P. M. Morse. *Vibration and Sound*, 2nd edn., McGraw-Hill, New York (1948) (see section 28).
28. T. F. W. Embleton and G. J. Thiessen. *J. acoust. Soc. Amer.*, **34**, 788 (1962).
29. C. L. Morfey. *Engineering*, **198**, 782 (1964).
30. K. W. Yeow. Univ. Southampton Inst. Sound and Vib. Res., Memo 143 (1966).

CHAPTER 11

The Reduction of Jet Noise

11.1 Introduction

Although the proportion of the energy of a jet engine converted into acoustic form is only of the order of one-tenth of one per cent, the experience of the public has been that this amount is usually too great and jet aircraft are too noisy. This problem of jet noise is, in general, restricted to the neighbourhoods of the airports used by these aircraft, though such areas may be several square miles in extent. This chapter deals with the suppression of exhaust noise, whether the aircraft is on the ground or in flight. The problem of rocket noise is also considered, but the specific topic of helicopter jet noise has been included in chapter 9.

11.2 Jet aircraft in flight

11.2.1 *Take-off flight plan*

The first means available to lessen the noise heard on the ground from a jet airliner is to use the best possible take-off flight plan, both with respect to aircraft direction and engine power. By choosing the most suitably-oriented runway and subsequent path, advantage can be taken of the great directivity of jet noise (chapter 7) to give appreciable alleviation to the regions of high population usually found close to the airport and which the latter serves.

Take-off over water from Kennedy Airport, New York, is an example of the application of this idea of preferential runways. Fig. 11.1 illustrates this point with acoustically poor and somewhat improved paths shown, the better flight path being curved so that residential areas may be least affected. The contours shown enclose all points which, during the course of the aircraft take-off, are subject to noise levels in excess of the quoted level. These sets are only shown for comparative purposes as atmospheric conditions such as prevalent wind directions influence the precise noise pattern below an aircraft as well as affecting the flight performance. Whilst the aircraft is still fairly low, say less than 300 ft, and subtending a small angle to the observer, the levels may be lower due to additional attenuation.[1] These general contours also help in the assessment

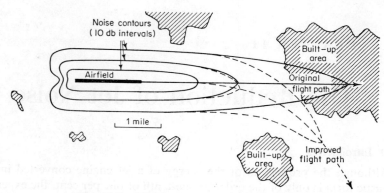

Fig. 11.1   Noise contours under flight paths.

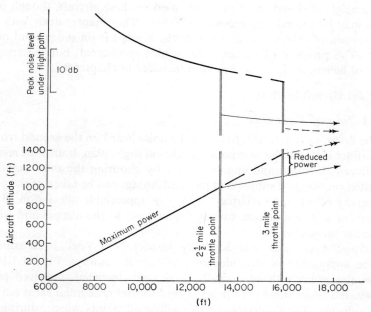

Fig. 11.2   Typical jet airliner take-off flight plans and associated noise.

of likely noise levels when new sites near to the airport are considered for building development.

Secondly, whilst aircraft must take off and do their initial climb at maximum thrust, throttling to a rating of the order of 50–75% of maximum thrust is acceptable at a later stage in the climb. This reduction in power which gives a noise reduction of the order of 10 dB should be made just before an aircraft passes over the first built-up area, say, 1,500 ft after the start of take-off roll. Lower power is a greater aid to noise alleviation at these close-in areas than is the use of full power to obtain greater altitude or higher forward speeds.

On the other hand, the lower rates of climb so obtained imply a spreading of the noise nuisance well beyond these areas. With the growth of engine power and output of acoustic energy, this wide spread of noise is beginning to counteract the advantages of throttling over the monitoring points and there are now many advocates of the full throttle take-off and maximum climb angle. Among these are pilots who have expressed concern at the deterioration in safety standards involved in the low power extended climb. Since the magnitude of the noise nuisance is related almost equally to the number of operations, the noise level, and the density of population, the optimum throttling technique must depend on the siting of the airport and the population around it.

Fig. 11.2 shows the result of throttling at both the $2\frac{1}{2}$ and 3 mile positions. North[2] and Richards[3, 4] also consider take-off techniques from the noise point of view.

11.2.2 *Engine design*

These operational methods are helpful, but in general they are insufficient to remove the noise problem. The only remaining methods involve tackling the noise-producing mechanism itself, which in turn involves an appreciation of the nature of jet mixing and manner of generation of noise.

Chapters 5 and 7 show that the dominating parameter is the effective jet velocity—the difference between the jet velocity and that of the surrounding air. Since the net thrust of a given engine is proportional to the *square* of the relative velocity while the noise emitted is approximately related to the *eighth* power of the velocity, from the acoustic point of view it is clearly better to have a larger jet pipe producing a low efflux velocity, than to obtain the same amount of thrust with a smaller jet working at a higher jet velocity. This is what we have with the now well-known by-pass engine, developed on both sides of the Atlantic. In this a proportion of the inlet air after going through the low pressure compressor 'by-passes' the high pressure compressor and meets the other gas again just prior to discharge. Lower noise levels are obtainable by using a (mechanical) mixer to ensure that the two streams are well-mixed before leaving the propelling nozzle. The exhaust noise will decrease as the proportion of by-pass air is increased, but a larger by-pass ratio implies a larger compressor and hence greater intake noise during approach to land (see chapter 10). Plotted against thrust, the noise levels of present-day commercial by-pass engines are 8 to 10 dB lower than those of straight turbo-jets. However, their design thrusts are generally much higher, and so at the same aircraft altitude a by-pass engine operating at full power can produce more noise than the straight jet. A discussion of straight jet, ducted fan and by-pass engine noise is given by Irving,[5] Pearson[6] and Greatrex.[7]

11.2.3 *Use of ejectors*

A simple method of reducing the exhaust velocity of an engine is to place some form of sleeve behind the propelling nozzle (see Fig. 11.3).

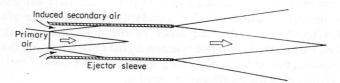

Fig. 11.3   Ejector mixing processes.

The ejector allows the entrainment of secondary air which mixes with the exhaust gases to produce a bigger jet of lower velocity. Such a scheme also includes the possibility of thrust augmentation at take-off but weight and drag problems may exist at cruise conditions. Unfortunately, with present aircraft, any permissible ejector would be so short, even with a retractable design, that little mixing between the gas streams would occur and little noise reduction would ensue.

The introduction of narrow-delta type supersonic airliners makes the installation of ejectors of adequate length more feasible. With this application in mind a comprehensive investigation using a model jet was undertaken at the University of Southampton. It was found that together with the white noise component, two types of discrete frequency sounds could be present in the noise from these ejectors and that each type might be of sufficient magnitude to swamp the reduction achieved in the broad-band noise. The first type was obtained with the flow in the ejector partly supersonic. This is the phenomenon of shock noise and is considered more fully in section 11.3.1 of this chapter. Secondly, discrete frequencies caused by acoustic cycles in the ejector were present at sub-critical pressure ratios. Both these are discussed in greater detail by Middleton.[8] Thickening of the turbulent boundary layer of the efflux gas tended to bring some amelioration without altering the broad-band noise. Until recently there has been little evidence of such discrete-frequency noise in practice, and it may well be that full scale conditions are too rough to allow the development of such phenomena. The attenuations produced by these ejectors have there-

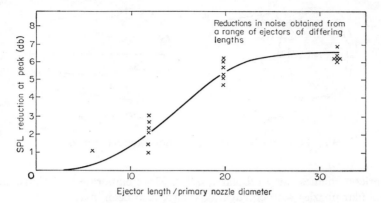

Fig. 11.4   The attenuation of ejectors, Southampton University Results.[8]

fore been plotted in Fig. 11.4 after the contributions of these pure tone sounds have been filtered out. It may be seen that excessive length is required to give a reasonable attenuation, unless some additional method of mixing is included.

11.2.4 *Modification of jet profile*

Early attempts to reduce the noise from a jet were made merely by a simple modification of the velocity profile. Powell[9] shows that a reduction in noise level is only obtained at the expense of considerable thrust loss. In fact, for a given thrust and jet diameter the usual square velocity profile gives less noise than a pipe-flow profile. This is due to the higher jet velocity in the centre of the jet. If a large jet pipe is used, the noise is reduced.

11.2.5 *Modification of jet shape*

The final possibility to commend itself is a change in the shape of the jet itself, and this is accomplished by some modification of the final exhaust nozzle. Unfortunately, every deviation from the normal circular jet is costly in propulsive efficiency and as jet airliners usually fly near their range limits, this has presented in many cases an unacceptable economic burden. However, by a mainly experimental approach, commercially-acceptable flight suppressors giving significant, though modest, noise reductions have been evolved.

The first innovation made was the introduction of vortex generators in the nozzle, and then the use of teeth was tried (Fig. 11.5a). The next step was to use corrugations (Fig. 11.5b) around the edge of the nozzle. Today forms of corrugated nozzles have been developed on British engines whilst

9

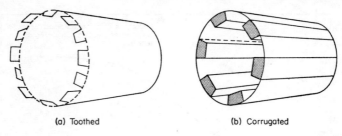

(a) Toothed   (b) Corrugated

Fig. 11.5   Early experimental noise-reducing nozzles.

American engines carry tubular nozzles. More elaborate variations, such as tubular nozzles with corrugations, have also been investigated.

It must be remembered that the replacement of a circular jet by a series of smaller ones, or by an irregularly shaped one, tends to produce a greater proportion of high frequency noise, the components most disliked by the human ear. Thus subjectively the noise reduction may not be as great as is indicated by objective noise meters (see chapter 6). This point is illustrated in Fig. 11.6 which shows the form of the spectra generated by a circular nozzle and by a suppressor nozzle, at the instant of emission of the peak overall noise with the parent aircraft in level flight at about 500 ft altitude. (The precise shape these spectra take naturally depends on many factors, especially the engine setting and aircraft forward speed.) The difference in overall sound pressure level is about 7 dB, but in subjective units it is less than 5 PN dB. However, when the aircraft altitude increases to 1,500 feet, the attenuation of the suppressor nozzle over the standard nozzle increases by about a further 2 PN dB. This is due to the greater atmospheric absorption which occurs in the higher frequencies and which therefore lessens the importance of the upper frequency sound in determining the subjective measure of a sound. These new spectra are also shown in Fig. 11.6 and include the 9 dB fall in levels due to the inverse square law as well as a typical allowance for the additional absorption factors in the higher frequencies.

A further point in favour of these suppressor nozzles is that they generally have a better 'time-history'. This can be seen in Fig. 11.7 which shows the noise received on the ground as an aircraft flies over with a plain nozzle and with a suppressor. The time the noise from the suppressor nozzle lies within say 5 dB of the peak noise generated by that nozzle is appreciably less than the corresponding time for the ordinary nozzle. Since proposals are now being made to include duration in subjective units, this aspect may grow in importance in the future.

References 10, 11, 12 and 13 describe some of the work involved in the

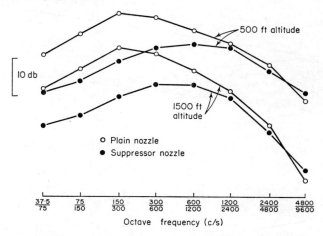

Fig. 11.6   Examples of conical and suppressor nozzle spectra, emitted
at instant of peak air-to-ground noise.

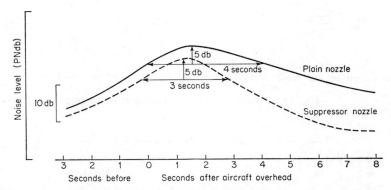

Fig. 11.7   Typical trace of overall noise against time for aircraft in
level steady flight, 500 ft altitude; plain and suppressor nozzles fitted.

development of these nozzles. In order to understand the mechanism of
noise suppression involved, let us consider the turbulence patterns in such
jets.

In the mixing region of a circular jet (Fig. 11.8) the wavelength of the
noise generated a certain distance downstream is typically associated with
the scale of the turbulence and this is related to the width of the mixing
region there. If this jet is now replaced by n equal tubes whose total efflux
area is the same, then the mixing width is decreased by a factor $n^{1/2}$, and
there is a corresponding rise in the frequency. Similar remarks apply to a

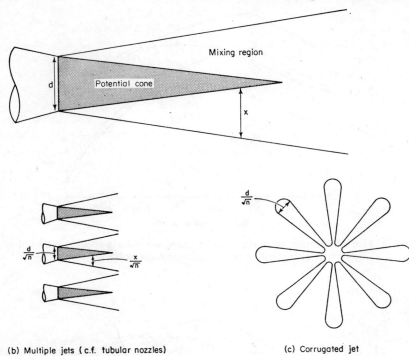

(b) Multiple jets (c.f. tubular nozzles)   (c) Corrugated jet

Fig. 11.8   Alteration of jet mixing parameters.

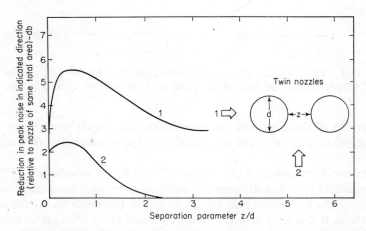

Fig. 11.9   The noise pattern of two jets in proximity (after Greatrex
and Brown[14]).

corrugated nozzle, the width of a corrugation being used instead of the diameter of the tube in this similarity argument.

This spectrum shift is not the only change when a single circular nozzle is replaced by several smaller nozzles. The screening of one jet by another when observed from a certain angle also occurs. If two parallel similar co-planar jets are sufficiently far apart to avoid any aerodynamic interference with each other (i.e. z/d is large in Fig. 11.9), it is found that the peak noise measured in the plane of the axis of the jets is less than that measured perpendicular to it.[14] This change can only be due to refraction, partial reflection and high-frequency scattering.

The next important effect is found when the distance between the jets is reduced sufficiently for them to interfere by competing for the same secondary mixing air. As Fig. 11.9 shows, the noise is reduced further, and this must be due to altering the fundamental noise-producing parameters. (This effect may be present on the Comet 4 airliner whose four engines are mounted in pairs, the members of each pair being very close together.) Laurence and Benninghoff[15] have shown that the scales and intensities of turbulence of aerodynamically interfering jets are lower than when the jets do not interfere.

Coles[16] has suggested that for a set of nozzles, arranged in an $(m \times n)$ rectangular array, the reduction of the peak polar noise in the direction normal to a row of m nozzles is found by combining a 'shielding' effect of $10 \log_{10} n$ dB, with an interference effect, the magnitude of which can be deduced from the 'vertical' attenuation shown in Fig. 11.9. Thus, in the plane of their exhaust, the polar distribution of noise about a bank of nozzles will be similar to that shown in Fig. 11.10.

To examine the precise parameters which are affected by this interference

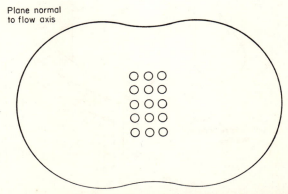

Plane normal
to flow axis

Fig. 11.10   The directional effect of noise from a bank of nozzles.

and increased mixing, a series of experiments was carried out at the University of Southampton.[17] A fuller knowledge of the mechanics of the problem may enable recommendations for the design of better noise-reducing nozzles to be made, as discussed by Richards and Williams.[18] The basic study was performed on a circular nozzle with a single 45° notch extending to the centre of the jet. This corrugation was designed to permit maximum secondary entrainment. Other configurations investigated include a six-corrugated nozzle and a pair of rectangular nozzles side-by-side to approximate to the two-dimensional case. These are all sketched in Fig. 11.11.

In the experiments the mean velocity, the longitudinal and radial intensities of the turbulent velocity fluctuations and the scales of the turbulence were all measured. In the case of the notched nozzle the results

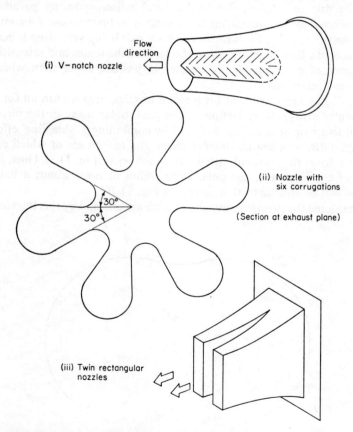

Fig. 11.11 Experimental nozzles, Southampton University.[17]

indicated that for the region immediately downstream of the notch a lower value of shear was formed due to the greater secondary airflow induced, and the initial value of intensity of the turbulence did not rise to the same high value as on the circular side. Values of turbulence scales were similar or slightly smaller on the notch side, both for the longitudinal and the transverse directions. This meant that a reduction in velocity shear is formed without a corresponding increase in turbulence eddy size. The volume over which turbulent fluctuations are correlated is important from the noise production point of view and the reduction found in the parameters implied less noise per unit volume of turbulence. A slight shift to higher frequencies in the spectral density of the longitudinal velocity fluctuations was also indicated. Figs. 11.12 and 11.13 show some typical results.[17]

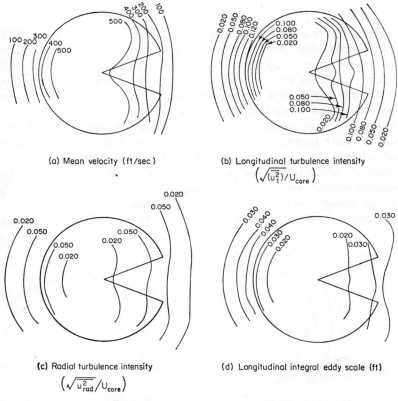

(a) Mean velocity (ft/sec)

(b) Longitudinal turbulence intensity

$$\left(\sqrt{(u_1^2)}/U_{core}\right)$$

(c) Radial turbulence intensity

$$\left(\sqrt{u_{rad}^2}/U_{core}\right)$$

(d) Longitudinal integral eddy scale (ft)

Fig. 11.12   Some experimental results for V-notch nozzle: 2 inches downstream. Jet velocity = 500 ft/sec.[17]

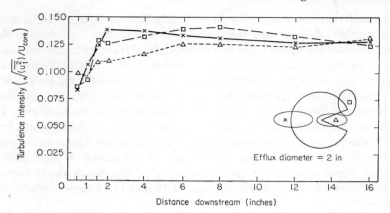

Fig. 11.13   Maximum turbulence intensity recorded in the indicated regions.[17]

All the effects of the notch appeared to have disappeared by about eight diameters downstream, and it is suggested[19, 20] that a limit exists to the noise reduction a suppressor can produce. This is because sufficiently far downstream the essential characteristics of the modified and unmodified jets will be the same so that no alteration can be made to whatever noise is generated downstream of such section.

Lighthill,[21, 22] Proudman,[23] Ribner,[24] Corcos,[25] Lilley,[26] and Cheng[27] have all made hypotheses about the structure of the noise-generating mechanism of turbulence, and these indicate, by implication, the possible manner by which suppressor nozzles work. For the notched nozzle the measured results were substituted in the expression due to Lilley and the acoustic power output per unit volume calculated for various stations downstream. A typical result is shown in Fig. 11.14. These results indicate

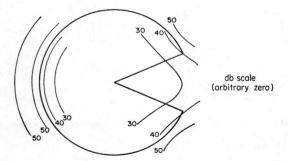

Fig. 11.14   Acoustic power output/unit volume, 2 inches downstream.[17]

that a reduction in acoustic power output does not depend on one para-meter alone, but rather on how the parameters change with reference to each other. A reduction in shear alone is not sufficient if the turbulence intensity and eddy volume increase. In the case of the notched nozzle, the results indicated a general reduction in sound produced and that close to the nozzle this reduction was due to the decrease in shear, with a decrease in turbulence intensity, and no increase in correlation volume, these effects outweighing the greater mixing region volume produced. Further down-stream the slight reduction in eddy scale became important. The results from the other nozzles confirmed these findings, and a large reduction in turbulence intensity was noted for the interference region of the two rect-angular nozzles. Tyler, Sofrin and Davis[28] have presented some further noise results for rectangular nozzles, and have shown that the reduction is a function of the convergence geometry of the nozzles.

The main requirement for noise suppression thus appears to be the entrainment of the maximum quantity of secondary air as smoothly as possible with no rapid changes in nozzle shape to produce regions of high shear. Corrugated multi-nozzle configurations should be designed to do this without increasing the turbulence, and Potter[17] concludes that the best suppressor nozzle may be a four-lobed one with deep corrugations. It has been shown that failure to have deep corrugations leaves a large central core similar to that of an unmodified nozzle. An alternative and novel way suggested to deal with a large potential cone was the introduc-tion of convergent auxiliary nozzles with helium, and inclined so that their efflux could penetrate and break up the central region.

Some current types of suppressor are sketched in Fig. 11.15.

11.3 Further exhaust noises

11.3.1 *Shock-cell noise*

As section 11.2.3 has indicated, when a model conical nozzle operates at a pressure ratio above the critical, or supersonic flow issues from a convergent-divergent nozzle operating at a pressure other than its design Mach number, another source of noise may be added to the usual jet roar.

This is because shock-cells are set up in the jet, and any disturbances at the orifice convected through such shocks emit sound waves which on passing back to the nozzle lip create new disturbances. Thus resonances are produced, the frequency of the note being related to the shock spacing and Mach number. The geometry of the loop is such that the frequency of the fundamental falls as the pressure ratio is increased, but discontinuous jumps in frequency can occur due to changes in the cell structure.

Fig. 11.16 shows the form of relationship which exists between the

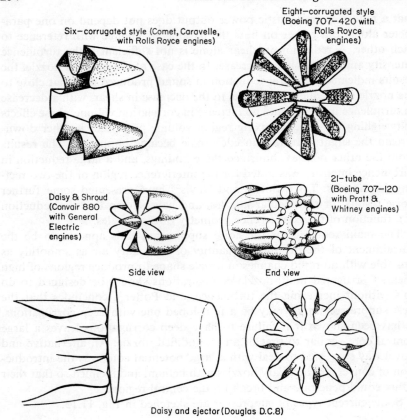

Six−corrugated style (Comet, Caravelle, with Rolls Royce engines)

Eight−corrugated style (Boeing 707−420 with Rolls Royce engines)

Daisy & Shroud (Convair 880 with General Electric engines)

21−tube (Boeing 707−120 with Pratt & Whitney engines)

Side view   End view

Daisy and ejector (Douglas D.C.8)

Fig. 11.15   Some current commercial flight noise suppressors.

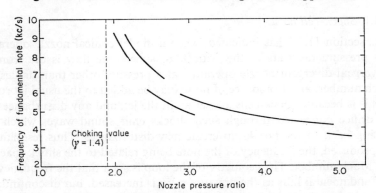

Choking value (γ = 1.4)

Fig. 11.16   Variation of frequency of shock noise with pressure ratio, 1 inch diameter circular nozzle.

frequency of the fundamental note and the pressure ratio (after Middleton[8]). However, it has been found that different experiments have obtained results which cannot collapse using Strouhal number as the only non-dimensional parameter, and it seems that the precise relation may be a function of nozzle design. Hysteresis may be present in any jumps, and the notes are strongest at the high-pressure end of any stage. This screech noise has also been found in tests on nozzles not circular in shape.

The important criterion[30] in the production of these notes is the condition of flow at the edge of the nozzle. Roughening the lip in some way, for example by the introduction of notches, has been found adequate to eliminate them on model jets. Although aircraft engines now operate at pressure ratios above choking, the roughness of their exhaust units and of the emergent flow appears to be sufficient to prevent such a resonance mechanism at sea level. However, at high altitudes where the pressure ratio is greater, evidence is now strong that such a mechanism is present and causes tailplane acoustic fatigue failures.

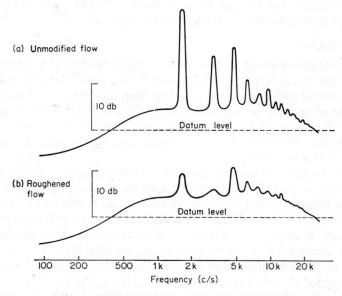

Fig. 11.17   Effect of roughening the flow from choked nozzle (narrow band spectra).[8]

Fig. 11.17 shows a typical spectrum measured at Southampton University with an unmodified nozzle, and with some degree of roughness. A reduction in the overall level of about 5 dB was achieved, bringing the noise to within 2 dB of the broad band component.

11.3.2 *Reheat noise*

Until now, the use of after-burning, which is the combustion of further
fuel aft of the turbine to raise the jet temperature, and hence its velocity
and thrust, has been restricted to military aeroplanes. If this practice is to
be developed for civil use the resulting noise picture must be carefully considered. The noise due to combustion, which contains discrete frequency
components related to the normal modes of the cavity, must be added to
the increased noise levels due to the higher velocities. The more thorough
the burning, the lower will be this combustion noise.[31] Further information
is given by Elias.[32]

Under after-burning conditions, the directivity of the radiated noise
changes, the peak occurring at a somewhat greater angle to the exhaust.

11.3.3 *Noise from aft fan engines*

With aft fan engines, discrete noise from the blades has been detected
in the exhaust noise. Gordon[13] reports the development of an exhaust
system incorporating acoustic filters to remove these discrete notes, and
these may have some effect on the broad-band noise.

11.4 Rocket noise

Mention must be made of the special problems produced by the exhaust
noise of rockets. A feature of rocket noise is that such a large proportion
of the acoustic output is in the very low frequency range, and therefore
not easily attenuated. In fact, even though some of the noise may be sub-
audible, doors can shake and windows rattle, giving the impression of
ground-borne vibration. The combustion noise mentioned in section
11.3.2 can also be a problem.

Unlike aircraft, it is scarcely possible to alter the flight path of rockets,
but some alleviation of the nuisance may be obtained by choosing the
optimum take-off *time*. Investigations have shown[33] that due to various
meteorological factors such as wind and temperature gradients, severe
focusing of the noise may occur at considerable distances from the launch-
ing area. Attempts have therefore been made, allowing for these parameters, to calculate on a digital computer the likely areas of excessive
noise at expected launch times. In this way, it may be possible to avoid
such occasions when launches will be liable to cause distress in vulnerable
localities.

Many tests require static firing of the rocket, and it becomes necessary
to provide a blast deflector to protect the region beneath the rocket. The
design of this deflector has a considerable effect on the polar noise distribution.[34] A study of the design of deflectors to determine their optimum
form from the acoustic point of view is clearly a necessity.

11.5 Jet aircraft on the ground

11.5.1 *Ear protection*

At most airports the essential maintenance and overhaul work entails the running of engines for appreciable periods to check their performance. It may be possible to provide airport personnel with adequate ear protectors (for example, the Liquid Seal type which gives up to about 40 dB attenuation), but passengers and visitors can scarcely be expected to adopt these. In any case, the noise problem associated with these static tests may extend well beyond the confines of the aerodrome.

11.5.2 *Orientation and screening*

The sharp directivity of jet noise implies that there is an optimum orientation for the aircraft to give the smallest annoyance in the area. At least one open-air test bed has been realigned to minimize the community disturbance. Typical noise contours around both an unsuppressed and suppressed jet aircraft are shown in Fig. 11.18 to illustrate the improvements which are possible by this means.

Aircraft may also take advantage of the shadow effect produced by testing behind airport buildings. Alternatively, banks or walls may be specifically built to produce a screening effect. Redfearn,[35] for example, has considered theoretically the noise reduction that a barrier can be expected to produce. More recently, Magyar[36] has produced results showing that leakage of sound round the ends of a wall is negligible if the length is more than six times the height. The actual attenuation is obviously a function of both the distance of the source and that of the receiver from the barrier, as well as of the barrier height itself and the wavelength of the sound (Fig. 11.19). The wall should be made as massive as possible, so that sound attempting the direct path through the wall may be attenuated by the mass law (see chapter 23).

Because of additional atmospheric effects, including wind and temperature gradients, turbulence and reflections, the sound reduction is often not as great as can be expected. On the other hand, if the noise source is at an appreciable distance (say more than a half-mile) from the nearest population, the attenuation can be somewhat greater than that computed from the inverse square law and molecular absorption data. This is because of the frequency-dependent 'ground effect' associated with grazing incidence and discussed by Parkin and Scholes.[37] Results of further measurements are given by Hayhurst.[38]

11.5.3 *Ground mufflers*

Although optimum orientation and the use of shadow zones may each give benefits of the order of 10 dB or more, these simple expedients may

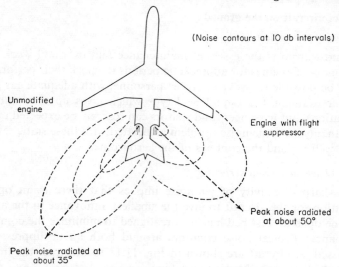

(Noise contours at 10 db intervals)

Unmodified
engine

Engine with flight
suppressor

Peak noise radiated
at about 50°

Peak noise radiated at
about 35°

Fig. 11.18   Directivity effects. The noise field around a stationary jet
aircraft with and without flight suppressors.

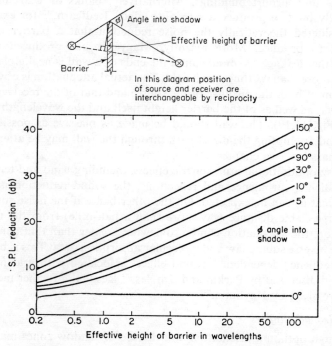

ϕ) Angle into shadow

Effective height of barrier

Barrier

In this diagram position
of source and receiver are
interchangeable by reciprocity

S.P.L. reduction (db)

Effective height of barrier in wavelengths

150°
120°
90°
30°
10°
5°

ϕ angle into
shadow

0°

Fig. 11.19   The noise reduction of a barrier (after A. J. King).

be insufficient to reduce the noise levels adequately and ground suppressors must be used. These vary in complexity from a simple pipe to a building enclosing the whole aircraft, like the Caravelle 'hush-house' developed at Stockholm.

Fortunately, the overriding importance in the flight case of negligible thrust loss from any suppressor does not apply on ground mufflers, and mufflers may be designed for any aircraft type, whether civil or military, with or without reheat. The principal condition is that the muffler produces no back pressure on the engine. This is essential in order that the true engine running characteristics may be checked, and that no design limitation on operating conditions be exceeded.

Clearly there are other desirable features such as ease of setting up. Portability of the muffler is advantageous as this avoids the necessity of manoeuvring the aircraft into the correct position relative to a fixed installation. Portability implies that the muffler must be lightweight and small, but these terms are only relative, as full scale fixed mufflers can weigh over 50 tons and be over 100 ft in length. A really light-weight structure in position would require some anchoring, as the muffler must withstand the momentum of the exhaust gases, and the design must be such that no significant noise is radiated through the muffler casing. Similarly the suppressor must be capable of use under a full range of engine operating conditions which, amongst other things, entails allowing for any differences in height of the tailpipe due to different engine thrusts. If after-burning is likely to be used, the equipment must be able to withstand the higher temperatures.

The following principles may be applied in the design of the mufflers themselves:

(i) reduction of the amount of noise generated in the mixing region
(ii) absorption of as much as possible of the noise remaining after (i)
(iii) use of the directional effect of the residual noise to give least disturbance.

The simplest piece of noise-reducing equipment is the straight-through pipe. Working on the ejector principle outlined in section 11.2.3, the pipe should be long enough to promote complete mixing and therefore a lower exhaust velocity and lower noise level. Reductions of about 10 dB can be achieved in this way.

This method simply converts the jet velocity to a lower value and employs only concept (i). Most of the remaining sound travels down the pipe from the high speed mixing region. If the second principle is now applied then we must add absorptive chambers, in which mineral wool or similar substance is placed (Fig. 11.20a). Clearly the greater the area of

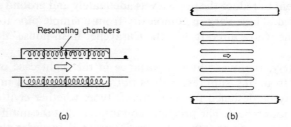

(a)   (b)

Fig. 11.20 (a) 'Straight through' muffler.
(b) Conventional splitter arrangement.

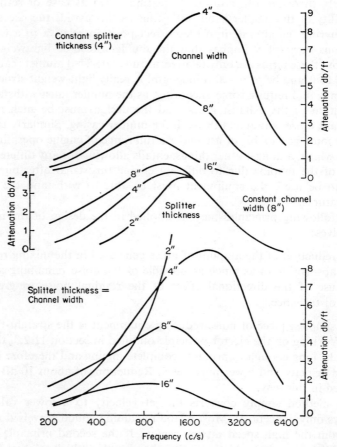

Fig. 11.21 Attenuation of sound in absorbent channels (after Fleming
and Copeland[39]).

absorbent material presented to the exhaust gases, the greater the reduction, and this is best achieved by splitting the flow into several narrow channels, as in Fig. 11.20b.

The attenuation produced by these annular splitter panels increases with length, but this attenuation is also a function of the geometry of the installation and is frequency-dependent. The effect of varying some of these geometrical parameters for a perforated metal sheet splitter with sound absorbent filling is shown in Fig. 11.21 from reference 39. Thus the precise design of a muffler can be finalized only when the characteristics of the noise to be absorbed are known. The high-frequency noise reduction can be improved by introducing bends in the duct or tortuous paths through the splitters.[40, 41]

Noise generation may be reduced by lowering the gas velocity within the muffler by, for example, a 'pepper-pot' arrangement with diffusing sections. The designs of the pepper-pot and diffuser are such that the gas flows fully in the chamber with as uniform a velocity as possible. A similar innovation employs a series of gauzes or holed screens near the efflux, the velocity falling rapidly as the gas expands through these. The small size of the resulting jets implies a shift in frequency scale upwards, and higher frequency sound is always easier to absorb.

When the temperature and velocities are as high as those encountered with reheat conditions, unprotected screens and similar devices cannot be placed near the nozzle. Water injection into the exhaust downstream from the nozzle is a normal method to use. It is effective, but the necessary temperature and velocity reductions are only achieved with a considerable mass of water, at least equal to the mass of the gas flow and preferably much greater.

Some mufflers bleed off the exhaust gases using a series of annular ducts or holes. A more common device is to discharge the efflux upwards. This takes advantage of the low noise level radiated at 90° to the jet. Using the results of section 11.2.5, a further reduction can be obtained by making the plane of the exhaust rectangular.

A generic muffler scheme is shown in Fig. 11.22, and Fig. 11.23 illustrates a muffler designed for a specific installational requirement. A

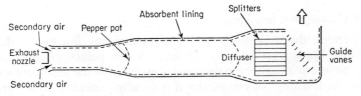

Fig. 11.22   Generic ground muffler.

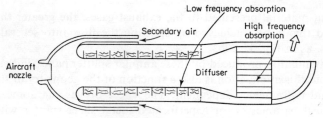

Fig. 11.23   SAAB silencer system.[42]

step-by-step account of its development is described by Olesten.[42] It was possible to reduce the noise at the inlet of this suppressor by coupling the muffler to the aircraft engine, and inducing the secondary air through separately treated ducts.

An attractive way of combining noise-reducing principles into a lightweight form of muffler is discussed by Richards.[43] Reductions of the order of 25 to 30 dB have been measured in model tests on a non-reheat version of the design. Fig. 11.24 (after Middleton[44]) shows that the noise in all frequencies was appreciably reduced.

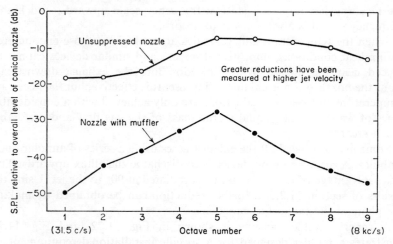

Fig. 11.24   Attenuation measured on model based on Reference 44.

11.5.4 *Ground pens and hangars*

Quietening the exhaust by use of a muffler may not be a sufficient palliative. The engine intake noise may require reduction and separate intake suppressors have been designed, usually using the absorbent splitter system. At this stage of complexity it is usually preferable to use ground pens as these give a greater degree of suppression of the noise sources.

In such a pen, solid walls form three sides of a rectangle, and the aircraft is placed into this. The mouths of the exhaust mufflers break the far wall while the main parts of the mufflers lie outside the pen. Some sort of sliding partition may seal the front of the pen. In this way the mufflers deal with the exhaust noise in the usual manner, whilst other noise is retained within the cell by the screening of the walls.

The final refinement is the construction of a full run-up hangar, in which all air-gaps are acoustically treated. It is also necessary to fit adequate sound absorptive materials inside to suppress reverberant build-up. This would otherwise create both risk of damage to the aircraft through acoustic fatigue and risk of auditory trouble to any unprotected personnel.

The attenuations that can be expected at 250 ft radius from various

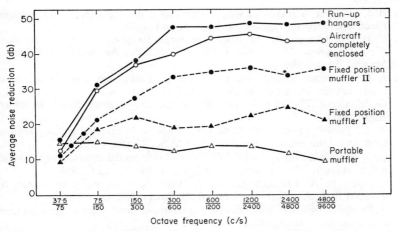

Fig. 11.25 Noise reductions achieved by various ground installations.[45]

ground installations are given in Fig. 11.25 (after von Gierke[45]). Attenuations of 50 dB are usually sufficient to bring noise levels down to background levels.

References

1. P. H. Parkin and W. E. Scholes. Oblique air-to-ground sound propagation over buildings, *Acustica*, 8, 99 (1958).
2. W. J. North. Effect of climb technique on jet-transport noise, *N.A.C.A. Tech. Note.*, 3582 (1956).
3. E. J. Richards. The noise at airports, *Proc. Inst. civ. Eng.*, 20, 533 (1961).
4. E. J. Richards. Aircraft Noise—Mitigating the Nuisance, *Aeronautics and Astronautics*, 5, 1, 34 (1967).

5. H. B. Irving. Jet aircraft noise and the by-pass engine, *The Aeroplane*, **94**, 262 (1958).

6. H. Pearson. Trends in aircraft propulsion, *J. Roy. Aero. Soc.*, **65**, 373 (1961).

7. F. B. Greatrex. The economics of aircraft noise suppression, *I.C.A.S. Preprint* (1966).

8. D. Middleton. The noise of ejectors, *Aero. Res. Council (London)*, R & M 3389 (1963).

9. A. Powell. The influence of the exit velocity profile on the noise of a jet, *Aero. Res. Council (London) Report No.*, A.R.C. 16156 (1953).

10. E. N. Sidor. Aircraft noise, Boeing Aircraft Company, Seattle; A.R.C. 19029 (1957).

11. F. B. Greatrex. Noise suppressors for Avon and Conway engines, *A.S.M.E. Preprint* (1959).

12. E. E. Callaghan. Noise suppressors for jet engines, *Noise Control*, **5**, 1, 18 (1959).

13. B. J. Gordon. A review of work in jet engine noise control at the General Electric Company, *Noise Control*, **7**, 3, 14 (1961).

14. F. B. Greatrex and D. M. Brown. Progress in jet engine noise reduction, First Congress International Council of the Aeronautical Sciences, Madrid, September 1958.

15. J. C. Laurence and J. M. Benninghoff. Turbulence measurements in multiple interfering air jets, *N.A.C.A. Tech. Note*, 4029.

16. G. M. Coles. The jet noise of multi-nozzle arrays, Rolls-Royce Brochure AP9 (1961).

17. R. C. Potter. Measurements of turbulence in interfering mixing regions from jets, University of Southampton M.Sc. Thesis, 1963.

18. E. J. Richards and J. E. Ff. Williams. Some recent developments in jet noise research, *Univ. Southampton Rep.*, *A.A.S.U.* 118 (1959).

19. I. Dyer, P. A. Franken and P. J. Westervelt. Jet noise reduction by induced flow, *J. acoust. Soc. Amer.*, **30**, 761 (1958).

20. A. Powell. Considerations concerning an upper limit to jet noise reduction, *J. acoust. Soc. Amer.*, **31**, 1138 (1959).

21. M. J. Lighthill. On sound generated aerodynamically: I—General theory, *Proc. Roy. Soc.*, (A), **211**, 564 (1952).

22. M. J. Lighthill. On sound generated aerodynamically: II—Turbulence as a source of sound, *Proc. Roy. Soc.*, (A), **222**, 1 (1954).

23. I. Proudman. The generation of noise by isotropic turbulence, *Proc. Roy. Soc.*, (A) **214**, 119 (1952).

24. H. S. Ribner. Aerodynamic sound from fluid dilations; a theory of the sound from jets and other flows, *Univ. Toronto Inst. Aerophysics Report*, 86 (1962).

25. G. M. Corcos. Some measurements bearing on the principle of operation of jet silencing devices, *Aero. Res. Council (London) Report No.*, A.R.C. 20213.

26. G. M. Lilley. On the noise from air jets, *Aero. Res. Council (London) Report No.*, A.R.C. 20376 (1958).

27. S-I. Cheng. The aerodynamic noise of a turbulent jet, *J. aerospace Sci.*, **28**, 321 (1961).

28. J. M. Tyler, T. G. Sofrin and J. W. Davis. Rectangular nozzles for jet noise suppression, *S.A.E. Preprint* 57T (1959); also: *J. Soc. Automotive Engineers*, **67**, 82 (1959).

29. A. Powell. On the mechanism of choked jet noise, *Proc. Phys. Soc.*, **66B**, 1039 (1953).

30. A. Powell. The reduction of choked jet noise, *Proc. Phys. Soc.*, **67B**, 313 (1954).

31. E. J. Richards. A note on combustion noise, *Aero. Res. Council (London) Report No.*, A.R.C. 23103 (1961).

32. I. Elias. Sonic combustion control, *Sound* **2**, 6, 8 (1963).

33. R. N. Tedrick and co-workers. Studies in far-field acoustic propagation, *N.A.S.A. TN D*-1277 (1962).

34. W. D. Dorland. Characteristics of the Saturn space booster as a noise source, Paper L25, Fourth International Congress on Acoustics (1962).

35. S. W. Redfearn. Some acoustical source-observer problems, *Phil. Mag.*, **30**, 233 (1940).

36. P. E. Magyar. Étude sur maquette de la réduction du bruit par l'interposition d'une barrière, Paper L46, Fourth International Congress on Acoustics (1962).

37. P. H. Parkin and W. E. Scholes. The horizontal propagation of sound from a jet engine close to the ground, at Radlett, *J. Sound Vib.*, **1**, 1 (1964).

38. J. D. Hayhurst. Acoustic screening by an experimental running-up pen, *J. Roy. Aeron. Soc.*, **57**, 3 (1953).

39. N. Fleming and W. C. Copeland. Principles of noise suppression, *Ann. Occup. Hyg.*, **1**, 28 (1958).

40. B. G. Watters, S. Labate and L. L. Beranek. Acoustical behaviour of some engine test cell structures, *J. acoust. Soc. Amer.*, **27**, 449 (1955).

41. I. Dyer. Noise attenuation of dissipative mufflers, *Noise Control*, **2**, 3, 50 (May 1956).

42. N. O. Olesten. SAAB silencer system, *Noise Control*, **5**, 215 (1959).

43. E. J. Richards. An optimum form of ground muffler, *Aero. Res. Council (London) Report No.*, A.R.C. 20724 (1959).

44. D. Middleton. The testing and development of a ground muffler for jet engine exhaust noise, *Aero. Res. Council (London) Current Paper No.*, CP. 610 (1962).

45. H. E. von Gierke. Recent advances and problems in aviation acoustics, Proceedings of the Third International Congress on Acoustics, Stuttgart, 1959; Vol. II Applications (L. Cremer, editor, Elsevier (1961)).

Further References

C. M. Harris (editor). *Handbook of noise control*, McGraw-Hill Book Co., New York (1957); chapters 33, 34.

L. L. Beranek (editor). *Noise reduction*, McGraw-Hill Book Co., New York (1960): chapters 16, 17, 24.

—— The control of noise, N.P.L. Symposium No. 12, Papers D1, D2, D3, D4, H.M. Stationery Office (1962).

D. J. Maglieri, D. A. Hilton and H. H. Hubbard. Noise considerations in the design and operation of V/STOL aircraft, *N.A.S.A. TN D*-736; A.R.C. 22986 (1961).

CHAPTER 12

Sonic Bangs

12.1 Introduction

The sonic bang is a phenomenon which nowadays is well understood from the physical standpoint. It looms large as a problem at the present time because of the bearing that it has on the operation of future supersonic civil aircraft. The problem centres around the level of bang that people generally will accept as tolerable from the annoyance point of view, for it is now fairly well-established that this level is well below the level that will cause serious damage to property. Work is currently going on to try and determine what the acceptable level is likely to be. A discussion of this subjective problem is outside the scope of this paper, however. What will be done here is to run over the theory of the sonic bang, and to present an ordered account of its physical aspects. By theory is meant not just the mathematics, for most of the mathematical results will merely be quoted, but rather the essential physics of the phenomenon, primarily in descriptive terms, for once this is understood the actual details can always be obtained from the literature. Nothing new will be presented, but it is hoped that a survey of the subject will prove useful.

12.2 The nature of a sonic bang

Sonic bangs are caused by the shock waves which an aircraft generates when it travels at supersonic speed. The bangs are the subjective manifestation of the passage of these shock waves, which cause sudden rises in pressure, past an observer. Near the aircraft the shock wave pattern is rather complex, but well away from the aircraft the disturbance from ambient conditions takes the form of what is called an *N-wave*, as illustrated in Fig. 12.1. This figure represents how the pressure experienced by an observer would vary with time in the usual ideal case in which the air is treated as an inviscid fluid. It will be noticed that there is a sudden rise in pressure at the *bow shock* to some value greater than ambient, then an almost linear fall in pressure to a value that is less than ambient by the same amount, followed by a sudden rise again at the *stern shock*, and finally a decay to ambient conditions once more. Clearly the ideal *N-wave*

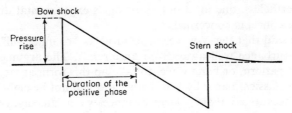

Fig. 12.1   The pressure in an ideal *N*-wave.

can be described in its main essentials by the *pressure rise* at the bow shock, and the so-called *duration of the positive phase*.

In a real fluid the effect of viscosity and other dissipative properties can be to 'round off' the sharp peaks in the ideal *N*-wave, as illustrated in Fig. 12.2. The additional parameter that would seem to describe this effect

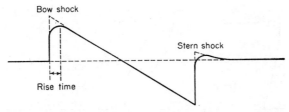

Fig. 12.2   The pressure in a real *N*-wave.

in its main essential is the so-called *rise time*. The effect and importance of this rounding off is not fully understood. Since it effectively reduces the high frequency content in the spectral representation of the waveform one would expect it to have the effect, perhaps, of making the impression of that of a 'boom' rather than a 'bang'. Further, the amount of rounding-off is a balance between the natural tendency of the compressive part of a waveform to steepen, and the attenuative effects of the fluid's dissipative properties. The latter are expected to dominate for the weaker shocks. However, little has been published on this aspect. At the level of sonic bang that is of interest (around 1 lb/ft^2) the rise time is less than a tenth of the duration of the positive phase. Nevertheless, the rise time is important from the subjective point-of-view, and it is one of the matters requiring further study. In passing, care should be taken to distinguish between the rise time, defined as the time taken for the pressure to rise to its maximum value, and the response time of any instrument that one might be using to measure the waveform. Clearly the fact that physical instruments have a finite response time makes it difficult in practice to measure short rise times. Nevertheless, one must not allow one's experimental difficulties to obscure the concepts concerned.

We have said that the *N*-wave, be it 'ideal' or 'rounded-off', character-izes the disturbance well away from an aircraft. The arguments for the disturbance pattern, or bang wave, to take up this form at large distances are put in a classical paper by Whitham,[1] and will not be elaborated upon here. It suffices to say that by large distances we mean, say, one hundred aircraft lengths away, or something of the order of 10,000 ft or so. This condition is usually met in sonic bang situations. However, occasionally one may be interested in situations closer to the aircraft, and an example of the more complex waveform that can be encountered in such cases is illustrated in Fig. 12.3, which shows the waveform that might be experi-enced by an observer who was sufficiently near, that the bow and stern shocks from the wing had not merged with the bow and stern shocks

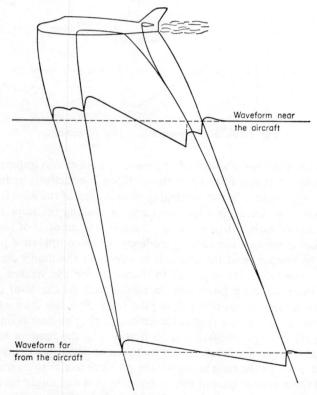

Fig. 12.3   The *N*-wave pressure distribution near to, and far from an aeroplane.

respectively from the fuselage, as they would eventually according to Whitham,[1] except in rare instances. Our methods for calculating sonic bangs are capable of dealing with this case, as will be explained when we come to the matter, although the calculations become a little more complicated. An interesting thing to observe at this stage is that the waveform in the example shown is characterized by four shocks in all. This raises the whole question of how many bangs one should hear. The only answer to this question is that it all depends, and on many things at that. Ideally one could say that one would expect to hear two bangs corresponding to the two shocks of the *N*-wave, but one would hasten to add the following qualifications:

   (i) One might hear more bangs than two—perhaps three or four—if the aircraft was sufficiently close that the disturbance had not attained the ideal *N*-wave form.

  (ii) One might hear less bangs than there were shocks if the time interval between any two shocks was less than about one hundredth of a second, for evidence suggests that the human ear would not then be able to distinguish the separate shocks.[2]

 (iii) One might not hear a shock as a bang if it were too weak or too 'rounded off'.

We shall have occasion to add other qualifications to our list later. It should be emphasized, however, that we shall be primarily concerned with the pressure–time record, or waveform, and not the human ear's interpretation of it, on which (ii) and (iii) above depend.

In the sections which follow we shall develop the basic framework of the theory of sonic bangs. It will be considered in three stages as follows:

   (i) Geometric acoustics. This introduces the concept of bang rays, and enables the development of the wave patterns in the large to be studied. It deals with the occurrence of sonic bangs, but not with their intensity.

  (ii) Whitham's technique. This enables the broad characteristics of the bang wave to be determined, including the formation of shocks. It gives the two main essentials concerning the intensity of a sonic bang—pressure rise and duration of the positive phase.

 (iii) Final refinement. This means allowing for the softening of the shock waves associated with the various dissipative processes involved—viscosity, humidity and turbulence.

12.3 Propagation of the bang waves

The bang waves, which become *N*-waves, may be assumed to be propagated along what are called *bang rays*, a bundle of bang rays forming a *bang tube.* For the purpose of constructing the bang rays an aircraft is treated as a point that moves along the flight path. Suppose that the speed of the aircraft at a given point is V, and that the ambient speed of sound at that point is a_a. The bang rays that emanate from this point make an angle $\frac{1}{2}\pi - \mu, = \sec^{-1}(V/a_a)$, with the direction of the flight path at the point. Locally these rays form part of a cone. Unless the atmosphere is homogeneous the bang rays are not straight. In particular, when there are ambient temperature variations in the atmosphere, leading to variations in the ambient speed of sound, or when there are winds, the bang rays from a point are curved and form a *bang conoid*, as from the point P in Fig. 12.4. Each point on the flight path has such a bang conoid associated with it.

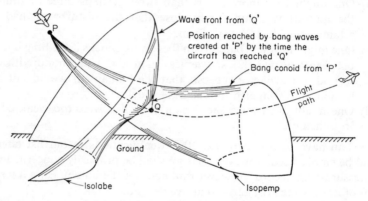

Fig. 12.4   Diagram illustrating the bang conoid.

In the present context the bang waves, which represent the disturbances created by the aircraft, are, like the aircraft, treated as points. The bang waves created at a given point of the flight path are propagated along the rays of the bang conoid through that point. At any given time the positions of all the bang waves that have been created up to that time form a *wavefront*, which is a conoid through the point at which the aircraft is at that time, as for the point Q in Fig. 12.4. For an atmosphere at rest the bang conoids and the wavefronts form two families or orthogonal surfaces.

The bang conoids may sometimes have an envelope, which is the surface where the phenomenon of focusing occurs. There is a curve on each bang conoid which is its line of contact with the envelope. Each ray of a bang conoid meets this curve at the *point of focus* for that ray. Each wavefront also intersects the envelope in a curve. Along this curve the wavefront has a cusped edge, the wavefront folding back upon itself, as shown in Fig. 12.5.

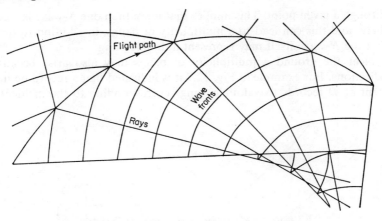

Fig. 12.5   Diagram illustrating focusing effects.

A bang conoid intersects the ground in a curve called an *isopemp* (see Fig. 12.6). An isopemp clearly connects those points on the ground that receives bangs created at the same instant of time. A wavefront intersects

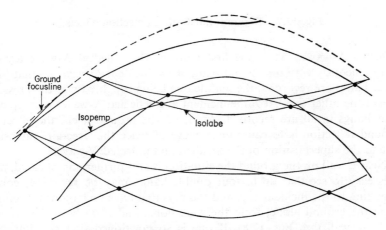

Fig. 12.6   Diagram illustrating isopemps and isolabes.

the ground in a curve called an *isolabe* (see Fig. 12.6). An isolabe clearly connects those points on the ground at which bangs are heard at the same instant of time. The envelope, or *focus surface*, intersects the ground in the *ground focusline* (see Fig. 12.6). The ground focusline is an envelope of the isopemps, and a cusp-locus of the isolabes. We observe from Fig. 12.6 that more than one isolabe, and also more than one isopemp, may pass through a given point. This implies that more than one N-wave is received there, and this can lead to a multiplicity of bangs, in addition to the fact that the N-wave itself may represent a double bang.

Near the ground a multiplicity of bangs may also arise because of reflections. For example in Fig. 12.7 it is indicated that a recording instrument at D records two double bangs corresponding to the incident and

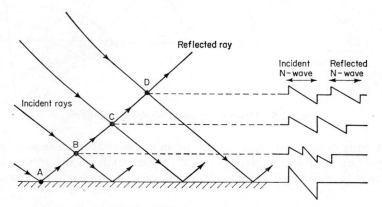

Fig. 12.7   Diagram illustrating reflection effects.

reflected N-waves. At C the first bang of the reflected N-wave arrives simultaneously with the second bang of the incident N-wave, so that only three bangs are recorded, the middle one being roughly twice the pressure rise of the other two. At B the incident and reflected N-waves overlap and four bangs in all are recorded. At A, on the ground itself, the effect of ground reflection is to cause an N-wave of twice the 'free-air' intensity, due to the superposition of the incident and reflected N-waves.

Near a wall on the ground the situation is still further complicated, and it is easy to see that up to four double bangs may be recorded, corresponding to the incident ray, and the rays that have undergone reflections from the ground alone, from the wall alone, and from both the ground and the wall (see Fig. 12.8). If one is so positioned that simultaneous arrival of different shocks occurs, then pressure rises of up to four times the free-air pressure rise could be recorded.

We are now in a position to add to the qualifications that we have already made in regard to the number of bangs that one could expect to hear, as follows:

(iv) One might hear more than two bangs if one happens to be situated on more than one bang ray.

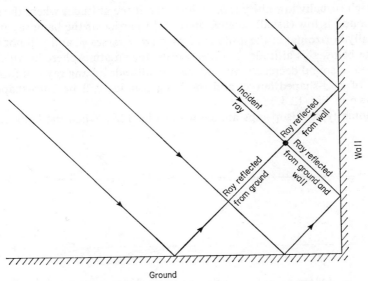

Fig. 12.8   Reflection of rays from the ground and a vertical wall.

(v) One might hear more than two bangs if one is near the ground, and
   more still if additionally one is near a wall.

Consider a bang ray that makes an angle φ with the horizontal at the
apex of the associated bang conoid, where the speed of sound is a_a. For an
atmosphere in which the speed of sound, a, is solely a function of altitude
the bang ray will lie in a vertical plane, and the law of refraction of the
bang ray is

$$D_{\bar{\omega}}R = \frac{a_a \sec \varphi}{a},$$

where R is the distance along the bang ray from the apex of the bang
conoid and $\bar{\omega}$ is the associated horizontal distance. Therefore, if $D_{\bar{\omega}}z$ is the
slope of the bang ray, we have

$$D_{\bar{\omega}}z = \frac{\sqrt{a_a{}^2 \sec^2 \varphi - a^2}}{a} \tag{12.1}$$

If the relationship expressing the speed of sound, a, in terms of the
altitude, z, is known, then equation 12.1 can be integrated[3] to yield the
equation of the bang ray. However, by inspection it is obvious that, no
matter what the relationship between the speed of sound and altitude, the
slope of the bang ray will be less at the altitudes where the speed of sound

is high (usually low altitudes), and greater at the altitudes where the speed of sound is low (usually high altitudes). In particular the bang ray will be locally horizontal at the altitude at which $a = a_a \sec \varphi$, and will not penetrate below this altitude. In other words, for an atmosphere in which the speed of sound decreases with increase in altitude a bang ray will form the arc of a U-shaped curve, and the bang conoids will be horn-shaped as shown in Fig. 12.4.

Some typical isopemps are shown in Fig. 12.9. When the bang conoid

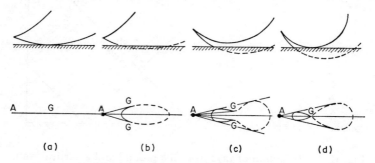

Fig. 12.9   Some typical isopemps.

just touches the ground the isopemp is merely the point of contact G (Fig. 12.9a). Fig. 12.9b shows the usual case when the bang conoid intersects the ground. The solid portion of the isopemp corresponds to downgoing rays. The broken portion corresponds to upgoing rays, which are not experienced in practice. The two portions join at the *points of graze*, G, which are the points of contact of the tangents from the projection of the apex of the bang conoid, A. The focus of the points of graze associated with each point of the flight path is called the *grazeline*. Figs. 12.9c and 12.9d are examples of what can occur in diving flight.

The pattern of isopemps for an aircraft doing a straight and level acceleration to supersonic speeds is shown in Fig. 12.10 and, for clarity, the associated pattern of isolabes is shown separately in Fig. 12.11. Now only part of each isopemp in Fig. 12.10 corresponds to downgoing rays, as represented by the solid portion of the curve lying to the left of the point of intersection with the grazeline GCBCG. Likewise only part of the ground focusline FCACF corresponds to downgoing rays: this part is the solid portion of the curve CAC. The grazeline, the focusline and the local isopemp (in this case that for $M = 1.22$) all touch at the point C. Further, the boundary GCACG of the area subjected to bangs (sometimes called the *cutoff line*) is given by the focusline from A up to the points C, and by the grazeline beyond the points C. Focusing occurs, therefore, to a limited

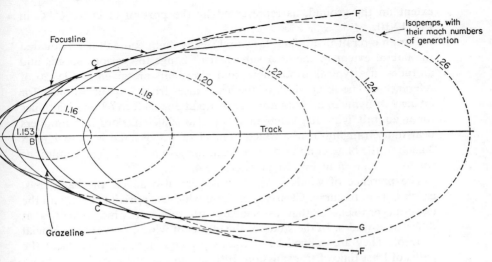

Fig. 12.10 The pattern of isopemps, grazeline and focusline for an aircraft during a straight and level acceleration to supersonic speed.

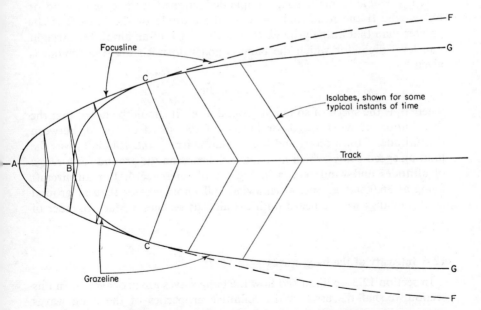

Fig. 12.11 The pattern of isolabes, grazeline and focusline for an aircraft during a straight and level acceleration to supersonic speed.

extent on the ground, as represented by the portion of curve CAC in Fig. 12.10.

For an aircraft in straight, level and steady flight Randall has evaluated the lateral extent of the area subjected to bangs for various speeds and altitudes in a typical atmosphere, and his results are given in Fig. 20 of reference 4. The total width of the area varies from 33 nautical miles for an aircraft flying at a Mach number of 1.5 at 36,000 ft to 63 nautical miles for an aircraft flying at a Mach number of 3 at 60,000 ft. For Mach numbers near unity, ranging from a Mach number of 1 at ground level to 1.15 at 36,000 ft, the bang conoids do not reach the ground at all. The conditions for this are given in Fig. 21 of reference 4.

The presence of winds in the atmosphere also affects the bang rays, causing them to curve. Clearly a uniform wind has a trivial effect, and the essential problem is that associated with different wind velocities at different altitudes. In the study of the effects of winds, rays are the crucial concept. They are defined by Fermat's principle that they represent the paths of least time of transmission. It turns out that the wavefronts are no longer orthogonal to the rays,[5] and the law of refraction is more complicated, but in principle the technique for tracing the rays is the same as that used when there are no winds.

It is possible to make some simple deductions for the case of head or tail winds. If the headwind speed at the altitude of the aircraft is Δw greater than that at the ground, then the speed V of an aircraft in straight and level flight for which the bang conoids do not reach the ground is given by

$$V < a_g + \Delta w,$$

where a_g is the speed of sound at ground level. It should be noted that the distribution of wind speed, or indeed of the speed of sound, between the altitude of the aircraft and the ground is irrelevant. Randall gives the limiting speed below which no bangs are heard on the ground for a range of altitudes and wind speeds in Fig. 21 of reference 4. For an aircraft flying at 36,000 ft against a headwind 100 knots greater than at ground level no bangs will be heard until the aircraft reaches a Mach number of 1.33.

12.4 Intensity of the bang waves

In section 12.3 we discussed how the bang waves are propagated. In this section we shall discuss how the definitive properties of the bang waves vary as they are propagated. As stated in section 12.2 these definitive properties are the pressure rise and the duration of the positive phase.

The determination of the definitive properties of the bang waves, which in effect means determining the disturbance field about the aircraft at large distances, can be accomplished by the methods of linearized theory improved by the technique of Whitham.[1] The original work was done by Rao[6,7] and Walkden.[8] It is not proposed to give the details here, for the matter is well covered in the literature, most recently in an article by Warren and Randall.[9]

Assuming first a homogeneous atmosphere, and negligible lift, linearized theory yields the result that the incremental pressure P at a distance r along a ray is given by*

$$\frac{\Delta P}{P_\infty} = \frac{\gamma M^{5/2} \mathscr{F}_v'(\eta)}{\left\{ 2(M^2 - 1)r \left[1 - \frac{rA_r}{(M^2 - 1)a_\infty^2} \right] \right\}^{1/2}}, \tag{12.2}$$

where γ is the ratio of the specific heats for air,
      P_∞ is the ambient pressure,
      a_∞ is the ambient speed of sound,
      M is the Mach number, $= V/a_\infty$,
      A_r is the resolute of the aircraft's acceleration along the ray,
and $\mathscr{F}_v(\eta)$ is a function of the aircraft's distribution of cross-sectional area, and is given by

$$\mathscr{F}_v(\eta) = \frac{1}{2\pi} \int_0^\eta \frac{S'(\xi)}{(\eta - \xi)^{1/2}} \, d\xi, \tag{12.3}$$

where $S(\xi)$ is the cross-sectional area at station ξ from the nose,
      η is a parameter which specifies the point on the waveform at which the incremental pressure ΔP is attained,
and a prime denotes a derivative with respect to the argument.

The resolute of the aircraft's acceleration along the ray, A_r, is given by

$$A_r = \frac{\dot{V}}{M} + \frac{(M^2 - 1)^{1/2} A_n \cos \theta}{M},$$

where $\dot{V}$ is the resolute of the aircraft's acceleration along the flight path,
      A_n is the resolute of the aircraft's acceleration normal to the flight path,
and θ is the angle that the instantaneous plane of motion makes with the plane containing the ray and the tangent to the flight path: θ is acute when the flight path is concave towards the ray.

Linearized theory does not predict a sharp N-wave, the waveform being

*See, for example, Warren and Randall[9] section 3.
10

in fact the form of the $\mathscr{F}'_v$-function, which is usually a smooth curve, without discontinuities corresponding to shocks. These come in when we pass to the improved theory.

The improved theory is concerned essentially with the parameter η. The other results of the linearized theory, and in particular equations 12.2 and 12.3, still apply. In essentials, the improved theory allows for the fact that disturbances are propagated, not at the ambient speed of sound a_∞, but at a speed $(a + u)$, where a is the local speed of sound and u is the local convective speed of the fluid. When this fact is allowed for, it is found that the 'crests' in the waveform move faster than the ambient speed of sound, and that the 'troughs' move more slowly. Clearly this will have the effect of steepening the compressive parts of the waveform and rendering less steep the rarefactive parts. It is found in fact that the compressive parts steepen sufficiently for it to be necessary to introduce shocks in order to maintain a single-valuedness in the pressure at a point, and much of the improved theory is concerned with this introduction of shocks.

As has already been stated, it usually happens that at large distances the disturbance takes the form of an N-wave. In such cases the pressure rise across the bow shock is given by equation 12.2 with η given by

$$\frac{\mathscr{F}_v(\eta)}{\{\mathscr{F}'_v(\eta)\}^2} = \frac{(\gamma + 1)M^{7/2}}{[8(M^2 - 1)]^{1/2}} \int_0^r \frac{\frac{1}{2}\,dr}{\left\{ r\left[1 - \frac{rA_r}{(M^2 - 1)a_\infty^2} \right] \right\}^{1/2}}. \qquad (12.4)$$

The duration of the positive phase Δt is given by

$$\Delta t = \frac{1}{Ma_\infty} \left\{ \eta_0 - \eta + \mathscr{F}'_v(\eta)\,\frac{(\gamma + 1)M^{7/2}}{[2(M^2 - 1)]^{1/2}} \right. $$

$$\left. \int_0^r \frac{\frac{1}{2}\,dr}{\left\{ r\left[1 - \frac{rA_r}{(M^2 - 1)a_\infty^2} \right] \right\}^{1/2}} \right\} \qquad (12.5)$$

where η_0 is the value of η for which $\mathscr{F}_v(\eta)$ is a maximum. These formulae are somewhat cumbersome, but they nevertheless complete the improved theory as far as the effects for negligible lift are concerned.

When there is lift the only change is in the $\mathscr{F}$-function, $\mathscr{F}_v(\eta)$ being replaced by a more general $\mathscr{F}$-function, $\mathscr{F}(\eta)$, defined by

$$\mathscr{F}(\eta) = \mathscr{F}_v(\eta) - (M^2 - 1)^{1/2} \cos \vartheta \mathscr{F}_i(\eta), \qquad (12.6)$$

with

$$\mathscr{F}_i(\eta) = \frac{1}{2\pi} \int_0^\eta \frac{L'(\xi)}{\gamma P_\infty M^2} \frac{d\xi}{(\eta - \xi)^{1/2}}, \tag{12.7}$$

where $L(\xi)$ is the lift from the nose to station ξ,
and ϑ is the angle that the plane containing the lift direction and the tangent to the flight path makes with the plane containing the ray and the tangent to the flight path: ϑ is acute when the lift has a positive resolute along the direction of the ray: in particular $\vartheta = \pi$ on the track for straight and level flight.

It will be noted that the more general $\mathscr{F}$-function, $\mathscr{F}(\eta)$, depends not only upon the geometry of and lift distribution of the aircraft, but in addition it varies along the flight path, since it depends upon M, and it also varies with the direction of the ray concerned, since it depends upon ϑ.

Let us now discuss some of the results so far obtained. Some typical $\mathscr{F}'$-functions are shown in Fig. 12.12. Now it can be shown that at large distances from the aircraft the parameter η giving the pressure rise across the bow shock tends to the value corresponding to the value of η for which $\mathscr{F}(\eta)$ is a maximum, or for which $\mathscr{F}'(\eta) = 0$, namely $\eta = \eta_0$. Hence the generalized version of equation 12.4 can be approximated by

$$\mathscr{F}'(\eta) = \left\{ \frac{\mathscr{F}(\eta_0)}{\frac{(\gamma + 1)M^{7/2}}{[8(M^2 - 1)]^{1/2}} \int_0^r \frac{\frac{1}{2} dr}{\left\{ r\left[1 - \frac{rA_r}{(M^2 - 1)a_\infty^2} \right] \right\}^{1/2}}} \right\}^{1/2}, \tag{12.8}$$

and the approximate generalized version of equation 12.2 for the pressure rise across the bow shock becomes

$$\frac{\Delta P}{P_\infty} = \frac{\frac{2^{1/2}\gamma M^{3/4}}{(\gamma + 1)^{1/2}(M^2 - 1)^{1/4}} \{\mathscr{F}(\eta_0)\}^{1/2}}{\left\{ r\left[1 - \frac{rA_r}{(M^2 - 1)a_\infty^2} \right] \int_0^r \frac{\frac{1}{2} dr}{\left\{ r\left[1 - \frac{rA_r}{(M^2 - 1)a_\infty^2} \right] \right\}^{1/2}} \right\}^{1/2}}. \tag{12.9}$$

Let us now interpret this result. The first point is that the effect of the aircraft's geometry, and of the lift that it generates, is contained in the term

$$\{\mathscr{F}(\eta_0)\}^{1/2}.$$

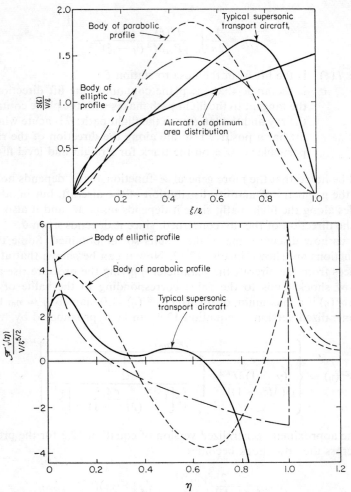

Fig. 12.12 Typical values of the function $\mathscr{F}'_v(\eta)$ and $S(\xi)/(v/l)$.

Secondly, the attenuation of the pressure rise across the bow shock as it moves along the ray is contained in the term

$$\left\{ r\left[1 - \frac{rA_r}{(M^2-1)a_\infty^2}\right] \int_0^r \frac{\frac{1}{2}\,dr}{\left\{r\left[1 - \frac{rA_r}{(M^2-1)a_\infty^2}\right]\right\}^{1/2}} \right\}^{1/2}.$$

Finally the effect of the aircraft's motion appears in both these terms, and also in the factor $M^{3/4}/(M^2-1)^{1/4}$. The dominant effect of the aircraft's motion, however, is through the resolute of the aircraft's accelera-tion along the ray, A_r. The aircraft's Mach number, M, is not a significant parameter: for example the factor $M^{3/4}/(M^2 - 1)^{1/4}$ varies roughly as $M^{1/4}$ for M not too close to unity.

If the aircraft has zero resolute of acceleration along a ray, then the attenuation with distance varies as $r^{-3/4}$. With acceleration the variation can be greater or less than this. Under certain circumstances the pressure may even increase with distance, the so-called focusing effect, and equa-tion 12.9 in fact predicts an infinite pressure rise when $r = (M^2 - 1)a_\infty^2/A_r$. This occurs at the point of focus of a ray, i.e. at a cusp of the wavefront. The present theory breaks down under these circumstances, primarily be-cause of the intersection of adjacent rays, and there is at present no way of handling this situation theoretically.

Consider the effect of aircraft size. We note by inspection of equation 12.6 that the $\mathscr{F}$-function is proportional to the square root of the linear dimensions of the aircraft, and therefore from equation 12.9 the pressure rise is proportional to the three-fourths power of the linear dimensions, or, in more popular terms, roughly as the fourth root of the all-up weight. The variation with size of aircraft is not, therefore, very great.

The next question that one might ask is whether there is anything that one can do in design that will affect the pressure rise appreciably. In other words, is there any way whereby one can affect the pressure rise by varying the distributions of volume and of lift, subject of course to the maintenance at least of a given overall length, total volume and total lift. This problem has been considered by Jones.[10] Considering only the con-tribution due to volume (i.e. the $\mathscr{F}_v$-function only), Jones has shown that the aircraft shape that yields the minimum pressure rise, for a given overall length l and total volume v, has a cross-sectional area distribution which tends to the shape given by

$$S(\xi) = \frac{3v}{2l}\left(\frac{\xi}{l}\right)^{1/2}.$$

Compared with aircraft of the same overall length and total volume having other cross-sectional area distributions, the ratios of the pressure rises are as follows:

optimum value	1
body of parabolic profile	1.52
body of elliptic profile	1.30
'typical supersonic transport aircraft'	1.25

The differences are not great, especially when set against the fact that one pressure rise has to be about 1.12 times that of another for the human ear to be able to distinguish it as greater. In any case, the marginal differences in pressure rise are as nothing compared with the differences in wave drag of the shapes concerned, and this is likely to be an overriding consideration in practice.

Similar statements can be made in regard to the lift distribution. Jones[10] has investigated the optimum lift distribution for minimum pressure rise in a number of circumstances, but the variations are, relatively, not great. The important thing to observe in regard to the lift contribution to the pressure rise is that it becomes increasingly more dominant as the altitude is increased. This follows from the presence of the P_∞ factor in equation 12.7. We also note that, because of the implied $(M^2 - 1)^{1/2}/M^2$ factor, the lift contribution is of greatest importance at Mach numbers around 1.4, for a given wing loading and altitude.

So far in this section we have treated the atmosphere as homogeneous. Comparatively little attention has been paid to the effects on the intensity of the sonic bangs of the pressure and temperature gradients that occur in the real atmosphere. Warren[11] proposed a simple method suggested by acoustic theory, to the effect that the variations of ambient pressure be allowed for by replacing the quantity P_∞ in equation 12.2 by the geometric mean of the ambient pressures at ground level and at the altitude of the aircraft for the particular ray concerned—i.e. by $(P_g P_a)^{1/2}$. The effect of variations in the ambient speed of sound is not so easily allowed for, for, as we have seen in section 12.3, it leads in the first place to changes in the bang rays themselves along which the bang waves are propagated. Recently Randall[12] has studied the intensities of sonic bangs in a typical atmosphere, and his results show that for an aircraft flying at a Mach number of 1.5 at 36,000 ft the simple method is low by a factor of 1.07, and that for an aircraft flying at a Mach number of 3 at 60,000 ft it is low by a factor of 1.12.

12.5 Discussion, with particular reference to focusing

Let us sum up the various stages in solving a problem about sonic bangs. The first stage is to determine when and where sonic bangs will occur. This is done by the methods of geometric acoustics, as outlined in section 12.3. The second stage is to determine the $\mathscr{F}$-function, which represents essentially the waveform according to linearized theory. The third stage is to derive the improved waveform by the technique of Whitham: this involves the introduction of shocks. In section 12.4 we have indicated the procedure for the determination of the pressure rise across the bow shock, which is the information usually sought, but the method for completely determining the waveform is described in the literature.[1]

The procedure just outlined yields what is called the 'free-air' waveform and associated pressure rise. Naturally in the vicinity of the ground, near buildings, and ultimately inside buildings, the waveform and pressure rise are quite different, for the reasons outlined in section 12.3, and depend upon the detailed situation. However, a situation that is of common occurrence is that called 'near the ground in the open', and here we expect the pressure rise to be double the free-air value owing to reflection, although if the rays are striking the ground at incidences very near grazing incidence, then there is evidence to suggest that regular reflection, and hence pressure doubling, does not occur. At grazing incidence itself the pressure rise near the ground in the open should be the free-air pressure rise.

The outstanding problem is the determination of what occurs in the vicinity of a point of focus, for which the theoretical procedure outlined above yields infinite pressures. At present there is no satisfactory and generally accepted extension of the theory to cater for this case, and recourse must be had at this stage to experiment. The most relevant information is that given by D. J. Maglieri.[13] An extensive series of ground-pressure measurements were made along the track for an aircraft in straight and level flight accelerating from a Mach number of 0.9 to 1.5. The pressure rises measured in a very localized region (of the order of the length of the aircraft) around the point of focus were about 2.5 times the value measured in the subsequent unfocused regions. Indeed, in the present state of the art, the pressure rise at a point of focus is usually reckoned to be about 2.5 times the pressure rise worked out assuming quasi-steady conditions—that is, by taking A_r as zero in equation 12.9.

The problem of the focus, however, is one on which more work is still required.

References

1. G. B. Whitham. The flow pattern of a supersonic projectile, *Commun. Pure Appl. Math.*, **5**, 3, 301-48 (August, 1952).
2. G. M. Lilley and coworkers. On some aspects of the noise propagation from supersonic aircraft, *J. Roy. Aeron. Soc.*, **57**, 510, June, 396–414 (1953).
3. C. H. E. Warren. The propagation of sonic bangs in a nonhomogeneous still atmosphere, *Internat. Council Aeronaut. Sci.*, Fourth Congress, Paris, Paper 64–547 (1964).
4. D. G. Randall. Methods for estimating distributions and intensities of sonic bangs, Aero. Res. Council Reports & Memoranda R. & M. 3113, H.M. Stationery Office, London (1959).
5. C. H. E. Warren. A note on the refraction of sound in a moving gas, *J. Sound Vib.*, **1**, 2, April, 175–178 (1964).
6. P. S. Rao. Supersonic bangs. Pt. 1, *Aeronaut. Qtly.*, **7**, Feb., 21–44 (1956).
7. P. S. Rao. Supersonic bangs. Pt. 2, *Aeronaut. Qtly.*, **7**, May, 135–55 (1956).

8. F. Walkden. The shock pattern of a wing-body combination, far from the flight path, *Aeronaut. Qtly.*, **9**, May, 164–94 (1958).

9. C. H. E. Warren and D. G. Randall. The theory of sonic bangs, *Prog. Aeronaut. Sci.*, **1**, 238–74 (1961); *Prog. Aeronaut. Sci.*, **5**, 295–302 (1964).

10. L. B. Jones. *J. Roy. Aeron. Soc.*, **65**, 606, June, 433–6 (1961).

11. C. H. E. Warren. An estimation of the occurrence and intensity of sonic bangs, *R.A.E. TN* Aero 2334.

12. D. G. Randall. Sonic bang intensities in a stratified, still atmosphere, *R.A.E. TR* 66002 (1966).

13. D. J. Maglieri. Some effects of airplane operations and the atmosphere on sonic boom signatures, *J. acoust. Soc. Amer.*, **39**, 5, S36–S42 (1966).

CHAPTER 13

Elements of Periodic Vibration Theory

13.1 The mass–spring–damper oscillator

In order to introduce or refresh ideas relating to the vibration of complicated systems, we consider first the simplest vibrating system which consists of a mass m, a spring of stiffness k, and a viscous damper of rate c. The displacement w of the free end of the spring relative to the unstrained position, when the mass is subjected to the time dependent force $F(t)$, is governed by the equation:

$$m \frac{d^2w}{dt^2} + c \frac{dw}{dt} + kw = F(t). \tag{13.1}$$

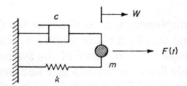

Fig. 13.1   Single degree of freedom system with viscous damper.

When $F(t)$ represents a harmonically varying force ($F_0 \cos \omega t$) it is convenient to represent it in the complex exponential form

$$F(t) = F_0 \cos \omega t = \mathcal{R} F_0 \, e^{i\omega t} \tag{13.2}$$

where $\mathcal{R}$ stands for the 'real part of'. For many purposes, we may drop the $\mathcal{R}$ and carry out the mathematics using the complex form

$$F(t) = F_0 \, e^{i\omega t} \tag{13.3}$$

it being understood, of course, that the real part must be recovered at the end of the analysis. In order to find how w is related to $F(t)$, we use the form of solution

$$w = w_0 \, e^{i\omega t} \tag{13.4}$$

285

in which w_0 is a complex number. Substituting equations 13.3 and 13.4 into equation 13.1 we find

$$w_0 \, e^{i\omega t}(-\omega^2 m + i\omega c + k) = F_0 \, e^{i\omega t} \tag{13.5}$$

from which we obtain

$$w = w_0 \, e^{i\omega t} = \frac{F_0 \, e^{i\omega t}}{(-\omega^2 m + i\omega c + k)}. \tag{13.6}$$

Now the complex number $(-\omega^2 m + i\omega c + k) = Z$ can be written in the form $|Z| \, e^{i\phi}$, where $|Z|$ is the modulus, and ϕ the argument of the number, i.e.

$$|Z| = \sqrt{(k - \omega^2 m)^2 + \omega^2 c^2}; \qquad \tan \phi = \frac{c}{(k - \omega^2 m)}. \tag{13.7}$$

Equation 13.6 now becomes

$$w = \frac{F_0 \, e^{i(\omega t - \phi)}}{|Z|}. \tag{13.8}$$

Restoring this equation to the real form (by extracting the real part from the right-hand side), we have

$$w = \frac{F_0 \cos(\omega t - \phi)}{|Z|}. \tag{13.9}$$

This states that the displacement w oscillates at the frequency ω, with an amplitude $F_0/|Z|$, and lags by a phase angle ϕ behind the exciting force.

We call Z the *complex obstructance* of the system. Its modulus is the amplitude of harmonic force required to give unit amplitude of harmonic displacement. Its argument ϕ is the angle by which the displacement lags behind the force.

The reciprocal of Z is the *complex receptance* of the system. Its modulus is the amplitude of harmonic displacement produced by unit amplitude of harmonic force.

Consider now the variation of the displacement amplitude per unit amplitude of exciting force (w/F_0) as the forcing frequency changes. From equation 13.9 we have

$$\frac{|w|}{F_0} = \frac{1}{\sqrt{(k - \omega^2 m)^2 + \omega^2 c^2}} = \frac{1}{|Z|}.$$

This may be written in the form

$$\frac{|w|}{F_0} = \frac{1}{k} \frac{1}{\{(1 - (\omega/\omega_r)^2)^2 + 4(\omega/\omega_r)^2 \zeta^2\}^{1/2}} \tag{13.10}$$

in which ω_r is the undamped natural frequency $(=\sqrt{k/m})$ and ζ is the 'damping ratio' or 'damping factor', i.e. $\zeta = c \div$ the critical damping coefficient $= c/2\sqrt{mk}$. (13.11)

When $\omega = 0$ (i.e. the force is steady), $w/F_0 = 1/k$. We call this the 'static displacement' w_s. As ω approaches the natural frequency, the displacement first increases sharply, and then passes through a maximum—*resonance* occurs. At this maximum, the displacement is Q times the static displacement. We call Q the *magnification factor*. It may easily be shown that $Q = 1/2\zeta$. Plotting w/F_0 against ω, we have the familiar frequency response curve, shown in Fig. 13.2.

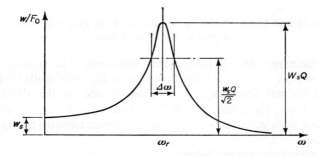

Fig. 13.2 Frequency response curve for a single degree of freedom system with viscous damping.

The 'width' of the frequency response curve is sometimes defined by the frequency $\Delta\omega$, which is the difference between the two frequencies at which the amplitude is $1/\sqrt{2}$ times the maximum amplitude. Provided the damping factor is no greater than, say, 0.3, we find

$$\frac{\Delta\omega}{\omega_r} = 2\zeta = \frac{1}{Q}.$$ (13.12)

We are often as interested in the amplitudes of the fluctuating velocity, $|\dot{w}|$, or acceleration $|\ddot{w}|$, as in the displacement. It is easily shown that

$$|\dot{w}| = \omega|w| \quad \text{and} \quad |\ddot{w}| = \omega^2|w|.$$

Using these expressions and equation 13.10, we may show that

$$\frac{|\dot{w}|}{F_0} = \frac{1}{c}\frac{2\zeta(\omega/\omega_r)}{\{(1-(\omega/\omega_r)^2)^2 + 4(\omega/\omega_r)^2\zeta^2\}^{1/2}} = \frac{1}{c}\frac{2\zeta(\omega/\omega_r)}{|Z|/k}$$ (13.13)

and

$$\frac{|\ddot{w}|}{F_0} = \frac{1}{m}\frac{(\omega/\omega_r)^2}{|Z|/k}.$$ (13.14)

As the frequency, ω, varies, each of these expressions varies in a way generally similar to that of $|w|/F_0$, but with certain significant differences, as shown in Figs. 13.3a and 13.3b.

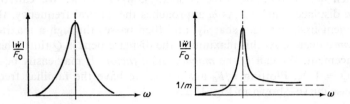

Fig. 13.3a and 13.3b   Velocity and acceleration amplitudes of a single degree of freedom system.

Each of the expressions is zero at zero frequency. The velocity function tends to zero as $\omega \to \infty$, but the acceleration function tends to $1/m$.

At the natural frequency, ω_r, the amplitude of the displacement, $|w|/F_0$, is $1/\omega_r c$, the amplitude of the velocity $|\dot{w}|/F_0$ is $1/c$, and of the acceleration is ω_r/c, i.e. it is the *damping*, c, which controls the amplitudes at resonance.

At very low frequencies (below ω_r),

$$\frac{|w|}{F_0} \doteqdot \frac{1}{k}; \qquad \frac{|\dot{w}|}{F_0} \doteqdot \frac{\omega}{k}; \qquad \frac{|\ddot{w}|}{F_0} \doteqdot \frac{\omega^2}{k}$$

i.e. the amplitudes are controlled by the *stiffness*, k, and the displacement tends to be independent of frequency.

At very high frequencies (above ω_r),

$$\frac{|w|}{F_0} = \frac{1}{m\omega^2}, \qquad \frac{|\dot{w}|}{F_0} = \frac{1}{m\omega}, \qquad \frac{|\ddot{w}|}{F_0} = \frac{1}{m}$$

i.e. the amplitudes are controlled by the *mass*, m, and the acceleration tends to be independent of frequency.

So far, we have used the concept of a viscous damper, which exerts a damping force which is in counterphase with and proportional to the velocity. In practical structures, this type of damping is seldom realized, and experimental evidence has suggested that a more realistic form, for harmonically vibrating systems, is the 'complex stiffness' type of damping. We represent the stiffness and damping forces together in the expression:

$$(k + ih)w.$$

This can only be used if we use the complex exponential form of solution and applies, strictly, only to harmonic motion. However, in the absence of any other satisfactory method of representation, it is usually used (with care) when dealing with random motion. The imaginary part of the stiffness implies the existence of a resistive force, which has an amplitude proportional to the displacement w, but which (by virtue of the i) is in counterphase with the velocity $\dot{w}$. If such 'structural' or 'hysteretic' damping replaces the viscous damping of equation 13.1, the expression for w (equation 13.6) becomes:

$$w = \frac{F_0\, e^{i\omega t}}{(-\omega^2 m + k + ih)};$$ (13.15)

also

$$|Z| = \sqrt{(k - \omega^2 m)^2 + h^2}$$ (13.16)

and

$$\tan \phi = \frac{h}{(k - \omega^2 m)}.$$ (13.17)

The undamped resonant frequency, ω_r, now coincides with the frequency for maximum displacement, at which

$$\left\{ \frac{|w|}{F_0} \right\}_{\max} = \frac{1}{h}.$$ (13.18)

This is independent of frequency.

In all the expressions we have derived which contain $(\omega/\omega_r)\zeta$, we may replace this term by

$$\zeta_H = \frac{h}{2k}.$$ (13.19)

We may call ζ_H the 'hysteretic damping factor'. The principal characteristic of the 'hysteretic' damping is that the energy dissipated per cycle of harmonic displacement is independent of the frequency, and is proportional to the square of the displacement amplitude. The energy dissipated by a viscous damper is proportional to the product of the frequency and to the square of the displacement amplitude.

Next, we consider the response of the system to a force which is no longer harmonic, but is periodic, repeating itself after every T secs. The force may be analysed into its Fourier (harmonic) components, and may be expressed in the form

$$F(t) = \sum_{n=1}^{\infty} (a_n \cos \omega_n t + b_n \sin \omega_n t)$$ (13.20)

where $\omega_n = 2n\pi/T$, i.e. the frequency having n cycles in the interval T. Equation 13.20 is equivalent to

$$F(t) = \mathscr{R} \sum_{n=1}^{\infty} (a_n - ib_n)\, e^{i\omega_n t}$$ (13.21)

or

$$\sum_{n=1}^{\infty} d_n \cos (\omega_n t + \epsilon_n).$$ (13.22)

The coefficients a_n and b_n are found by the usual methods to be

$$a_n = \frac{2}{T} \int_0^T F(t) \cos \omega_n t \, dt; \qquad b_n = \frac{2}{T} \int_0^T F(t) \sin \omega_n t \, dt$$

or   (13.23)

$$a_n - ib_n = \frac{2}{T} \int_0^T F(t) \, e^{-i\omega_n t} \, dt.$$

The total response of the system may now be found by superimposing the harmonic responses corresponding to the individual components of the analysed force. Using the form of equation 13.21, the nth component gives rise to the component of displacement (after transients have decayed):

$$w_n = \mathcal{R} \frac{(a_n - ib_n) \, e^{i\omega_n t}}{Z(\omega_n)}$$ (13.24)

(from equation 13.6). $Z(\omega_n)$ is the value of the complex obstructance at the frequency ω_n. The total displacement at any instant is therefore

$$w = \sum_{n=1}^{\infty} w_n = \mathcal{R} \sum_{n=1}^{\infty} \frac{(a_n - ib_n) \, e^{i\omega_n t}}{Z(\omega_n)}$$ (13.25)

or

$$w = \sum_{n=1}^{\infty} \frac{d_n \cos (\omega_n t + \epsilon_n - \phi_n)}{|Z(\omega_n)|}$$

where $\tan \phi_n = \arg Z(\omega_n)$. Notice that since ϕ is a function of frequency, each component of displacement lags behind its corresponding component of force by a different phase angle. Also, since $Z(\omega_n)$ has its minimum value when ω_n is close to ω_r, those components of the force closest to ω_r will cause the largest components of displacement.

13.2 The forced vibrations of continuous systems

A 'continuous system' implies that the mass and flexibility are distributed throughout the system. It is not a system of discrete masses connected by massless springs and links. As an example, a uniform simply supported beam will be considered.

The manner in which such a beam can vibrate *freely*, in the absence of damping, is well known. The fundamental mode of vibration is a simple sine wave, and the overtones are sine waves with different integral numbers of half-waves along the beam length. Now any deflected shape of the beam can be resolved into spatial harmonic components, by the methods of Fourier analysis. Symbolically, it follows that if the arbitrary deflected shape is $w(x)$, we may write

$$w(x) = \sum_{r=1}^{\infty} q_r \sin \frac{r\pi x}{l} \tag{13.26}$$

where the q_r's are the Fourier coefficients.

In this case, the terms of the Fourier series are identical with the natural modes of vibration of the beam. In the more general case, when the beam is non-uniform or is not simply-supported, the modes of vibration are not sine waves, but are more awkward functions of x. Denote these modes by $f_1(x), f_2(x), \ldots, f_r(x) \ldots$, etc., $f_1(x)$ being the fundamental mode, and the others being the overtones. Now it is characteristic of the natural modes of vibration that they may be readily used in a 'generalized' type of Fourier analysis, i.e. for a given beam, any possible deflected form of the beam may be represented by

$$w(x) = \sum_{r=1}^{\infty} q_r f_r(x). \tag{13.27}$$

This suggests an alternative form of describing the deflection of the beam. Instead of specifying the continuous function $w(x)$ for all points in the beam, we may specify the q_r's. If it was necessary to specify an infinite set of q_r's, there would be no advantage gained by this alternative method, but as with an ordinary Fourier series, a good approximation to $w(x)$ may be obtained by using the first few terms only of the infinite series. This feature effects a very great simplification in the calculations of the motion of a beam or plate, as follows:

If w is now a function of both space *and* time, equation 13.26 may be written

$$w(x, t) = \sum_{r=1}^{\infty} q_r(t) f_r(x). \tag{13.28}$$

The time dependence is contained entirely within the q_r's, and the spatial dependence entirely within the f_r's. We call the q_r's *generalized coordinates* corresponding to the modes of displacement f_r. The problem of finding the deflection $w(x, t)$ at any point of the beam or plate is now resolved into the problem of finding the q_r's. Now it may be shown that when the system is subjected to a time-dependent load, the displacement q_r of any one of the natural modes of vibration is governed by the same sort of equation as that of a simple mass–spring–damper oscillator. In fact, for most purposes we may write

$$M_r \ddot{q}_r + C_r \dot{q}_r + K_r q_r = L_r(t) \tag{13.29}$$

In this the terms M_r, C_r, K_r and $L_r(t)$ are known respectively as the *generalized mass*, the *generalized damping coefficient*, the *generalized stiff-*

ness and the *generalized force*, each corresponding to the *r*th natural (or normal) mode. The fact that the equation does not contain any other general coordinates q_s is due to the characteristic 'normal' property of the natural modes. Actually, a set of terms

$$\sum_{s=1(\neq r)}^{\infty} C_{rs}\dot{q}_s$$

should be included in the left-hand side since this 'normal' property does not always apply to the system of damping forces acting on the system, but these forces are usually small enough to be ignored—at least, we shall assume that they are in this book.

The *generalized mass*, M_r, has the same magnitude of that mass which when moving with velocity $\dot{q}_r$ has the same kinetic energy as the whole system when moving with the velocity $\dot{q}_r f_r(x)$. I.e.

$$\left. \begin{aligned} M_r \frac{\dot{q}_r^2}{2} &= \frac{\dot{q}_r^2}{2} \times \int_A \mu f_r^2(x)\,dA. \\[2mm] \text{Hence } M_r &= \int_A \mu f_r^2(x)\,dA. \end{aligned} \right\} \tag{13.30}$$

μ is the mass per unit area of the system (plate) and the integration is taken over the whole vibrating surface A.

The *generalized stiffness*, K_r, is the stiffness of the linear spring which, when displaced by q_r from its unstrained position, has the same potential energy stored within it as the actual system when displaced by $q_r f_r(x)$. For a beam, this is expressed by

i.e.
$$\left. \begin{aligned} K_r \frac{q_r^2}{2} &= \frac{q_r^2}{2} \int_l EI\left\{\frac{d^2 f_r(x)}{dx^2}\right\}^2 dx, \\[2mm] K_r &= \int_l EI\left\{\frac{d^2 f_r(x)}{dx^2}\right\}^2 dx. \end{aligned} \right\} \tag{13.31}$$

EI is the flexural stiffness of the beam, and l is the length.

The *generalized force*, $L_r(t)$, is that single force which when moved through the 'virtual' displacement δq_r, does the same amount of work as all the externally applied forces and pressures acting on the system when the system is moved through the virtual displacement $\delta q_r \cdot f_r(x)$. I.e.

or
$$\left. \begin{aligned} L_r(t)\delta q_r &= \int_A p(x,t)\delta q_r \cdot f_r(x)\,dA \\[2mm] L_r(t) &= \int_A p(x,t)f_r(x)\,dA. \end{aligned} \right\} \tag{13.32}$$

Here, $p(x, t)$ is the instantaneous pressure at the point x and time t. The integration again covers the whole surface.

The *generalized damping coefficient* C_r is the rate of that damper which, when extended at velocity $\dot{q}_r$, dissipates energy at the same rate as the whole system of damping forces and pressures on and within the system, when moving with velocity $\dot{q}_r f_r(x)$. I.e.

$$C_r \frac{\dot{q}_r^2}{2} = \int_A p_{dr}(x) \frac{\dot{q}_r^2}{2} f_r^2(x) \, dA \left.\right\}$$

or

$$C_r = \int_A p_{dr}(x) f_r^2(x) \, dA. \tag{13.33}$$

$p_{dr}(x)$ is the local viscous damping pressure per unit velocity in counter-phase with the velocity $\dot{q}_r f_r(x)$. If hysteretic damping stresses exist within the system, we may include a hysteretic damping force, $iH_r \cdot \dot{q}_r$, in equation 13.29. H_r is then the complex part of the stiffness of the linear spring which dissipates the same amount of energy in the harmonic displacement cycle of amplitude q_r as the actual system when displaced in the mode $f_r(x)$ through the cycle of amplitude q_r.

There is an equation of the type 13.29 for each of the normal modes of vibration of the structure. When the generalized force is periodic, q_r is given by an expression of the same form as equation 13.25, i.e.

$$q_r = \mathscr{R} \sum_{n=1}^{\infty} \frac{(a_n - ib_n)}{Z_r(\omega_n)} e^{i\omega_n t} \tag{13.34}$$

$Z_r(\omega_n)$ is the *generalized complex obstructance of the rth mode* at the frequency ω_n, and is given by

$$Z_r(\omega_n) = (K_r - M_r \omega_n^2) + i\omega_n C_n. \tag{13.35}$$

The term $1/Z_r(\omega)$ varies in exactly the same way with frequency as the term for the simple system. In particular, at or near the frequency

$$\omega_r = \sqrt{K_r/M_r}$$

resonance will occur in the mode $f_r(x)$. We may also speak of the 'damping factor' corresponding to the rth mode, i.e.

$$C_r/2\sqrt{M_r K_r.} = \zeta_r.$$

This term will govern the width and height of the response curve corresponding to the rth mode.

When the continuous system is excited by a harmonically varying pressure distribution giving rise to the rth generalized force $L_r \cos \omega t$, the response in the rth mode is (in the real form)

$$q_r = \frac{L_r \cos(\omega t - \phi_r)}{|Z_r|}. \tag{13.36}$$

The total displacement at any point, from equation 13.28, is clearly

$$w(x, t) = \sum_{r=1}^{\infty} \frac{L_r f_r(x)}{|Z_r|} \cos(\omega t - \phi_r), \tag{13.37}$$

and if only the first s modes are contributing appreciably,

$$w(x, t) \doteq \sum_{r=1}^{s} \frac{L_r f_r(x)}{|Z_r|} \cos(\omega t - \phi_r). \tag{13.38}$$

If we are now concerned with the stress at some point in the system, it is necessary to specify the stress at that point due to unit displacement in each of the significant modes. Let unit displacement in the rth mode cause a stress at the point x of $\sigma_r(x)$. Then the total stress is obviously

$$\sigma(x, t) = \sum_{r=1}^{\infty} \frac{L_r \sigma_r(x)}{|Z_r|} \cos(\omega t - \phi_r). \tag{13.39}$$

If the exciting pressure is periodic, but not harmonic, then the total stress is the sum of all such expressions as equation 13.39, each one corresponding to a different harmonic component, ω_n, of the exciting pressure. Hence

$$\sigma(x, t) = \sum_{n=1}^{\infty} \sum_{r=1}^{\infty} \frac{L_r \sigma_r(x)}{|Z_r(\omega_n)|} \cos(\omega_n t - \phi_{rn}). \tag{13.40}$$

We write 'ϕ_{rn}' since ϕ_r depends on the frequency, ω_n. Further, if L_r derives from an acoustic pressure field, it too will depend on the frequency.

13.3 Particular forms of the generalized force

We shall now examine the magnitude of the generalized force corresponding to different modes and pressure distributions, $p(x, t)$. Consider the normal modes of flexural vibrations of a uniform rectangular plate which is simply supported along each edge. The modes are represented by:

$$f_r(x) = \sin \frac{r\pi x}{l} \sin \frac{s\pi y}{b} \tag{13.41}$$

(r and s are integers, l and b are the plate length and breadth respectively). For simplicity, put $s = 1$. There are then r half waves along the length, and one across the width of the plate.

If the plate is excited by a uniform harmonic pressure over the surface (i.e. in phase at all points, and of uniform amplitude), $p(x, t)$ is of the form $p_0 \cos \omega t$. The generalized force (from equation 13.32) is then:

$$L_r(t) = \int_A p_0 \cos \omega t \sin \frac{r\pi x}{l} \sin \frac{\pi y}{b} \, dx \, dy$$

i.e.

$$L_r(t) = p_0 \cos \omega t \cdot \frac{4}{\pi} \times \frac{\text{plate area}}{r} \quad (r \text{ odd})$$

and

$$L_r(t) = 0 \quad (r \text{ even}).$$

That this should be zero for even r may be seen from Fig. 13.4.

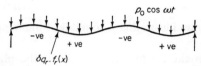

Fig. 13.4 Diagram illustrating the effective generalized force on a vibrating plate.

When the pressure $p_0 \cos \omega t$ is moved through a small displacement $\delta q_r \cdot f_r(x)$ as shown, the pressure does positive work over two of the half-wave sections, but an equal amount of negative work over the other two half-wave sections. Obviously, whenever r is even, the net work done is zero, and the corresponding generalized force is zero. When r is odd, a non-zero amount of work (negative or positive) is always done, and the generalized force is non-zero, but it gets progressively smaller as r increases since the area over which work is done without being cancelled gets smaller.

When the rth generalized force is zero, it follows from equation 13.29 that the rth coordinate and mode cannot be excited—always provided there is no significant damping coupling with other modes.

If the plate is excited by a harmonic pressure the amplitude of which varies sinusoidally over the plate (i.e. $p(x, t) = p_0 \cos \omega t \cdot \sin r\pi x/l$) we find that the generalized force is zero for every mode except the rth, provided that the half wavelength of the pressure on the plate (l/r) is an integral fraction of the plate length, i.e. r is an integer. When this is not so, an interesting and important effort occurs which is described later.

13.4 The forced vibration of a plate in a simple sound field

The simplest sound field we can consider is a field of plane harmonic waves. Suppose such a field impinges upon a plate, the wavefronts making an angle θ with the plate surface. Consider only the case when the sound wave motion has no component of velocity across the width of the plate, i.e. the wave travels in the direction of the *length* of the plate. (See Fig. 13.5.)

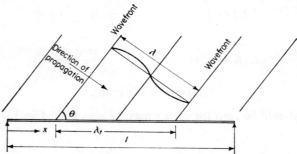

Fig. 13.5   Diagram illustrating a sound field impinging on the plate.

The wavefronts make an intercept of λ_t on the plate. We call λ_t the 'trace wavelength', and the speed at which the wavefront moves along the plate the 'trace velocity' (a_t). Clearly, $\lambda_t = \lambda \csc \theta$, and $a_t = a \csc \theta$ (a = speed of sound in air).

The instantaneous pressure at any point along the plate is

$$p(x, t) = p_0 \cos\left(\omega t - \frac{2\pi x}{\lambda_t} + \phi\right). \tag{13.42}$$

The generalized force (see equation 13.32) is given by:

$$L_r(t) = \int_0^b \int_0^l p_0 \cos\left(\omega t - \frac{2\pi x}{\lambda_t}\right) \sin\frac{r\pi x}{l} \sin\frac{\pi y}{b} \, dx \, dy$$

$$= p_0 \frac{2b}{\pi} \int_0^l \cos\left(\omega t - \frac{2\pi x}{\lambda_t}\right) \sin\frac{r\pi x}{l} \, dx.$$

Evaluating the integral, and replacing l/r by $\lambda_m/2$ (the half-wavelength of the mode of vibration) we find:

$$L_r(t) = p_0 l b \frac{4}{\pi^2} \left\{ \frac{\cos\left(\frac{\pi}{2} r\lambda_m/\lambda_t\right)}{r[(\lambda_m/\lambda_t)^2 - 1]} \right\} \cos(\omega t + \epsilon) \qquad (r \text{ odd})$$

and $\hspace{10cm} (13.43)$

$$p_0 l b \frac{4}{\pi^2} \left\{ \frac{\sin\left(\frac{\pi}{2} r\lambda_m/\lambda_t\right)}{r[(\lambda_m/\lambda_t)^2 - 1]} \right\} \cos(\omega t + \epsilon) \qquad (r \text{ even})$$

which may be written in the form

$$L_r(t) = p_0 l b j_r \cos(\omega t + \epsilon). \tag{13.44}$$

The total force acting on the plate at any instant when the pressure is distributed uniformly over the plate is $p_0 lb \cos(\omega t + \epsilon)$. j_r is always less than one, and is a factor which describes the proportion of this force which a particular mode of distortion can 'accept' and convert into the corresponding generalized force. Since it is a function of both the modal wavelength, λ_m, and the trace-wavelength, λ_t, we may call it the *joint acceptance* of the mode and pressure field. Figure 13.6 shows how j_r varies with both r and λ_m/λ_t. Apart from the curve for $r = 1$, it is characteristic of each curve that its maximum value occurs at $(\lambda_m/\lambda_t) = 1$, i.e. when the modal wavelength is the same as the trace-wavelength.

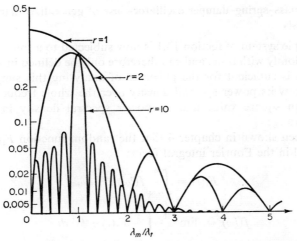

Fig. 13.6 The variation of the joint acceptance, j_r, with the wavelength ratio λ_m/λ_t.

The response of the plate in any of its modes to this pressure field is given (as before) by equation 13.34 or 13.36. Substituting from equation 13.44 into equation 13.36 we have:

$$q_r = p_0 lb \frac{j_r}{|Z_r|} \cos(\omega t + \epsilon - \phi_r). \tag{13.45}$$

It is now clear that a large response in this mode will occur if, at the resonant frequency ω_r (when $|Z_r|$ is a minimum) j_r is at the maximum value corresponding to $\lambda_m/\lambda_t = 1$. A dual coincidence then occurs, the sound field frequency being equal to the resonant frequency and the trace wavelength being equal to the modal wavelength. For this reason the term *coincidence effect* is often used to describe this particular phenomenon.

CHAPTER 14

Linear Dynamical Systems under Random Loading

14.1 The mass–spring–damper oscillator—use of generalized harmonic analysis

The simple system of section 13.1 is now subjected to a force $F(t)$ which varies randomly with time, and can therefore only be defined in a statistical manner. It is sufficient for the purposes of analysing this simple system that we know its power spectral density over the whole frequency range, or its mean square value and its power spectral density in a limited frequency range.

It has been shown in chapter 4 that the random function $F_1(t)$ may be represented in the Fourier integral form:

$$F_1(t) = \int_{-\infty}^{\infty} f(i\omega)\, e^{i\omega t}\, d\omega \qquad (14.1)$$

where

$$f(i\omega) = \lim_{t_0 \to \infty} \frac{1}{2\pi} \int_{-t_0}^{t_0} F_1(t)\, e^{-i\omega t}\, dt.$$

We may now replace $F(t)$ in equation 13.1 by the above form, giving

$$m\frac{d^2w}{dt^2} + c\frac{dw}{dt} + kw = \int_{-\infty}^{\infty} f(i\omega)\, e^{i\omega t}\, d\omega. \qquad (14.2)$$

The displacement of the system, w, which now varies randomly, may also be expressed in the Fourier integral form:

$$w(t) = \int_{-\infty}^{\infty} w(i\omega)\, e^{i\omega t}\, d\omega. \qquad (14.3)$$

Evidently $w(i\omega)$ is the Fourier spectrum of $w(t)$, and must be found before the power spectrum of $w(t)$ can be found. If $w(t)$ from equation 14.3 is substituted into the left-hand side of equation 14.2, we can differentiate with respect to t within the integral, yielding

$$\int_{-\infty}^{\infty} w(i\omega)[-m\omega^2 + ic\omega + k]\, e^{i\omega t}\, d\omega = \int_{-\infty}^{\infty} f(i\omega)\, e^{i\omega t}\, d\omega. \qquad (14.4)$$

If this is to be true for all values of t, it is necessary that the integrands on each side of this equation be equal. Hence

$$w(i\omega)[-m\omega^2 + ic\omega + k] = f(i\omega)$$

or

$$w(i\omega) = \frac{f(i\omega)}{(k - m\omega^2) + i\omega c} = \frac{f(i\omega)}{Z(\omega)}. \tag{14.5}$$

This important relationship states that the Fourier spectrum of the response is equal to the Fourier spectrum of the exciting force $\div$ the complex obstructance of the system.

The total response of the system, $w(t)$, is found from equation 14.3, i.e.

$$w(t) = \int_{-\infty}^{\infty} \frac{f(i\omega) e^{i\omega t}}{Z(\omega)} d\omega. \tag{14.6}$$

Now the power spectral density of $w(t)$, $S_w(\omega)$, has been shown to be related to the Fourier spectrum, $w(i\omega)$, by

$$S_w(\omega) = \lim_{t_0 \to \infty} \frac{\pi |w(i\omega)|^2}{t_0} \tag{14.7}$$

(see chapter 4). Substituting from equation 14.5, and remembering that $f(i\omega)$ above depends on t_0, we have

$$S_w(\omega) = \frac{S_f(\omega)}{|Z(\omega)|^2}, \tag{14.8}$$

where $S_f(\omega)$ is the power spectral density of the random force.

Equation 14.8 states the important fact that

$$\text{Power spectral density of response} = \frac{\text{power spectral density of force}}{\text{(modulus of obstructance)}^2}$$

The mean square value of w $(= \overline{w^2(t)})$ may now be found from

$$\overline{w^2(t)} = \int_{-\infty}^{\infty} S_w(\omega) \, d\omega = 2 \int_{0}^{\infty} \frac{S_f(\omega)}{|Z|^2} \, d\omega \tag{14.9}$$

since both $S_f(\omega)$ and $|Z|^2$ are even functions of ω.

Now consider typical graphs of $S_f(\omega)$ and $1/|Z|^2$, as shown in Fig. 14.1.

Provided the damping of the system is small, and the magnification factor Q is large, it is evident that the integral of the product of the two functions consists mainly of the area 'under' the resonant peak. The height of this peak (see section 13.1) is $Q^2/k^2 = 1/4\zeta^2 k^2$. The bandwidth of the peak, $\Delta\omega$, is $2\zeta\omega_r$ (from equation 13.12). The area under the peak must therefore be proportional to

$$S_f(\omega_r) \times 2\zeta\omega_r \times \frac{1}{4\zeta^2 k^2} = \frac{S_f(\omega_r)}{kc}.$$

Carrying out the integration analytically, and assuming that $S_f(\omega)$ does not vary appreciably in the region of the natural frequency, we find

$$\overline{w^2(t)} = \frac{\pi S_f(\omega_r)}{kc} = \frac{\pi}{2} S_f(\omega_r) \frac{\omega_r}{k^2 \zeta}. \tag{14.10}$$

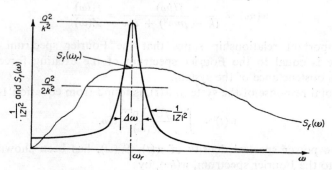

Fig. 14.1 Typical graphs of $S_f(\omega)$ and $1/|Z|^2$.

If we had used the hysteretic damping force, ihw, instead of the viscous damping in equation 14.2, the analysis would have been similar but the complex obstructance $Z(\omega)$ would have been

$$(k - m\omega^2) + ih.$$

The mean square displacement is found by replacing c in equation 14.10 by h/ω_r, giving

$$\frac{\pi S_f(\omega_r) \cdot \omega_r}{kh}.$$

This may be written in the alternative forms

$$\pi \frac{S_f(\omega_r)}{k^{1/2} m^{1/2} h} \quad \text{or} \quad \frac{\pi}{2} S_f(\omega_r) \frac{\omega_r}{k^2 \zeta}. \tag{14.11}$$

Notice that the second of these is identical to the last term in equation 14.10.

When the system is very lightly damped, most of the energy of vibration due to the random loading is contained within a narrow band centred on ω_r. The waveform of the random response may then be described as 'sinusoidal with a randomly varying amplitude'. In fact, since the narrow band has a finite width which is proportional to $\Delta\omega$ ($\propto \zeta\omega_r$), the frequency components on either side of ω_r tend to 'beat' with one another. This gives a crude modulation of the envelope of the response peaks. The frequency of this modulation is roughly proportional to the damping. These features are shown in Fig. 14.2.

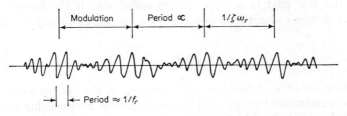

Fig. 14.2 Typical waveform of the response of a lightly damped oscillator to random excitation.

14.2 The mass–spring–damper oscillator—use of impulse response functions

In the previous section, the exciting force was (effectively) first of all analysed into its continuum of harmonic components, and the problem was solved knowing the response of the system to each component. An alternative method of solution is to analyse the exciting force into a succession of impulses, and then to study the total response of the system knowing its response to a single impulse. The two methods must obviously give the same final result, but the latter can be used conveniently to give some further insight into the nature of the response.

When an impulse I is imposed upon our simple oscillator (initially at rest) the instantaneous velocity increase is I/m, and the subsequent velocity varies periodically and decays exponentially in the form

$$\dot{w} = \frac{I}{m} e^{-\zeta \omega_r t} \cos (\sqrt{1 - \zeta^2} \omega_r t). \tag{14.12}$$

The corresponding displacement at the time t after the application of the impulse is

$$w = \frac{I}{k} \frac{\omega_r}{\sqrt{1 - \zeta^2}} e^{-\zeta \omega_r t} \sin (\sqrt{1 - \zeta^2} \omega_r t)$$

$$= \frac{I}{k} \mathcal{R} \left\{ -\frac{i \omega_r}{\sqrt{1 - \zeta^2}} e^{(-\zeta + i\sqrt{1 - \zeta^2}) \omega_r t} \right\}. \tag{14.13}$$

The real part of the complex term in the brackets is k times the response of the system to a unit impulse. We write

$$h(t) = -\frac{i \omega_r}{\sqrt{1 - \zeta^2}} e^{(-\zeta + i\sqrt{1 - \zeta^2}) \omega_r t} \tag{14.14}$$

$$= 0 \quad \text{for all } t < 0.$$

Although $h(t)$ as defined here is complex, its real part only is required in the operations carried out in this section. This should be remembered when the real part is not specifically called for. If the impulses were applied at time $t = \tau$, the response at time t would be simply the real part of

$$\frac{I}{k} h(t - \tau).$$

It is now possible to build up the response of the system to a general transient excitation by regarding the exciting force as consisting of a series of impulses, as in Fig. 14.3.

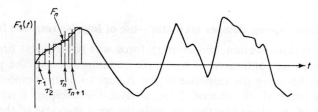

Fig. 14.3   Representation of the force $F_1(t)$ by a succession of impulses.

The exciting force is now represented by the sequence of impulses of magnitude $F_n \Delta \tau_n$ occurring at the time τ_n. The nth impulse gives rise to the increment of response

$$\frac{F_n \Delta \tau_n}{k} h(t - \tau_n)$$

at the time t.

For a *linear* system, we may superpose these incremental responses to obtain the overall response, i.e.

$$w(t) = \sum \frac{F_n \Delta \tau_n}{k} h(t - \tau_n).$$

In the limit, as $\Delta \tau_n \to 0$, the summation becomes a 'convolution integral'

$$w(t) = \int_0^t \frac{F(\tau)}{k} h(t - \tau) \, d\tau. \tag{14.15}$$

If the excitation did not start at $t = 0$ but had been continuing from $t = -\infty$, we should have

$$w(t) = \int_{-\infty}^t \frac{F(\tau)}{k} h(t - \tau) \, d\tau. \tag{14.16}$$

Changing the variable of integration from τ to $(t - \tau) = T$, we can write this in the alternative form

$$w(t) = \frac{1}{k} \int_0^\infty h(T)F(t - T) \, dT. \tag{14.17}$$

We shall now proceed to determine the auto-correlation function of the response $R_w(\tau)$. By definition

$$R_w(\tau) = \lim_{T \to \infty} \frac{1}{2T} \int_{-T}^T w(t)w(t + \tau) \, dt$$

$$= \frac{1}{k^2} \lim_{T \to \infty} \frac{1}{2T} \int_{-T}^T \int_0^\infty h(\tau_1)F(t - \tau_1) \, d\tau_1 \int_0^\infty h(\tau_2)F(t + \tau - \tau_2) \, d\tau_2 \, dt.$$

Interchanging the order of integration, we have

$$R_w(\tau) = \frac{1}{k^2} \int_0^\infty \int_0^\infty h(\tau_1)h(\tau_2) \lim_{T \to \infty} \frac{1}{2T} \int_{-T}^T F(t - \tau_1)F(t + \tau - \tau_2) \, dt \, d\tau_2 \, d\tau_1.$$

Contained within this integral is the auto-correlation function of the excitation with a time delay $\tau + \tau_1 - \tau_2$. Hence

$$R_w(\tau) = \frac{1}{k^2} \int_0^\infty \int_0^\infty h(\tau_1)h(\tau_2)R_F(\tau + \tau_1 - \tau_2) \, d\tau_1 \, d\tau_2. \tag{14.18}$$

The auto-correlation function of the response is therefore a double convolution of the auto-correlation function of the exciting force.

We shall now confine attention to a system with small damping ($\zeta^2 \ll 1$) excited by a force with a flat power spectrum. The real part of the impulse response function degenerates into

$$\omega_r \, e^{-\zeta \omega_r t} \sin \omega_r \tau.$$

If the power spectral density of the exciting force is constant and equal to S_0, say, then the auto-correlation function is given simply by $2\pi S_0 \delta(\tau)$. $\delta(\tau)$ is the Dirac delta function which is zero for $\tau \neq 0$, is indefinitely large for $\tau = 0$, but

$$\int_{-\infty}^\infty \delta(\tau) \, d\tau = 1.$$

The auto-correlation function of the force is therefore zero for all values of τ except $\tau = 0$. At $\tau = 0$, the function gives the mean square value, which in this case is infinite. This is precisely what would be expected from the specified 'white' spectrum, which is not in fact physically realizable.

The auto-correlation of the response is now

$$R_w(\tau) = \frac{2\pi}{k^2} \int_0^\infty \int_0^\infty h(\tau_1)h(\tau_2)S_0\delta(\tau + \tau_1 - \tau_2)\, d\tau_1\, d\tau_2$$

$$= \frac{S_0 2\pi}{k^2} \int_0^\infty h(\tau_1)h(\tau + \tau_1)\, d\tau_1$$

$$= \frac{S_0 2\pi}{k^2}\, \omega_r^2\, e^{-\zeta\omega_r\tau} \left[\frac{\cos \omega_r\tau}{4\zeta\omega_r} - \frac{\sin \omega_r\tau}{4\omega_r} \right]. \tag{14.19}$$

If the damping is small, this degenerates still further to give

$$R_w(\tau) \doteq \frac{2\pi S_0}{k^2} \frac{\omega_r}{4\zeta}\, e^{-\zeta\omega_r\tau} \cos \omega_r\tau. \tag{14.20}$$

The auto-correlation function of the response to white noise of a lightly damped single-degree-of-freedom system therefore approximates to a damped cosine wave. The rate of decay of the auto-correlogram with respect to time delay is identical to that of the free motion of the system with respect to real time. Ideally, therefore, the auto-correlogram may be used to obtain a measure of both the damping and the natural frequency of the system.

If we now put $\tau = 0$, we obtain the mean square value of the random response, i.e.

$$\frac{\pi}{2} \frac{S_0\omega_r}{k^2\zeta},$$

which is identical to that obtained by the method of section 14.1.

Reference 1 gives a more detailed exposition of the use of impulse response functions.

14.3 The mean square force due to random forces on a rigid beam

As a first step towards the consideration of the generalized force due to random pressures acting on a deformable structure, consider a rigid beam on which just two discrete random forces are acting, $P_1(t)$ and $P_2(t)$. These are illustrated in Fig. 14.4.

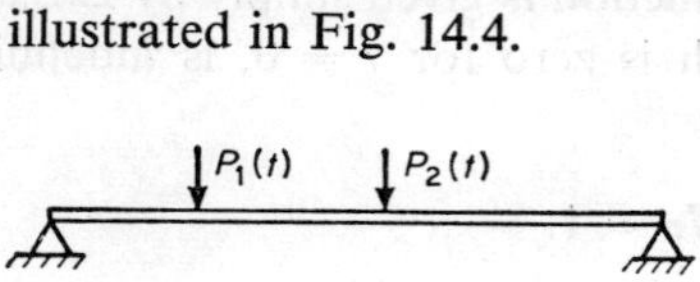

Fig. 14.4   A beam subjected to two random forces.

Each of the loads varies randomly with time about a mean value of zero, and at any instant may be positive or negative. The average properties of the total load will therefore depend upon the extent to which these loads are dependent, i.e. upon the correlation between the loads. The combined load at any instant is $(P_1 + P_2)$, the average value of which is zero. The mean square value of the total load is

$$\overline{(P_1 + P_2)^2} = \overline{P_1^2} + \overline{P_2^2} + \overline{2P_1 P_2},$$

and we have already defined $\overline{P_1 P_2}$ as the correlation between the two loads. If we now assume the loads to have the same mean square values (corresponds to a 'homogeneous' pressure field) then $\overline{P_1^2} = \overline{P_2^2}$, and

$$\overline{(P_1 + P_2)^2} = \overline{2P_1^2}[1 + \rho]$$

where ρ is the cross-correlation coefficient, $\overline{P_1 P_2}/\overline{P_2^2}$. As the value of ρ varies from -1 to $+1$, so the mean square value of the total load rises from zero to a maximum value.

Consider now a distributed pressure, $p(x, t)$, acting on the beam of span l; $p(x, t)$ is a random function of both space and time, having an average value of zero. Suppose the beam has unit width. The total load on the beam at any instant is

$$L(t) = \int_0^l p(x, t)\, dx. \tag{14.21}$$

The total value must also be a random function of time, having a zero mean value. To find the mean square value, we shall first find the auto-correlation function of this load, and then put τ (time delay) $= 0$. The auto-correlation function, $R_L(\tau)$, of $L(t)$ is

$$\lim_{T \to \infty} \frac{1}{2T} \int_{-T}^{+T} L(t) L(t + \tau)\, dt$$

$$= \lim_{T \to \infty} \frac{1}{2T} \int_{-T}^{+T} dt \int_0^l p(x_1, t)\, dx_1 \int_0^l p(x_2, t + \tau)\, dx_2$$

$$= \int_0^l \int_0^l dx_1\, dx_2 \lim_{T \to \infty} \frac{1}{2T} \int_{-T}^{+T} p(x_1, t) p(x_2, t + \tau)\, dt$$

$$= \int_0^l \int_0^l R_p(x_1, x_2; \tau)\, dx_1\, dx_2 \tag{14.22}$$

where $R_p(x_1, x_2; \tau)$ is the space–time correlation function of the distributed pressure.

Now if the pressure field is homogeneous, the mean square pressure at all points will be the same $(= \overline{p^2})$. Furthermore, the correlation function has the same shape everywhere and depends only on $(x_2 - x_1) = \xi$. We may then write

$$R_p(x_1, x_2; \tau) = \overline{p^2}\, \rho(x_1, x_2; \tau) \tag{14.23}$$

where $\rho(x_1, x_2; \tau)$ is the cross-correlation coefficient. $R_L(\tau)$, from equation 14.22, now becomes

$$R_L(\tau) = \overline{p^2} \int_0^l dx \int_{-x}^{l-x} \rho(\xi; \tau) \, d\xi. \qquad (14.24)$$

The mean square value of $L(t)$ is given by equation 14.23 when $\tau = 0$. I.e.

$$\overline{L^2(t)} = \overline{p^2} \int_0^l dx \int_{-x}^{l-x} \rho(\xi) \, d\xi. \qquad (14.25)$$

We now consider the extreme cases:

(i) The correlation coefficient is almost constant, and equal to unity over the whole beam. Then

$$\overline{L^2(t)} = \overline{p^2} l^2. \qquad (14.26)$$

(ii) The correlation coefficient drops very rapidly to zero, as in Fig. 14.5.

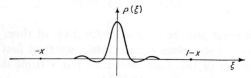

Fig. 14.5   The rapidly decreasing cross-correlation coefficient.

In this case

$$\int_{-x}^{l-x} \rho(\xi) \, d\xi$$

is scarcely dependent upon x at all, and having the dimensions of a length, may be replaced by the constant λ. We call this the 'correlation length' Integrating equation 14.25 we now have

$$\overline{L^2(t)} = \overline{p^2} \lambda l. \qquad (14.27)$$

If we put $l/\lambda = n$, the number of correlation lengths required to fill the span, we obtain

$$\overline{L^2(t)} = \overline{p^2} \frac{l^2}{n} \qquad (14.28)$$

showing clearly that as the correlation length becomes smaller, the mean square value of the load decreases.

14.4 The meaning and significance of the correlation length: types of excitation

In chapter 4 it was shown that auto-correlation in a curve is a property which follows from its continuity. If we assume that the pressure distribution on the beam is a continuous function of position along the beam, it follows that the pressure, as it fluctuates randomly, must be correlated in space. The correlation relative to a given point drops off at a rate which depends on how rapidly the load distribution can vary with position.

When we calculate the integral $\int_{-x}^{l-x} \rho(\xi) \, d\xi$, we are integrating along the beam weighting each position with the appropriate value of the space correlation coefficient. The result is an equivalent length over which the load is perfectly correlated, i.e. the area under the curve of Fig. 14.5 is equivalent to the area under the curve of Fig. 14.6.

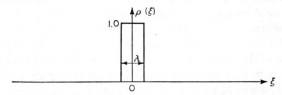

Fig. 14.6 A cross-correlation distribution equivalent to that of Fig. 14.5.

It should be noticed in equation 14.24 that the auto-correlation function depends on the integral

$$\int_{-x}^{l-x} \rho(\xi, \tau) \, d\xi,$$

i.e. it depends on τ as well as on ξ. Evaluation of this integral for different values of τ will give different correlation lengths, which have significance when we deal with deformable beams and plates.

Having seen that the mean square total load depends on the correlation of pressures at different points, and therefore on the way in which the curve of pressure distribution varies with time, it is of interest to consider how this curve can behave. There are two extreme cases which we can find:

(i) A curve, all points of which fluctuate randomly, the only condition being that the curve shall remain continuous. Such a curve has space correlation because of its continuity, but at large separation distances the pressures at two points are in no way correlated. The curve is therefore random in space and time. Such a pressure distribution exists in the near field of a jet.

(ii) Suppose we have a semi-infinite string and that we move the end so as to propagate waves along it (see Fig. 14.7).

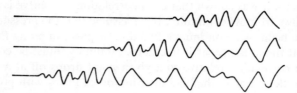

Fig. 14.7   Diagram illustrating the progress of a random wave motion along a taut string, from right to left.

A string propagates all waves at the same speed, and the random wave therefore retains its shape as it travels. Viewed instantaneously, it will appear as a random wave in space; viewed continuously, it will appear as a fixed waveform being translated at a fixed velocity, a. If we watch the motion of the string at a point x from the end, then we shall see there a reproduction of the motion of the end which occurred at a time x/a earlier. Thus, whereas in case (i) it was possible to find separation distances over which the pressures at two points were completely uncorrelated, we now have complete correlation at any separation distance ξ, provided the correct delay time, $\tau = \xi/a$, is used.

The two situations to which the model approximates are:

(a) An acoustic field in which the waves are effectively plane. This is very similar to the string wave motion.
(b) The hydrodynamic pressure fluctuations in a boundary layer. These are associated with 'bubbles' of turbulence which are convected downstream at a certain speed. Here, the above model only applies over short distances, for the turbulence in a given bubble decays as it is convected.

14.5 The generalized force due to random plane waves

We now consider the second of the two extreme cases of the last section, in order to determine the corresponding generalized force on a beam. The wavefronts are all inclined at an angle θ to the beam (as in section 13.4).

Since the fluctuations in time at a point may be resolved into a continuous frequency spectrum, the space waveform (pressure distribution) may be resolved into a continuous 'spectrum' of travelling waves, of all wavelengths. The velocity of the plane waves, perpendicular to the wavefronts, is the velocity of sound a, and the trace velocity on the beam is $a_t = a \csc \theta$.

We may represent the pressure at a point by

$$p(t) = \sum_{n=1}^{\infty} d_n \cos (\omega_n t + \epsilon_n)$$

and the pressure distribution on the beam by

$$p(t, x) = \sum_{n=1}^{\infty} d_n \cos \{\omega_n (t - [x/a] \sin \theta) + \epsilon_n\}.$$

Expressing this in the continuous, complex form of equation 14.1

$$p(t, x) = \int_{-\infty}^{\infty} f(i\omega)\, e^{i\omega(t - [x/a] \sin \theta)}\, d\omega. \tag{14.29}$$

Firstly, consider the *total* load acting on the beam, i.e.

$$\int_0^l p(t, x)\, dx = \int_0^l \int_{-\infty}^{+\infty} f(i\omega)\, e^{i\omega(t - [x/a] \sin \theta)}\, d\omega\, dx$$

$$= L(t). \tag{14.30}$$

The integration with respect to x may be carried out within the time integral, whence, since $\omega(x/a) \sin \theta = 2\pi x/\lambda_t$, we have

$$L(t) = \int_{-\infty}^{\infty} f(i\omega) \frac{i\lambda_t}{2\pi} (e^{-i2\pi l/\lambda_t} - 1)\, e^{i\omega t}\, d\omega \tag{14.31}$$

Now the Fourier spectrum of $p(t)$ is $f(i\omega)$. The Fourier spectrum, $l(i\omega)$, of $L(t)$ is related to $L(t)$ by

$$L(t) = \int_{-\infty}^{\infty} l(i\omega)\, e^{i\omega t}\, d\omega.$$

Comparing equations 14.31 and 14.30, we see

$$l(i\omega) = f(i\omega)i \frac{\lambda_t}{2\pi} (e^{-i2\pi l/\lambda_t} - 1).$$

Now the power spectrum, $S_L(\omega)$, of $L(t)$ is

$$\lim_{T \to \infty} \frac{\pi}{T} |l(i\omega)|^2.$$

Evaluating $|l(i\omega)|^2$ from equation 14.31, we find

$$S_L(\omega) = \lim_{T \to \infty} \frac{\pi}{T} |f(i\omega)|^2 \left\{ \frac{\sin (\pi l/\lambda_t)}{\pi l/\lambda_t} \right\}^2 \cdot l^2.$$

We recognize

$$\lim_{T \to \infty} \frac{\pi}{T} |f(i\omega)|^2$$

11

as the power spectrum of $p(t)$, i.e. $S_p(\omega)$. Hence

$$S_L(\omega) = S_p(\omega) \cdot l^2 \cdot \left\{ \frac{\sin{(\pi l/\lambda_t)}}{\pi l/\lambda_t} \right\}^2 \tag{14.32}$$

The function $\{\sin{(\pi l/\lambda_t)}/(\pi l/\lambda_t)\}^2$ is shown in Fig. 14.8.

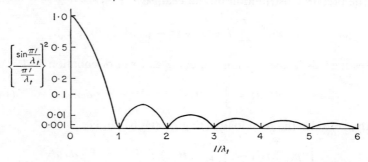

Fig. 14.8   The function $\{\sin{(\pi l/\lambda_t)}/(\pi l/\lambda_t)\}^2$ plotted against l/λ_t.

Equation 14.32 demonstrates a very important point, viz. that by virtue of its length the beam acts as a 'wavelength filter'. Whenever the frequency ω is such that $\pi l/\lambda_t \; (=\omega l \sin{\theta}/2a)$ is an integer, then the corresponding power spectral density of the total force is zero. The principal part of the generalized force comes from those components, the wavelengths of which are long compared with the length of the beam. That this must be so can be seen from simple physical considerations; a wave component whose trace wavelength is exactly equal to the span of the beam provides a positive pressure over one half of the span and a negative pressure over the other half. The corresponding net force is therefore zero. The statement also ties in with the fact that the generalized force decreases as the correlation length decreases, for with short wavelength components, although the correlation coefficient oscillates and does not define a length as in sections 14.3 and 14.4, yet we can take its 'period' of oscillation as a new 'correlation length'.

A similar calculation may be carried out to determine the power spectrum of the generalized force corresponding to the mode of displacement $f_r(x) = \sin{(\pi r x/l)}$. The generalized force, $L_r(t)$, is given by

$$L_r(t) = \int_0^l p(t, x) \sin{\frac{\pi r x}{l}} \, dx.$$

Substituting for $p(t, x)$ from equation 14.29, and carrying out the integration in the same way as above, we find that the power spectrum of $L_r(t)$ is

$$S_L(\omega) = S_p(\omega) l^2 \frac{4}{\pi^2} \left\{ \frac{\sin{[(\pi/2)r\lambda_m/\lambda_t + r\pi/2]}}{r[(\lambda_m/\lambda_t)^2 - 1]} \right\}^2 \tag{14.33}$$

in which $\lambda_m/2 = l/r$.

If we had considered a plate, vibrating in the mode $\sin (\pi rx/l) \sin (\pi y/b)$, this expression would have become*

$$S_L(\omega) = S_p(\omega)(lb)^2 \frac{16}{\pi^4} \left\{ \frac{\sin \left[(\pi/2)r\lambda_m/\lambda_t + r\pi/2 \right]}{r[(\lambda_m/\lambda_t)^2 - 1]} \right\}^2. \tag{14.34}$$

which by comparison with equations 13.43 and 13.44 is seen to be

$$S_L(\omega) = S_p(\omega)(lb)^2 j_r^2. \tag{14.35}$$

The wavelength 'filter' characteristics of the plate are again seen by the manner in which j_r (and hence j_r^2) varies with l/λ_t.

Equation 14.35 states simply that the power spectral density of the generalized force is equal to the product of the power spectral density of the (homogeneous) pressure, the square of the plate area and the square of the joint acceptance.

14.6 The generalized force corresponding to an entirely random, homogeneous pressure field

We now remove the restriction of the former paragraph that the sound waves must be plane and moving down the length of the plate. Further, we shall consider a general mode shape, $f_r(x, y)$ which is not necessarily a sinusoidal mode.

The generalized force is

$$L_r(t) = \int_A f_r(x, y) \cdot p(x, y; t)\, dA$$

where the integration is now over the surface. The Fourier spectrum of this is

$$l_r(i\omega) = \lim_{t_0 \to \infty} \frac{1}{2\pi} \int_{-t_0}^{t_0} \int_A f_r(x, y)p(x, y; t)\, e^{i\omega t}\, dA\, dt$$

$$= \int_A f(i\omega)f_r(x, y)\, dA,$$

in which $f(i\omega)$ is the Fourier spectrum of $p(x, y; t)$. The power spectrum of the generalized force, $S_{Lr}(\omega)$, is given by

$$S_{Lr}(\omega) = \lim_{t_0 \to \infty} \frac{\pi}{t_0} |l_r(i\omega)|^2 = \lim_{t_0 \to \infty} \frac{\pi}{t_0} l_r(i\omega)l_r^*(i\omega)$$

$$= \lim_{t_0 \to \infty} \frac{1}{t_0} \frac{1}{4\pi} \int_A \int_A f_r(x_1 y_1)f_r(x_2 y_2) \int_{-t_0}^{+t_0} \int_{-t_0}^{+t_0} p(x_1 y_1 t_1)p(x_2 y_2 t_2)$$

$$e^{i\omega(t_1 - t_2)}\, dt_1\, dt_2\, dA_1\, dA_2. \tag{14.36}$$

---

* Equation 14.33 applies to a beam of unit width, whereas equation 14.34 applies to a plate of width b. It is this fact which makes the dimensions of the two equations appear to be different. The asterisk in equation 14.36 implies 'complex conjugate'.

Now denote the point $x_1 y_1$ by $\bar{x}_1$

$\qquad\qquad\qquad\qquad x_2 y_2$ by $\bar{x}_2$

$\qquad\qquad$ and $t_2 - t_1$ by τ.

Equation 14.36 becomes

$$S_{L_r}(\omega) = \int_A \int_A f_r(\bar{x}_1) f_r(\bar{x}_2) \lim_{t_0 \to \infty} \frac{1}{t_0} \frac{1}{4\pi} \int_{-t_0}^{+t_0} \int_{-t_0}^{+t_0} p(\bar{x}_1, t_1) p(\bar{x}_2, t_1 + \tau)$$
$$e^{-i\omega\tau} \, dt_1 \, d\tau \, dA_1 \, dA_2 \qquad (14.37)$$

But

$$\lim_{t_0 \to \infty} \frac{1}{2t_0} \int_{-t_0}^{+t_0} p(\bar{x}_1, t_1) p(\bar{x}_2, t_1 + \tau) \, dt_1 = R_p(\bar{x}_1, \bar{x}_2, \tau) \qquad (14.38)$$

which is the cross-correlation function of the pressures at $\bar{x}_1$ and $\bar{x}_2$ and is sometimes known as the 'space–time correlation function'.

The limit of the time integral contained within the right-hand side of equation 14.37 now becomes

$$\frac{1}{2\pi} \int_{-\infty}^{\infty} R_p(\bar{x}_1, \bar{x}_2, \tau) \, e^{-i\omega\tau} \, d\tau = S_p(\bar{x}_1, \bar{x}_2, \omega) \qquad (14.39)$$

which is the 'cross power spectral density' of the pressures at $\bar{x}_1$ and $\bar{x}_2$. Since $R_p(\bar{x}_1, \bar{x}_2, \tau)$ is *not* an even function of τ, the cross-power spectral density will be a complex quantity.

Equation 14.37 can now be written in the form

$$S_{L_r}(\omega) = \int_A f_r(\bar{x}_1) \int_A f_r(\bar{x}_2) S_p(\bar{x}_1, \bar{x}_2, \omega) \, dA_2 \, dA_1. \qquad (14.40)$$

When this integral is examined in detail, it will be found that the imaginary component vanishes identically (as it must, since $S_{L_r}(\omega)$ is essentially real). For the purposes of evaluating the integral of equation 14.40, we therefore need only the real part of $S_p(\bar{x}_1, \bar{x}_2, \omega)$, i.e.

$$\frac{1}{2\pi} \int_{-\infty}^{\infty} R_p(\bar{x}_1, \bar{x}_2, \tau) \cos \omega\tau \, d\tau.$$

This is identical to the 'narrow-band' space correlation of the pressures at the two points.

The cross-power spectral density may be non-dimensionalized by dividing it by the power spectral density, $S_p(\omega)$, of the (homogeneous) pressure field. Equation 14.40 becomes

$$S_{L_r}(\omega) = S_p(\omega) \int_A f_r(\bar{x}_1) \int_A f_r(\bar{x}_2) \rho(\bar{x}_1, \bar{x}_2, \omega) \, dA_2 \, dA_1 \qquad (14.41)$$

$\rho(\bar{x}_1, \bar{x}_2, \omega)$ has the form of a correlation coefficient, and being a function of ω, may be regarded as the spectrum of the correlation coefficient.

The inner integral,

$$\int_A f_r(\bar{x}_2)\rho(\bar{x}_1, \bar{x}_2, \omega) \, dA_2$$

has the dimensions of an area, and may be regarded as a correlation area. In a homogeneous field, $\rho(\bar{x}_1, \bar{x}_2, \omega)$ is a function of the separation ξ, of the points $\bar{x}_1$ and $\bar{x}_2$ and not on $\bar{x}_1$ alone. The inner integral is therefore the area under the product of two curves, as in Fig. 14.9.

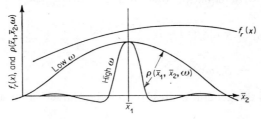

Fig. 14.9   Diagram showing the relationship between a mode $f_r(x)$ and two correlation curves.

At high frequencies the distance to the first zero crossing of $\rho(\bar{x}_1, \bar{x}_2, \omega)$ is small compared with the wavelength of $f(x)$, and the integral is approximately equal to

$$f_r(\bar{x}_1) \int_A \rho(\bar{x}_1, \bar{x}_2, \omega) \, dA_2,$$

which may be written in the form

$$f_r(\bar{x}_1)\lambda^2(\omega).$$

$\lambda^2(\omega)$ is the *correlation area* at the frequency ω and is independent of either $\bar{x}_1$ or $\bar{x}_2$.

Equation 14.41 now becomes

$$S_{L_r}(\omega) \doteqdot S_p(\omega)\lambda^2(\omega) \int_A f_r^2(\bar{x}) \, dA \qquad (14.42)$$

which is the approximate expression for the power spectrum of the generalized force at high frequencies.

At low frequencies, when $\rho(\bar{x}_1, \bar{x}_2, \omega)$ is nearly constant (≈ 1) all over the surface, equation 14.41 becomes

$$S_{L_r}(\omega) \doteqdot S_p(\omega) \left[\int_A f_r(\bar{x}) \, dA \right]^2. \qquad (14.43)$$

Now when f_r represents a sinusoidal mode of r half-waves over the surface area, the integral in equation 14.43 will be zero when r is even, and when

r is odd it will get increasingly smaller as r increases. The spectral density at low frequencies may therefore have zero values, and will be very small for the high order modes. The integral in equation 14.42 can never be zero as the integrand is squared. For sinusoidal modes, the integral is independent of r, and is equal to one-quarter of the plate area. At high frequencies, therefore, the spectral density of the generalized force in all the modes is proportional to the magnitude of the correlation area.

The above statements and deductions follow immediately from the consideration of the generalized force in sections 13.3, 14.2 and 14.3.

Finally, we write equation 14.41 in the form

$$S_{L_r}(\omega) = S_p(\omega)A^2 \left\{ \frac{\int_A \int_A f_r(\bar{x}_1)f_r(\bar{x}_2)\rho(\bar{x}_1, \bar{x}_2, \omega)\, dA_1\, dA_2}{A^2} \right\}$$

$$= S_p(\omega)A^2 j_{rr}^2 \tag{14.44}$$

which defines the general expression for the 'direct' joint acceptance, j_{rr}.

14.7 Calculation of the total response

We have now seen how to calculate the generalized force in certain special cases, and also how to calculate the response in just one mode (as for the simple oscillator, section 14.1). At first sight one might think that the total mean square response could be obtained by summing the mean square responses of the individual contributory modes, but this may only be done if the responses in the various modes are statistically independent. This is not so here, for a particular frequency component of the pressure fluctuations can excite vibrations in several modes, albeit by small amounts in some of them. The corresponding frequency components in the fluctuating response of these modes will be correlated to a certain extent.

To calculate the total response, we return to equation 13.6 which says that in the interval $-t_0 < t < t_0$, the amplitude of motion in the rth mode is

$$q_r(t) = \int_{-\infty}^{\infty} \frac{l_r(i\omega)\, e^{i\omega t}}{Z_r(\omega)}\, d\omega \tag{14.45}$$

in which we have replaced the Fourier spectrum, $f(i\omega)$, of the force $F(t)$ by the Fourier spectrum, $l_r(i\omega)$, of the rth generalized force, $L_r(t)$. Z_r is the complex generalized obstructance of the rth mode.

The displacement in the rth mode at any point x on the vibrating structure is

$$q_r(t)f_r(x) = \int_{-\infty}^{\infty} \frac{l_r(i\omega)f_r(x)\, e^{i\omega t}}{Z_r(\omega)}\, d\omega, \tag{14.46}$$

having the Fourier spectrum $l_r(i\omega)f_r(x)/Z_r$. The Fourier spectrum of the total displacement arising from all the modes is

$$\sum_{r=1}^{\infty} \frac{l_r(i\omega)f_r(x)}{Z_r(\omega)} \qquad (14.47)$$

and the corresponding power spectrum is

$$S_w(\omega) = \lim_{t_0 \to \infty} \frac{\pi}{t_0} \left| \sum_{r=1}^{\infty} \frac{l_r(i\omega)f_r(x)}{Z_r(\omega)} \right|^2. \qquad (14.48)$$

This is the same as

$$S_w(\omega) = \lim_{t_0 \to \infty} \frac{\pi}{t_0} \sum_{r=1}^{\infty} \frac{l_r(i\omega)f_r(x_1)}{Z_r(\omega)} \sum_{r=1}^{\infty} \frac{l_r^*(i\omega)f_r(x_1)}{Z_r^*(\omega)}$$

On multiplying this out and proceeding to the limit of $t_0 \to \infty$ (as we did in the last section) we find

$$S_w(\omega) = \sum_{r=1}^{\infty} \frac{f_r^2(x)}{|Z_r(\omega)|^2} \int_A f_r(\bar{x}_1) \int_A f_r(\bar{x}_2) S_p(\bar{x}_1, \bar{x}_2, \omega) \, dA_2 \, dA_1$$

$$+ \sum_{r=1}^{\infty} \sum_{\substack{s=1 \\ r \neq s}}^{\infty} \frac{f_r(x)f_s(x)}{Z_r(\omega)Z_s^*(\omega)} \int_A f_r(\bar{x}_1) \int_A f_s(\bar{x}_2) S_p(\bar{x}_1, \bar{x}_2, \omega) \, dA_2 \, dA_1 \qquad (14.49)$$

The first series gives the sum of the spectra of the responses in the individual modes. The second (double) series is the 'correction' term which is necessary due to the correlation between the responses in different modes, and is said to be the response due to 'modal frequency overlap'.

It may be shown that both the real and imaginary parts of $S_p(\bar{x}_1, \bar{x}_2, \omega)$ contribute to the second series. The final value of this is, of course, entirely real. This is in contrast to the first series, to which only the real part contributes, as was shown in the last section.

The double integral in the first series is the power spectrum of the rth generalized force, and may be written in the form of equation 14.44, involving the joint acceptance. It then becomes

$$A^2 S_p(\omega) \sum_{r=1}^{\infty} \frac{f_r^2(x)}{|Z_r(\omega)|^2} j_{rr}^2.$$

The second series may be manipulated in a similar way[4] to yield

$$A^2 S_p(\omega) \sum_{r=1}^{\infty} \sum_{s=1}^{\infty} \frac{f_r(x)f_s(x)}{|Z_r||Z_s|} j_{rsr}^2.$$

Notice that the moduli of the obstructances now appear in the denominator. Their arguments (i.e. phase angles) have been converted into time delays at the appropriate frequencies, and the difference between these two time delays is added to the time delay in the cross-correlation function which is contained within j_{rst}^2. In fact, the value of j_{rst}^2 is derived using the cross power spectral density

$$S_p(\bar{x}_1, \bar{x}_2, \omega) = \frac{1}{2\pi} \int_{-\infty}^{\infty} R_p(\bar{x}_1, \bar{x}_2, \tau - \tau_r + \tau_s)\, e^{-i\omega\tau}\, d\tau$$

where τ_r and τ_s are the time delays referred to.

Powell's final result is then

$$S_w(\omega) = A^2 S_p(\omega)\left\{ \sum_{r=1}^{\infty} \frac{f_r^2(x)j_{rr}^2}{|Z_r(\omega)|^2} + \sum_{r=1}^{\infty} \sum_{\substack{s=1 \\ r \neq s}}^{\infty} \frac{f_r(x)f_s(x)j_{rst}^2}{|Z_r(\omega)|\,|Z_s(\omega)|} \right\} \quad (14.50)$$

It is considered, however, that the equation 14.49 gives a better physical understanding of the phenomena.

The second series is extremely difficult to evaluate, and can probably only be dealt with statistically.[3,4] For some simple problems, however, it can be ignored. This is possible when only two or three modes participate in the response and their natural frequencies are widely separated.

To find the mean square response of the system, equation 14.50 must be integrated over the frequency range in the usual way. Suppose the mean square *stress* is required in the system. Denote by $\sigma_1(x)$ and $\sigma_2(x)$ the stresses due to unit displacements in modes 1 and 2, which are the only modes participating and which are well separated in frequency. The stress spectrum is given by

$$S_\sigma(\omega) \doteq \frac{A^2 S_p(\omega)\sigma_1^2(x)j_{11}^2}{|Z_1(\omega)|^2} + \frac{A^2 S_p(\omega)\sigma_2^2(x)j_{22}^2}{|Z_2(\omega)|^2} \quad (14.51)$$

Integrating this from $\omega = -\infty$ to $\omega = +\infty$, we find the mean square stress to be

$$\overline{\sigma^2(t)} \doteq \frac{\pi}{2} A^2\left\{ \frac{\omega_1 S_p(\omega_1)\sigma_1^2(x)j_{11}^2}{K_1^2\zeta_1} + \frac{\omega_2 S_p(\omega_2)\sigma_2^2(x)j_{22}^2}{K_2^2\zeta_2} \right\}. \quad (14.52)$$

References

1. S. H. Crandall. *Random Vibration* (S. H. Crandall, ed.), chaps. 1 and 4. Cambridge, Massachusetts: M.I.T. Press.
2. Y. K. Lin. Stresses in continuous skin stiffener panels under random loading, *J. Roy. aerospace Soc.*, **29**, 67 (1962).
3. A. Powell. On the approximation to the 'infinite' solution by the method of normal modes for random vibrations, *J. acoust. Soc. Amer.*, **30**, 1136 (1958).
4. A. Powell. On the fatigue failure of structures due to vibrations excited by random pressure fields, *J. acoust. Soc. Amer.*, **30**, 1130 (1958).

CHAPTER 15

Acoustically Excited Modes
of Vibration

15.1 Introduction

The general random vibration theory of chapter 14 may be used, in principle, to estimate the response of any acoustically excited structure. The practical problems associated with its use are, of course, the determination of the modes of vibration that will be excited and the determination of the pressure field characteristics. In this chapter we shall consider the type of mode of vibration that can be excited, and also some of the interaction effects that exist between the structure and the surrounding (or enclosed) medium. We shall also consider in a simple way the effect on structural stresses of increasing the thickness of the structural material.

The mean square stress in the structure in just one mode of vibration has been shown to be given by:

$$\int_{-\infty}^{\infty} \frac{S_p(\omega)A^2 j_{rr}^2 \sigma^2(x)}{|Z_r(\omega)|^2}\, d\omega \tag{15.1}$$

This will have a large value for those modes which have a large value of both the joint acceptance, j_{rr}, and the pressure spectral density $S_p(\omega)$ in the region of the modal natural frequency. For this reason, usually we need consider only those modes having frequencies between 100 and 1000 c/s, and very often the frequency range is narrower still. Within this range a typical structure has many different natural modes of vibration. For convenience, we may consider two groups (a) The 'overall modes' and (b) The 'local modes'. The distinction really relates to the methods of analysis that may be used to calculate them, rather than to essential differences in their natures.

15.2 The modes of vibration

15.2.1 *The overall modes of vibration of a reinforced cylinder*

The modes of vibration of a real fuselage structure are influenced by such non-uniformities as the stringer spacing, frame spacing, taper, skin-thickness, etc. In this section we shall consider only a uniform cylinder, with uniformly spaced stringers and frames of the same section throughout. Such a model gives considerable insight into the modes which can be excited by a jet-noise field.

The so-called 'overall modes of vibration' of this model involve distortion of the whole cross-section and length. The distortion is periodic around the circumference and along the length, as shown in Fig. 15.1. The frames, stringers and skin all take part in the distortion.

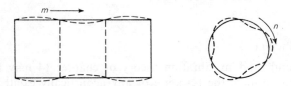

Fig. 15.1   Diagram illustrating an 'overall' mode of vibration of a cylinder. (m = number of half-waves along the length; n = number of full-waves around the circumference.)

Natural modes of vibration can occur (ideally) with any combination of values of n and m. Fig. 15.2 shows the frequencies at which they occur for a particular stiffened cylinder, 30 ft long and 10 ft in diameter.[1,2] The

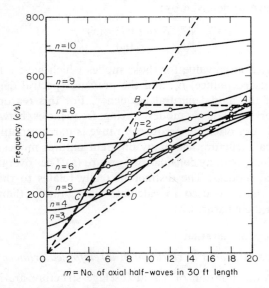

Fig. 15.2   The natural frequencies for a particular stiffened cylinder.

intersection of a constant n line with the vertical m lines gives the corresponding natural frequency.

An important feature of these modes is the unusual variation of the natural frequency with n for most of the values of m that are shown. For example, for $m = 10$, as n increases from 2 to 4, the natural frequency decreases; thereafter, the frequency increases with increasing n. This implies that with very large circumferential wavelengths, the frequency decreases with decreasing wavelength. With small circumferential wavelengths, the frequency increases with decreasing wavelength in the usual way. This is a consequence of the strain energy of distortion being predominantly extensional for the long wavelength modes, and predominantly flexural for the short wavelength modes.

These overall modes are susceptible to excitation by the acoustic coincidence effect. If plane harmonic sound waves impinge on the side of the cylinder, the wavefronts being inclined at, say, 20° to the cylinder side, then the trace-wavelength of the sound waves for any given frequency is given by the line OB on the last diagram. If the wave-fronts are inclined at 45°, then the corresponding line is OA. Now in a general random incident field, there will be a randomness in the inclination of the wavefronts, as well as in the pressure fluctuation. If a particular sound field has inclinations of all values between 20° and 45°, and frequencies of all values between 200 c/s and 500 c/s, then all the modes contained within the region ABCD of the diagram can be excited by this field by coincidence. For each of these modes, there is a component of the incident field which has the same frequency as the modal natural frequency and the trace-wavelength equal to the axial wavelength of the mode.

It is seen that there are a large number of these modes in this region. Some early acoustic fatigue failures seemed to be due to vibration in these modes, and remedies based on this supposition were quite effective. However, investigations on another aeroplane at a later date (see chapter 16) did not indicate their importance. The sound field considered in the above argument is very idealized, and is only partially realized in the medium-to-far field of a jet. The further away from the jet, the narrower is the range of inclination angles, so that the angle AOB on the diagram closes up with increasing distance from the jet. Close to the jet, the concept of the trace-wavelength must be replaced by the narrow-band correlation wavelength, determined from filtered correlation measurements centred on the natural frequency of the mode. This correlation wavelength becomes significantly dependent on the jet velocity in the near field of the jet, due to the convection with the jet of the hydrodynamic pressure fluctuations.

Another important feature of these modes is their ability to couple readily with the standing waves of the air within the cylinder. This has already been mentioned in chapter 3. Its effect is not likely to be important in well soundproofed interiors, but in unsoundproofed regions the associated internal noise levels could be very damaging to, say, electronic equipment.

15.2.2 *The 'local' modes of a reinforced cylinder*

We consider now a section of the fuselage between two adjacent frames. In the local modes, the frames do not distort apart from some twisting of their flanges. Consequently, nodal lines exist around the cylinder at the frame positions. The section of skin between the frames is reinforced by stringers which can bend and/or twist as the skin vibrates.

Consider first a long section of skin plating (assumed to be flat), which is simply supported across its width at constant intervals (see Fig. 15.3). The long sides may also be assumed to be simply supported.

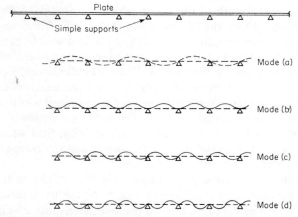

Fig. 15.3   Diagram illustrating some 'local' modes of a multi-supported plate.

The modes of vibration of such a plate are well known. The fundamental mode is sinusoidal with a longitudinal half-wavelength equal to the support spacing as shown in Fig. 15.3, mode (a).

The lateral half-wavelength (across the plate) is equal to the plate width. Each bay vibrates in anti-phase with its neighbour, and as the displacement pattern in one bay is symmetrical about the panel centre, we call this a 'symmetric mode'. Another symmetric mode occurs with all bays vibrating in-phase; as in Fig. 15.3, mode (b).

Each of these modes has its anti-symmetric overtone mode, as shown in Fig. 15.3, modes (c) and (d).

When the simple supports are replaced by flexible stringers which per-mit deflection in the direction perpendicular to the plate, and restrain (to some extent) the rotation of the plate, similar modes to these still occur. Lin[3] has discussed them, and concludes that rotation and flexure of each stringer do not occur together. Modes (a) and (b) are now as shown in Fig. 15.4 where they are denoted by A and B respectively.

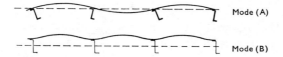

Fig. 15.4 Two local modes of a plate with constraints at the supports.

The first of these we call the *symmetrical stringer torsion mode*, and the second we call the *symmetrical stringer bending mode*. Despite this, it is possible that under certain extreme conditions there will be very little stringer torsion or bending, and most of the energy of vibration will derive from the plate deflection.

The natural frequencies of modes (A) and (B) may be greater or less than the frequencies of the corresponding modes (a) and (b), depending on whether the natural frequencies of the pure torsional and flexural modes of the stringers are greater or less than those of modes (a) and (b).

Modes (c) and (d) also have their counterparts in the stiffened plate, but they do not appear to be at all important in connection with acoustic fatigue.

Mode (A) may be described as having a semi-wavelength equal to the stringer spacing. There exists another set of modes having greater 'wave-lengths' than this, and which are 'borderline' modes between the 'local' and 'overall' modes. For instance, a three-bay plate has the following lowest order modes:

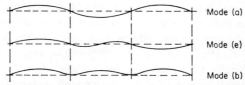

Fig. 15.5 Some local modes of a three-bay plate.

The modes of the stringer-reinforced plate which correspond to modes (a) and (b) are those of Fig. 15.4, (A) and (B). Corresponding to (e) we now have the mode of Fig. 15.6.

Stringers 1, 4 and 7 are twisting, stringers 2, 3, 5 and 6 are twisting *and* bending. The half-wavelength may be said to be equal to $1\frac{1}{2}$ bay widths.

The frequency of this mode is found to be in between those of (A) and (B). Further modes exist with half-wavelengths of 2, 2½, 3, etc. bay widths, but the corresponding natural frequencies still lie between those of (A) and (B). Modes (A) and (B) are therefore called the 'bounding modes', giving the lower and upper bounds respectively to the natural frequencies of these numerous modes of vibration. These other modes may be called 'intermediate modes'.

Fig. 15.6   An 'intermediate mode' of a stiffened plate.

The bounding frequencies are obviously quite critically dependent upon the torsional and flexural stiffnesses of the stringers. If the torsional stiffness is low, the lower bounding frequency may not be much different from that of mode (a). On the other hand, if the torsional stiffness is very high, rotation of the plate along the stringer line will be restrained or largely prevented and the mode will be of the form shown in Fig. 15.7.

Fig. 15.7   The stringer torsion mode for a plate with stringers of high torsional stiffness.

The frequency of this mode is very close to that of mode (b).

From these simple considerations, we deduce that the bounding frequencies will be widely separated if the stringer torsional stiffness is low, but they will be close together if this stiffness is high. In the latter case, the numerous intermediate modes will be closely packed together in a narrow frequency band.

So far, we have considered only flat reinforced plates. If the plates are curved (as on a real fuselage), the same types of local and intermediate mode occur. The principal effect of curvature is to increase the natural frequencies above those of flat plates. The effect on the stringer torsion mode is not very marked, but the frequency of the stringer bending mode increases quite rapidly with increasing curvature.

Just as the overall modes of the cylinder were susceptible to acoustic excitation by the coincidence effect, so also are the intermediate modes described above. Suppose that acoustic plane waves impinge on the plate at the incidence angle θ_i, moving along the plate from left to right, and downwards at the same time. The trace velocity of the acoustic field along the plate is $a \operatorname{cosec} \theta_i$, and the trace wavelength is $(a \operatorname{cosec} \theta_i)/f$, where f is the frequency. (See chapter 3.)

Now the intermediate modes each have their characteristic half-wavelength, and their own natural frequency, f_i. Twice the half-wavelength will be called the 'pseudo-wavelength' of the mode, and will be denoted by λ_{pi}. The word 'pseudo' is used in order to distinguish the wavelength from that of a pure sine mode. It is, in fact, a periodic distance.

If the plate were infinite in extent, the intermediate mode would become a travelling wave having the same pseudo-wavelength at the same frequency. Its wave velocity would be $(\lambda_{pi}f_i)$. If this 'pseudo wave velocity' is equal to the acoustic trace velocity, then the wave is readily excited by the acoustic field. The condition for this is

$$\frac{a \operatorname{cosec} \theta_i}{f_i} = \lambda_{pi}f_i$$

i.e.

$$\lambda_{pi}f_i^2 = a \operatorname{cosec} \theta_i.$$

Now $a \operatorname{cosec} \theta_i$ varies from a to infinity as θ_i varies from 0 to $\pi/2$. The lowest value of $\lambda_{pi}f_i^2$ is usually that corresponding to the stringer torsion mode for which $\lambda_{pi} = 2l$ (l = bay length) and $f_i = f_t$, the lower bounding frequency. Hence

$$(\lambda_{pi}f_i^2)_{\min} = 2lf_t^2.$$

The highest value of $\lambda_{pi}f_i^2$ is that of the stringer bending mode, for λ_{pi} for this mode is very large. Indeed it can be argued that for an infinite beam, λ_{pi} for this mode is infinite. The term $\lambda_{pi}f_i^2$ is then infinite. Hence $\lambda_{pi}f_i^2$ varies between the limits $2lf_i^2$ and infinity, for the different pseudo-waves, or intermediate modes.

It follows, then, that above the frequency given by

$$\lambda_{pi}f_i^2 = a$$

all the pseudo-waves can be excited by coincidence by acoustic waves at the appropriate incidence. If θ_i is fixed, and the plate is finite, then the intermediate mode which will dominate the response is that for which $\lambda_{pi}f_i^2$ is closest to $a \operatorname{cosec} \theta_i$. This feature has been identified experimentally.[4]

15.2.3 *Modes involving rib distortion*

The ribs of tailplane and control surface structures play an important part in some of the modes of vibration of the skin plate surfaces. This has been discussed briefly by Clarkson[5] who has considered a simple system of coupled beams arranged in the form shown in Fig. 15.8.

The top and bottom beams represent the top and bottom skins of the structure and the vertical beams represent the ribs. The fundamental mode (f) of the system is similar to mode (A) of Fig. 15.4, but the stringers are now replaced by the ribs, which must bend to accommodate the top and bottom skin distortion.

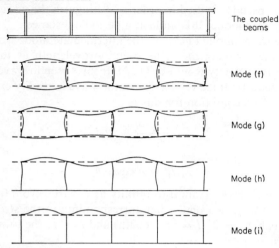

Fig. 15.8   Modes of a skin and rib type of structure.

If the top and bottom skins have different stiffnesses, there may be much less bending deflection in one skin than in the other, and the rib deflection will be similar to that of a fixed-pinned beam, as shown by mode (g).

If the ribs are relatively much stiffer than the plates, then the mode of skin vibration becomes similar to the type A mode (section 15.2.2) with torsionally stiff stringers, as shown by mode (h).

Modes similar to type (b) still occur, with adjacent bays vibrating in phase. Now, however, there is no rib distortion, and no plate normal displacement at the rib intersection (see mode (i)).

15.2.4 *Modes involving distortion of stringer cross-section*

Under the action of the inertia forces of the local modes of section 15.2.2 the cross-section of a stringer must distort to a certain extent. When a stringer vibrates in a torsion mode, the associated inertia forces must bend the cross-section in the manner shown in Fig. 15.9 (a).

This is only of importance for open section stringers, and can cause considerable bending stresses at the root of the stringer web. When the stringer vibrates in a flexural mode, the inertia forces are in the other (normal) direction; shown in Fig. 15.9 (b).

It is extremely difficult to analyse these modes, as it is necessary to consider them in close conjunction with all the local and intermediate modes of section 15.3. The flexural stiffness of the flange of the stringer in the direction parallel with the plate surface is an important parameter, and adds to the number of variables which must be included in a general analysis.

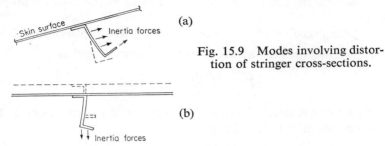

(a)

Fig. 15.9   Modes involving distortion of stringer cross-sections.

(b)

15.3 The effect on randomly excited stresses of increasing skin thickness, etc.

We shall assume in this section that when the skin thickness is increased so also are the dimensions of the reinforcing members. The flexural stiffness of the skin is proportional to (skin thickness, t)^3. We shall assume that the stiffnesses of the reinforcing are also increased such that the generalized stiffness of the modes increases in proportion to t^3. Furthermore, we shall assume that the mass of the whole structure is proportional to t (these last two assumptions are not, in fact, exactly reconcilable).

Now if just one mode of vibration is excited by the noise, the mean square bending stress in the plate is given by equation 15.1. Provided the product $S_p(\omega)j_{rr}^2$ (which is proportional to the spectrum of the generalized force) does not vary rapidly in the region of resonance, the integral may be evaluated as in section 14.1. The root mean square stress is therefore proportional to

$$\left\{ \frac{S_p(\omega_r)j_{rr}^2(\omega_r)\omega_r\sigma_r^2(x)}{K^2\zeta} \right\}^{1/2}. \tag{15.2}$$

Now the stress per unit displacement in the rth mode, $\sigma_r(x)$, is proportional to t, and K is proportional to t^3. Further, ω_r is proportional to t. The r.m.s. stress is therefore proportional to

$$[S_p(\omega_r)j_{rr}^2(\omega_r)]^{1/2}\, t^{-3/2}\zeta^{-1/2}. \tag{15.3}$$

It is not known in general how the damping ratio, ζ, is likely to vary with t. The results quoted in chapter 18 related to the effect of increasing the thickness of the skin plating on a substructure of reinforcing members which were not stiffened as the skin thickness increased, and so are not strictly applicable here. In any case they showed that the damping could increase or decrease with thickness. We shall therefore assume in this crude investigation that ζ does not change with t. It is then found that the plate stress is inversely proportional to $\zeta^{1/2}t^{3/2}$.

Consider now the inertia forces exerted on the reinforcing members by the vibrating skin. The power spectrum of these inertia forces is proportional to

(skin mass)^2 × ω^4 × power spectrum of the displacement.

The mean square inertia force is therefore proportional to

$$\int_{-\infty}^{+\infty} \frac{S_p(\omega)j_{rr}^2 A^2 \omega^4 m^2}{|Z_r(\omega)|^2}\, d\omega. \tag{15.4}$$

If the damping is small, this may be evaluated in the same way as the integral of equation 15.1, yielding

$$S_p(\omega_r)j_{rr}^2 A^2 \frac{\omega_r^5 m^2}{K^2\zeta} = S_p(\omega_r)j_{rr}^2 A^2 \frac{\omega_r}{\zeta}. \tag{15.5}$$

Since ω_r is proportional to t, this is seen to be proportional to t (unless ζ varies with t). The root mean square inertia loads are therefore proportional to $t^{1/2}$, and *increase* with increasing skin thickness.

The inertia force exerted on the reinforcing by the plate is transmitted by the attachment rivets, which therefore carry an increasing load as the thickness of the skin increases. Any moment transferred by the skin to the stringers (in a stringer torsion mode, say) is also proportional to this inertia force.

The above derivations are based on the assumption that the spectrum of the generalized force ($S_p(\omega)j_{rr}^2 A^2$) does not vary rapidly in the region of the modal resonant frequency. If high order modes are excited by the coincidence effect, we know that this assumption is not satisfied, for j_{rr}^2 can vary *very* rapidly. (See section 13.4.) It is not possible to derive simple expressions for the mean square stress when the joint acceptances are of this general form, but we can examine a limiting case which occurs with very high order modes. Consider the curves of j_{rr}^2 and $1/Z(\omega)^2$, as illustrated in Fig. 15.10.

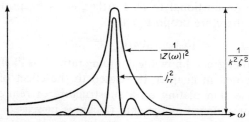

Fig. 15.10   Curves of the functions $1/|Z(\omega)|^2$ and j_{rr}.

In such a case as this, the important peak in the j_{rr}^2 curve is 'contained within' the resonant peak of the $1/Z(\omega)^2$ curve, and the integral of equation 15.1 can be approximated by

$$\frac{S_p(\omega_r)A^2\sigma_r^2(x)}{|Z(\omega_r)|^2} \int j_{rr}^2 \, d\omega$$

$$= S_p(\omega_r)A^2 \int j_{rr}^2 \, d\omega \, \frac{\sigma_r^2(x)}{K^2\zeta^2}, \tag{15.6}$$

in which it has been assumed that $1/|Z(\omega)|^2$ does not vary rapidly in the region of the principal peak of j_{rr}^2. From this, we see that the r.m.s. stress is inversely proportional to $\zeta.t^2$. Increasing both the damping, ζ, and the skin thickness t is now more effective in reducing the stress than in the former case when the spectrum of the generalized force did not vary rapidly.

Under the same conditions the r.m.s. inertia force is found to be inversely proportional to ζ, and is independent of the skin thickness.

In the intermediate cases between the two extremes considered above the stresses and inertia forces will be very roughly inversely proportional to $\zeta^a t^b$, where a lies between $\frac{1}{2}$ and 1, and b lies between 3/2 and 2 for the stresses and between 1 and 0 for the inertia forces.

15.4 The effect of the surrounding medium on the modes of vibration

It is generally assumed that the displacement patterns of the natural modes are unaffected by the surrounding and enclosed media. However, the mechanical impedances of the modes can be drastically changed, as shown in chapter 3 when a very simple system was considered. If the enclosed region contains a great deal of sound-absorbing material, then the effect of the enclosed medium will be primarily resistive, i.e. it increases the damping of the modes. If there is no sound absorption, then the medium will add to the stiffness or mass of the modes.

It is possible that the additional stiffness will be *infinite* at frequencies at which certain standing waves occur in the enclosure. These standing wave patterns have nodes at the structural surface. If they are excited acoustically from the outside under harmonic conditions, the acoustic pressure on the inside surface of the structure is the same as on the outside surface, but, acting in the opposite direction, effectively cancels the effect of the external pressure and no structural motion results.

It is also possible that the additional stiffness will be *zero*. Under these conditions, the standing wave excited within the enclosure has velocity anti-nodes and pressure nodes at the structural internal surface. The acoustic pressure at some other points inside the enclosures will certainly be very high.

Such 'cavity resonances' can occur within cylinders (see section 3.6) or within tailplane and control surface structures, e.g. corresponding to the tailplane modes (f), (g) and (h), etc. (section 15.2.3). There exist standing waves within the structure which will cause high internal noise levels at certain frequencies. There are also other standing waves with nodes at the surfaces which will inhibit the skin motion at other frequencies, yet still cause large sound pressures to exist within the structure. Under some conditions, the cavity resonance effects can cause the modal frequency to be considerably different from that of the structural mode in vacuo.

Outside the structure, the motion of the structure causes sound waves to be radiated away from the surface. Since, in general, these waves are not plane waves, the pressure at the skin surface due to this radiation has components in phase with the velocity, and also in quadrature with the velocity. The former components constitute the acoustic damping pressure and are discussed in greater detail in chapter 18. The latter components are in anti-phase with the structural acceleration, and constitute the 'virtual inertia' of the medium in conjunction with the structure. This virtual inertia is more important for long wavelength modes.

15.5 The effect of the structure on the noise field

The remarks of section 15.4 apply equally well to boundary layer induced vibration as to jet noise induced vibration. In this section we shall only consider acoustic noise fields such as those around jets or rockets.

We saw in chapter 3 that when plane waves impinge on a wall, the effective exciting pressure on the wall was *twice* that of the incident field, due to the effect of total reflection of the wave. A similar effect takes place when sound waves impinge on a fuselage or a tailplane. Those pressure components having wavelengths less than the characteristic dimensions of the body (diameter, or chord) are effectively doubled by the reflection effects. If the wavelength is much greater than these dimensions, 'scattering' of the incident wave occurs due to non-uniform reflection. The effective pressure is still increased, but not by as much as a factor of two. However, the wavelengths of the frequency components of greatest interest are usually much shorter than the structural dimensions. The noise pressures used for response calculations should therefore be twice those measured in free-field conditions.

The effect of structural vibration has been stated (section 15.4) to radiate acoustic waves away from the surface. When the associated pressures are added into the doubled incident pressure, it is found that the resultant pressure is *less* than the double incident pressure. (See also section 3.6.) This accounts for 'reflection factors' of less than two which have been quoted as a result of measurements of jet noise close to an aeroplane structure. The actual effective exciting pressure is still twice the incident pressure.

References

1. D. J. Mead. The effect of a damping compound on jet-efflux excited vibrations: Part II, *Aircraft Engineering*, **32**, 106 (1960).
2. P. R. Miller. The free vibrations of a stiffened cylindrical shell, Aeronautical Research Council, London, R & M 3154 (1958).
3. Y. K. Lin. Free vibrations of continuous skin stringer panels, *J. appl. Mech.*, **27**, 669 (1960).
4. D. J. Mead. The damping of stiffened plate structures, chapter 26 of *Acoustical Fatigue in Aerospace Structures* (Walter J. Trapp and Donald M. Forney, Jr., eds.), Syracuse Univ. Press, 1965.
5. B. L. Clarkson. The design of structures to resist jet noise, *J. Roy. Aeron. Soc.*, **66**, 603 (1962).

CHAPTER 16

Response of Practical Structures to Noise

16.1 Introduction

Chapter 14 has presented a method of analysing the response of a continuous structure to random acoustic loading, the total response having been expressed as a sum of responses in the many normal modes. If all the normal modes were known in detail the method could be used directly to calculate stresses. Unfortunately, however, it is not yet practicable to calculate all the significant modes, nor their natural frequencies and dampings. Approximate methods of analysis must therefore be sought.

There are two principal approximate methods which may be considered. In the first, the number of allowed normal modes of the structure is reduced to a minimum. The problem now is to know which are the significant modes to retain. This has been investigated by experimental studies on full-scale aeroplane structures in jet-noise fields, and a major portion of this chapter is devoted to describing these studies. From the evidence presented, it is shown that the response of some structures can be calculated to an acceptable degree of accuracy by restricting the motion to just one normal mode of vibration. Other simplifications may also be made in the analysis, and these are described.

The second approximate method goes almost to the other extreme, and assumes that so many modes contribute to the response that a study of individual modal responses is completely out of the question. The vibration field is assumed to be so complex that it is best described in terms of its associated energies (kinetic, potential and dissipative). Furthermore, the total energy is assumed to be shared out equally amongst the modes. This method obviously requires different (or additional) concepts from those already presented, and is described in the last sections of this chapter.

16.2 Purpose and description of experiments

The total random response of a jet-excited structure is multimodal. At any frequency, however (particularly at a resonant frequency), the response

330

may be dominated by a single mode. In order to verify such a feature, it is essential to examine the phase-relationships between the deflections or stresses at different parts of the structure at the predominant resonant frequencies. Firstly, the resonant frequency must be identified and then the corresponding vibration pattern must be analysed in an appropriate manner. The knowledge so obtained can then be used to assess the validity of the assumptions made in the approximate methods of calculating response, natural frequencies and modes. It also gives insight into the most effective method of improving the structure if the stress levels are too high, whether by adding damping or in other ways.

Experimental measurements are described which have been made on three different conventional structures:

ITEM 1 A rear fuselage, having a thin skin, Z section stringers and relatively stiff frames.

ITEM 2 A tailplane, having a thick skin, closed-section stringers and relatively thin ribs.

ITEM 3 An elevator (stabilizer) having a thin skin and relatively stiff ribs.

The analysed results have shown predominant modes of vibration of the skin panels. A rather less complete picture has been obtained of the stringers, frames and ribs although some inference has been made about the predominant modes of vibration which occur. Details of the test structures, gauge positions and test results are given in two reports.[1,2] In this chapter only the basic method, the general trend of results and the conclusions are outlined.

The experimental work has been carried out on two test installations. Items 1 and 3 were investigated on the Caravelle test rig built by Sud Aviation at Toulouse and the tailplane measurements were made on a tailplane test section mounted behind a jet engine at the R.A.E. Farnborough.

In the Caravelle test rig a section of the aircraft consisting of most of the structure aft of the pressure dome was mounted at the correct height above the ground and one engine was fitted. Structure aft of the jet nozzle included part of the rear fuselage, and also the fin, rudder, tailplanes and elevators. The first part of the testing was directed towards obtaining information about the modes of vibration of the fuselage skin panels, and therefore two rows of skin panels between frames were strain gauged. The stringers and frames were also strain gauged but with the limited instrumentation it was not possible to fully investigate the form of vibration of this support structure. On the elevator it was not possible to attach any gauges to the internal structure and therefore attention was again concentrated on the study of skin panel vibrations. Gauges were placed at the centres of panels forming two rows on the upper surface of the elevator (thinner skin than the lower surface and thus higher stresses) and also on the skin above the ribs.

To carry out the tests the Rolls-Royce 'Avon' engine was run at full take-off thrust and the signals from the strain gauges were recorded on twin-track magnetic tape in pairs. As will be discussed later the determination of mode shape involves the comparison of simultaneous signals from gauges on adjacent pieces of structure.

The tailplane test section consisted of a typical piece of conventional tailplane structure of about 8 ft chord (including the elevator) and 5 ft span made as a parallel sided section. The main tailplane structure consists of a leading edge diaphragm, front spar and a rear spar with ribs at approximately 11 in pitch. The stringers are of the closed section top hat type and are bonded to the skin at approximately 6 in pitch. This test specimen was mounted vertically behind a de Havilland 'Ghost' jet engine at the R.A.E., Farnborough. It was positioned close to the nozzle in order to obtain as high a noise level on the structure as possible with the relatively low thrust engine.

Skin panels were strain gauged in much more detail in order that the mode shape of individual panels might be determined. Some gauges were placed on the stringers and on the rib bend radii.

16.3 Methods of analysis

In order to analyse the form of vibration taking place in these structures it is first necessary to define the strain at significant parts of the structure. The strains in different parts can then be related to one another to pick out the mode shapes and the extent of coherent vibration. Random strains of the type excited by jet noise can only effectively be described by their power spectra and amplitude distributions. A considerable amount of work has been done by Sud Aviation to measure the strain amplitude distribution at many points in the Caravelle structure. These results show that the distribution approximates closely to Gaussian. For most applications therefore it will be adequate to assume a Gaussian distribution.

The power spectrum of the strains of interest can be found either by the use of electronic filters or by transformation of the autocorrelogram as discussed in detail in chapter 4. It should be borne in mind here that the damping of the structure is very low (of the order of 2% critical) and therefore it is necessary to use narrow band filters of band width less than 2% if a reasonable representation of the true spectrum is to be obtained.

The phase measurements can be made by using a reference strain gauge (say, in the centre of a panel) and other gauges for comparison on adjacent panels and support structures. To make the comparison between any two strain signals the signals must first be filtered so that only one of the resonances is studied at a time.

The phase relationship between these two filtered signals will not, in general, be constant, but will fluctuate randomly about a mean value. The measurement of this mean value requires statistical methods and this is performed by correlation techniques. The normalized correlation function (i.e. the correlation coefficient) is the cosine of the mean phase angle. If two filtered strain signals are exactly in phase *over the whole of the filter bandwidth*, then the correlation coefficient is exactly $+1$. If the phase relationship varies through the bandwidth, or if the phase difference is constant but neither zero, nor $180°$, then the correlation coefficient is between $+1$ and -1.

The filtering of the signals before correlating presents some difficulty as the peaks in the spectra are narrow and some may be 1% relatively close together. A filter bandwidth of the order of 1% is therefore required. Unfortunately the two signals must be passed through identical filters to avoid spurious correlation due to relative phase shifts in the filters. Normal narrow band filters will not be matched in this way. To overcome this difficulty two methods are possible.

(1) Use of the same filter in both channels.

By using two tape recorders it is possible to play back the original signals, one of which passes through the filter before being re-recorded on the second recorder. The next step is to play back from the second recorder, this time filtering the second signal before re-recording on the first recorder. This system works successfully if the signals to be compared have peaks in the spectrum occurring at the same frequencies. If there is any frequency difference the differential phase shift in the side band of the filter will lead to spurious correlation readings.

(2) Transformation of the cross-correlation coefficient.

This process, described in chapter 4, gives the cross power spectrum which can be normalized by use of the individual power spectra to give a correlation spectrum. This is in effect a narrow band plot of correlation coefficient against frequency. This method eliminates the differential phase shift of the electronic filter but again long time delays may be needed if high accuracy is to be achieved.

An estimate of the damping of the modes may be possible where the peaks in the spectra are separated in frequency. It has been shown in chapter 14 that the autocorrelogram of the output of a single degree of freedom system excited by white noise takes the form of a damped cosine wave. The damping of the autocorrelogram gives directly the damping present in the system. In the practical case if the resonant frequencies are separated in frequency it may be possible to isolate the predominant ones with a relatively wide filter (say ⅓ octave) and measure the decay of the autocorrelogram of the filter output.

16.4 Stress spectra and cross correlations

The results for the three specific structures described earlier are now discussed, as it is considered that they give an insight into the type of vibration which occurs in structures of conventional stiffened skin design. Even in the latest designs such as the supersonic transport aircraft, as large a part of the structure as possible will be of conventional design for ease and cheapness of manufacture and maintenance.

16.4.1 *Fuselage*

The rear fuselage skin, 0.048 in thick, has a radius of curvature of approximately 40 in at the measuring section. The open Z section stringers are at approximately 4 in pitch and the channel section frame pitch is 7 in. The response of the skin panels has been studied in some detail but it is only possible to indicate general trends for the stresses in the support structure.

Skin panels

The stress spectra for the row of skin panels between two frames all have a similar form—a typical spectrum is shown in Fig. 16.1. There is

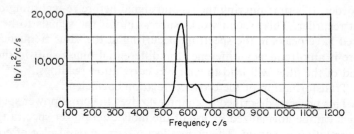

Fig. 16.1   Fuselage skin stress spectrum.

very little stress in the frequency range up to about 500 c/s and then there is a main panel resonance in the frequency range 550 c/s to 750 c/s. At higher frequencies the spectra differ from one another. Some have a relatively flat spectrum from 700 to 1,000 c/s at a power spectral density level about half to one-third of the main peak value, suggesting a combination of several closely spaced resonant frequencies. In other cases however there is very little stress at these higher frequencies. The frequency at which the main resonance occurs varies from panel to panel due to variation in dimensions in the tapering rear fuselage, and the stiffness of the end fixing. Thus no more than three adjacent panels have been found to have the same frequency. Cross-correlation measurements have been made where the panels to be compared show the same resonant frequency. A typical cor-

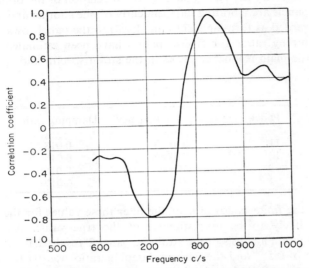

Fig. 16.2   Correlation spectrum of adjacent panel strains.

relation spectrum between two adjacent gauges is shown in Fig. 16.2. The results for the row of eight panels can be summarized as shown in Table 16.1.

Table 16.1

Panel	Width (in)	Main resonant frequencies (c/s)			Phase relationship	
1	4.33			715	out of phase at 715	in phase at 715
2	3.55			715	
3	4.41		610	715	out of phase at 610	
4	3.55		610		
5	4.33	580	610		out of phase at 580	in phase at 580
6	3.55	580			
7	3.35	580	625		
8	4.33		625		

From these results we can see that the fundamental panel mode is one in which the stringers are twisting such that adjacent panels vibrate out of phase. But because of physical differences in the panels typical modes of vibration show only two or three panels having appreciable motion at any one resonant frequency.

A single check on adjacent panels across a frame shows that at the main resonant frequency the panels have a low correlation of the order of 0.05. Thus the panels are vibrating independently of one another and the frames can be regarded as forming rigid boundaries to the panel rows.

The damping ratios for typical panels have been estimated from the width of the main resonance peaks in the strain spectra and are given in Table 16.2.

Table 16.2

Panel	Resonant frequency	Damping ratio
11	590 c/s	0.018
10	720 c/s	0.018
9	705 c/s	0.018
2	715 c/s	0·017

Due to the finite bandwidth of the filter these values for the damping ratio are likely to be overestimates of the true value. An extended autocorrelogram analysis of the strain in panel 2 resulted in an effective bandwidth of 0.5% and the derived damping ratio was 0.013. Thus the over-estimation of the damping in Table 16.2 is likely to be of the order of 0.005. Measurements by Douglas[3] and other companies quote values of the same order as those given above but again the accuracy of their results is likely to be dependent on the bandwidth of the filter used in the power spectral density analysis. As is discussed in section 16.5 even the values obtained after corrections have been made for the filter bandwidth seem high for the types of mode being excited. The difficulty is probably due to the fact that one peak in the power spectrum does not necessarily arise from only one mode but could be due to several modes close together in frequency appearing as a single broad peak instead of three closely spaced narrow peaks. Methods of resolution in this case are discussed in section 16.5.

Since it has been shown that the fuselage panels are vibrating in a relatively simple manner it was thought worthwhile to calculate the resonant frequencies. As outlined in chapter 15, Lin[4] has suggested a method of dealing with certain modes of vibration of a row of panels coupled to their supporting stringers and simply supported on two rigid frames. The solution gives the frequencies and mode shapes for the mode in which adjacent panels vibrate out of phase (i.e. stringers twisting only) and for the mode in which adjacent panels vibrate in phase (i.e. stringers bending only). These two modes form respectively the lower and upper limits of a band of modes which have wavelengths of two or more panel widths. The calculated frequency of the lower bounding mode which corresponds with the predominant mode identified in the experimental investigation is:

Panel width	*Calculated frequency, c/s*
4.33	585
3.55	626

The calculation is for the case of simple supports at the frames whereas in practice there must be some fixing at these edge supports. The effect of fixing will be to increase the frequency of the mode and therefore the computed frequencies are seen to be in reasonable agreement with the experimental results. The next simple mode shape is one in which the stringers bend and adjacent panels vibrate in phase. But because of the difference between torsional and bending stiffness of the stringers in this structure the stringer bending mode has a frequency of the order of 1,100 c/s. There was no evidence of a marked resonance in the measured spectrum at this frequency but cross-correlation measurements show that adjacent panels are in phase around this frequency. In structures with closed section stringers these two frequencies become much closer together.

Stringers

The stringers in this case are of J cross-section and it was only possible to attach gauges to the flat web. Fig. 16.3 shows a typical stringer stress

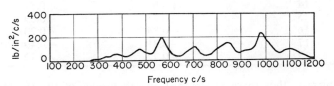

Fig. 16.3   Fuselage stringer stress spectrum.

spectrum but as the main panel mode causes the stringer to twist the direct stresses measured on the stringer web are very low. There are some peaks in the 1,000 c/s region which may be associated with the stronger bending mode but from the limited measurements no precise conclusions can be drawn.

Frames

Stress spectra have been measured at positions on one frame corresponding to the measurements on the row of eight panels. The frame spectra (Fig. 16.4) all show to a greater or lesser extent two resonant frequencies, one at 180 c/s and the other at 250 c/s. In the higher frequency range

600–700 c/s there is little response which can be associated with the inertia loading from the predominant panel mode. This is because in this mode adjacent panels are vibrating out of phase and therefore the net inertia

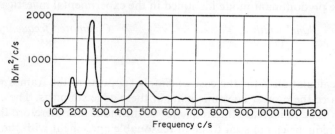

Fig. 16.4   Fuselage frame stress system.

force on the supporting frames will be small. Correlation measurements summarized in Table 16.3 show that the frame stresses are in phase over at least five stringer bays at 180 c/s and four bays at 250 c/s.

Table 16.3

Gauge no.	Correlation coefficient	
	$\frac{1}{3}$ octave: 160 c/s	$\frac{1}{3}$ octave: 250 c/s
12 (reference)	+ 1.0	+ 1.0
13	+ 0.93	+ 0.70
14	+ 0.95	+ 0.21
15	+ 0.75	+ 0.09
16	+ 0.34	− 0.20
17	− 0.31	− 0.49
18	− 0.72	− 0.72

This suggests that these modes are overall modes of the fuselage itself. Other resonant frequencies appear in only a few of the spectra and are therefore likely to be associated with very localized vibrations.

16.4.2 *Tailplane*

The general description of the tailplane structure has been given in section 16.2 and more details are given in reference 2. The main object of the tests in this case was to study in some detail the skin panel modes of vibration for a structure with a different type of stiffener from that used in the curved fuselage structure discussed in the previous section. The tailplane stiffener was of the closed section top hat type which has a very high torsional rigidity compared with the open section Z stiffeners in the Caravelle fuselage.

Skin panels

The stress spectra for points on the surface of the skin panels showed only one peak which occurred at about 380 c/s as shown in Fig. 16.5. A group of four panels divided into two pairs by one of the ribs vibrated at the same frequency. The panel mode shape, as later measured by a photographic method, was found to be that of a plate fully fixed at the ribs and along the stringer flanges at a position approximately along the centreline of the flange. Two pairs of panels across the dividing rib vibrated in phase with each other as would be expected because the spanwise pressure correlation length is large compared with the panel dimensions. Adjacent panels across stringers vibrated out of phase. As the stringers were very

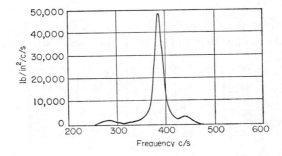

Fig. 16.5 Tailplane skin stress spectrum.

stiff, offering approximately the same resistance to both bending and twisting, the calculated frequencies for adjacent panels out of phase, and adjacent panels in phase, were approximately equal. Since the chordwise pressure correlation length is large when compared with the panel width, it would be expected that two panels across a stringer would be in phase and the stringer would bend. This form of vibration was detected in discrete frequency tests in the laboratory, and occurred at a slightly higher frequency (440 c/s) but it was not excited in the jet noise tests. The reason why these panels vibrated out of phase across the stringers instead of in the apparently more easily excited in phase mode is difficult to understand. The damping of the two modes appears to be approximately equal although the acoustic damping is likely to be lower in the mode in which adjacent panels vibrate out of phase with one another. The excitation levels and pressure correlation patterns are very similar at the two frequencies. It may be that the out of phase vibration represents a minimum energy condition for the relatively large cross-section stringer.

Calculations of the frequency of vibration from the measured or an assumed fully fixed mode shape give qualitative agreement with experimental values as shown:

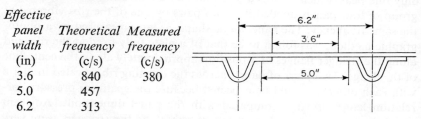

Effective panel width (in)	*Theoretical frequency* (c/s)	*Measured frequency* (c/s)
3.6	840	380
5.0	457	
6.2	313	

The main difficulty is the representation of the closed section stringers which have dimensions comparable to the panel width.

Stringers

The spectra of the stringer crown stress show two or three marked resonant peaks as illustrated in Fig. 16.6. The overall stress levels are very

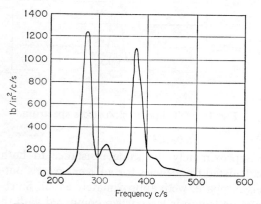

Fig. 16.6   Tailplane stringer stress spectrum.

low (200–300 lb/in² r.m.s.) and it is therefore not possible to draw any detailed conclusions. It appears however that one of the resonant peaks corresponds to the skin panel resonant frequency of 380 c/s. At the lower resonant frequency of 280 c/s the stresses at all the points measured are in phase, indicating either a stringer fundamental fully fixed mode or an overall vibration of the structure. Simple calculations suggest that the fundamental stringer mode with fixed ends should have a frequency of the order of 2,000 c/s. It is therefore reasonable to conclude that the 280 c/s resonance is an overall mode of the test section and not, therefore, necessarily representative of the overall modes which would occur on the full tailplane.

Rib bend radii

The stress levels in the reinforced rib bend radii are also low (720 lb/in² r.m.s. overall) and the spectrum, shown in Fig. 16.7, has a single pre-

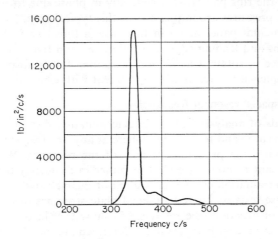

Fig. 16.7 Tailplane rib bend radius stress spectrum.

dominant resonant peak at 340 c/s. This does not correspond to any panel or stringer inertia load effect (380 c/s). It has now been confirmed from more detailed tests on a different tailplane of similar construction that this phenomena is due to overall vibrations of the tailplane as a whole. The top and bottom stiffened skins and the ribs are vibrating in coupled modes. In any particular mode the ratio of the rib deflection to the skin deflection depends on the relative stiffness of the skin and ribs. Thus with a relatively stiff skin and a relatively light rib the lowest mode is one in which the rib vibrates in an almost fully fixed mode with only a very small motion of the skin taking place. This is the type of mode which is indicated in the measurements of Fig. 16.7 and is the type which does the most fatigue damage as the stress concentration at the skin attachment line makes this region the weakest from a fatigue point of view.

16.4.3 *Elevator skin panels*

The particular structure investigated in this case had a thin skin and relatively stiff ribs with a light longeron sub-dividing the skin in a direction normal to the ribs. The stress spectra for mid-panel positions generally had only one predominant peak but the frequency of this peak varied from panel to panel because of the variation in panel dimensions. There was some coupling between panels in the fore and aft direction

12

across the longeron which twisted. Stress levels measured across the ribs were of the same order of magnitude as the levels at the panel centres suggesting fully fixed modes of vibration with the ribs acting as rigid supports. Across the ribs panels were generally in phase due to the fact that the excitation pressures are in phase over this typical area.

The fundamental panel mode in this case is thus one in which three edges (two ribs and leading edge) are fully fixed and the rear edge at the longeron is free to rotate. Calculations assuming full fixation on all four sides give frequencies in reasonable agreement with experimental values.

16.5 Closely spaced resonant frequencies

The methods of analysis and mode determination discussed in the preceding sections are valid for the case where at any one frequency only one mode contributes the predominant part of the response. Where modes having significant response are close together in frequency the results of the mode determination can be misleading or inconclusive.*

The response tests on curved stiffened plate structure of the type used for fuselage construction (see reference 5) show that in between the two bounding panel modes there are many intermediate modes all very close together in frequency. In such a situation amplitude response measurements are not adequate to identify the intermediate modes because the responses in the modes overlap in frequency. For example, one might excite one of the intermediate modes at resonance but at this frequency other modes would have significant response—some forced below and some above their resonant frequency.

To separate out the components of each individual mode it is necessary to use the vector analysis method as suggested by Kennedy and Pancu.[6] The basis of the method is the polar plot of the response vector relative to the vector of the forcing pressure. For a lightly damped single mode this polar plot is a complete circle whose diameter is inversely proportional to the damping of the mode. In the case of the panel modes being considered here the damping is very low and therefore the circumference is traversed in a relatively small frequency interval.

Consider now the case of the response vector for the strain in the centre of one of the panels in the test specimen. As the frequency of the forcing pressure is increased through one resonance the tip of the response vector describes an arc approximating to part of the circle of the single mode response. Identification of these arcs gives the normal modes which might otherwise be hidden if only the amplitude response were measured. An example of a polar diagram for one response vector is shown in Fig. 16.8.

From this figure it can be seen that further complication arises when two

* As in the case of the Comet tailplane discussed in section 16.4.2.

modes are very close together in frequency such that their arcs join to-gether giving the appearance of a single mode. In this case it is necessary to go one step further to isolate the two modes. In the simple single degree of freedom case the tip of the response vector travels the greatest distance per cycle per second at resonance. Therefore, if the rate of change of arc length with frequency, ds/df, is plotted against frequency a maximum occurs at resonance. In this way it is possible to identify two closely spaced resonances even when the arcs on the vector diagram form a continuous curve.

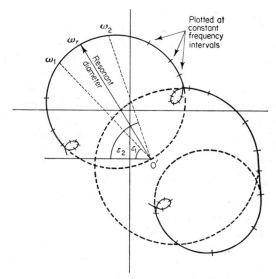

Fig. 16.8   Typical polar diagram.

Once a resonant frequency has been established the arc may be com-pleted to give the single degree of freedom resonant circle and the resonant diameter drawn. The origin of this diameter represents the displaced origin of the single degree of freedom system, the length represents the response in the mode and the argument represents the phase relative to the exciting pressure. The damping in the mode may be determined from the resonant circle by measuring the phase angle at two frequencies close to resonance.[7]

The damping ratio, ζ, is given by

$$\zeta = \frac{\omega_2 - \omega_1}{\omega_n} \frac{1}{\epsilon_2 - \epsilon_1} \tag{16.1}$$

as illustrated in Fig. 16.8.

This method gives finer resolution of the modes and providing that a reasonable length of arc for each resonant circle is available an estimate of the damping can be made. Typical figures for the damping ratio of the Lin type of stiffened skin modes are:

$$\text{Lower bound: } 0.005$$
$$\text{Upper bound: } 0.008$$

The acoustic damping is likely to be higher in the upper bounding mode and may thus account for the majority of the difference in the measured values.

The Kennedy–Pancu method, as just outlined, relates to discrete frequency response tests. Similar information can be obtained in the case of random excitation from the cross power spectrum of the strain and the pressure excitation as measured by a microphone. The real part of the cross power spectral density gives the real component of the response vector and the imaginary part of the cross power spectrum gives the imaginary component of the response vector. The main problems now are due to the bandwidth of the analysis—a sufficiently long cross correlogram must be measured to make the equivalent bandwidth (see chapter 4) small compared with the structural 'bandwidth'. This method is discussed in some detail in reference 8.

16.6 Theoretical estimation of response levels

In order to obtain an estimate of the response of practical structures to jet noise, considerable simplifications to both the structure and the noise field must be made. One simplification is to assume that the structure is responding predominantly in one mode only. This assumption has formed the basis of many of the early design charts and also of the recent method of estimation put forward by the Royal Aeronautical Society in the Data Sheet Series.[9] The other simplification which has formed the basis of much recent work has been that of assuming a uniform distribution of vibrational energy amongst the modes in the frequency range of excitation. This method has been developed mainly by Bolt Beranek and Newman Inc.[10,11,12] The two methods are now outlined in more detail.

16.6.1 *Single degree of freedom assumption*

The response of a complete structure to a random noise field can be fully described by Powell's equation 14.49 but unfortunately this equation cannot be used because of the prohibitively lengthy computation required. To enable estimates to be obtained equation 14.49 is simplified in the following way:

(1) Assume that the predominant form of skin vibration is that of each panel vibrating independently in a single mode at its fundamental

natural frequency. This reduces the series of equation 14.49 to a single term, or the two terms of equation 14.52 to one term.

(2) Assume that the mode of vibration is identical with the mode of deflection of the panel when subjected to a uniform static pressure. The panel is usually assumed to be fully fixed at its edges.

(3) Assume that the pressure is exactly in phase over the whole panel. The double integral in equation 14.49 then reduces to the square of the integral of the mode. This assumption is reasonable for jet-excitation of panels of typical size, but may be inadequate for boundary layer excitation.

(4) Assume that the power spectral density of the pressure is constant over the frequency range near the natural frequency of the panel. This is justifiable if the panel damping is light.

Under these assumptions, it can be shown that the mean square stress at any point on the panel is given by

$$\overline{\sigma^2(t)} \approx \frac{\pi}{4\zeta} f_r G_p(f_r) \sigma_{sp}^2 \tag{16.2}$$

In this f_r = the natural frequency of the fundamental mode in c/s

$G_p(f_r)$ = spectral density of the acoustic pressure at the frequency f_r

σ_{sp} = static stress at the point of interest due to a unit static pressure over the whole surface of the panel (corresponding to the mode of assumption (2), above)

ζ = damping ratio of the fundamental mode, usually having a value between 0.01 and 0.02.

[Equation 16.2 can be derived from equation 14.52 using the above assumptions. Note that in equation 16.2 we are using the 'single-sided' power spectrum, $G_p(f_r)$, whereas in equation 14.52 the 'double-sided' spectrum, $S_p(\omega)$, was used. From the definitions given in chapter 4, it may be seen that

$$G_p(f_r) = 4\pi S_p(\omega).]$$

Equation 16.2 may be used to find the stress at, say, the critical point along a rivet attachment line. In this case, the two predominant assumptions and simplifications (1) and (2) may tend to have a self-cancelling effect. The assumption of a single mode of response is not realistic in general and should result in an under-estimate of the actual stress. The assumption of full fixation at the stiffeners on the other hand gives an over-estimate of the stress at the elastic edge supports. Comparison of the estimated and measured stresses on a wide range of stiffened skin structures is required before this simple approach can be fully justified. Such a comparison of estimates and experimental results is shown in Fig. 16.9.

From this it can be seen that the simple theory gives a reasonably good estimate of measured stresses.

In the case of control surfaces, two skins are attached together by ribs and thus both skins and ribs vibrate because of the mechanical coupling between them. This type of structure is not so amenable to the simple form of analysis outlined above. In the case of tailplanes, elevators and flaps, the sound pressure will be greater on one side of the control surface due to acoustic shielding. In this case the incident sound energy minus the re-radiated sound energy is absorbed by two vibrating skins and ribs. Thus

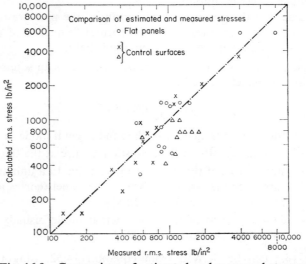

Fig. 16.9   Comparison of estimated and measured stresses.

the stress level should be about one-third of that which would have been induced in a single plate. Where there is incidental sound energy on both skins, as in the case of a fin and rudder, the overall excitation level is increased by approximately 3 dB. Thus the stresses would be approximately 1.4 times higher than for the one engine case. In comparisons shown in Fig. 16.9 a factor of one-third has been used in the calculation of skin stress.

16.7 The statistical energy method

When a complex structure is excited by forces having broad frequency spectra the time averaged stress at any point can theoretically be determined by the addition of the responses in the many normal modes excited, due account being taken of average phase relationships. However, the large number of modes involved and their sensitivity to small changes in boundary conditions make attempts to calculate point values laborious and of doubtful accuracy. Hence an approach to vibration analysis has been developed in which quantities averaged over space as well as in time are the basic ingredients, rather than point quantities of the classical approach. Such averaged quantities are found to be less sensitive to variation in boundary conditions and calculations require less detailed information about the structure than has in the past been thought necessary for analysis. This approach, which was originally developed for structural vibration analysis by Bolt Beranek and Newman, Inc., has been called the 'Statistical energy method'. It has been found to be most successful in the analysis of high frequency vibration of coupled structural or acoustic-structural systems but it has not yet been fully developed as a standard engineering calculation procedure.

The classical approach involves setting up differential equations of motion in terms of forces on, and within, a system, which are generally solved in terms of the natural (undamped) modes of vibration. On the other hand the energy approach considers the energy flow into and out of a vibrating system and employs such concepts as modal energy and power flow. Vibrating systems, whether they are mechanical or fluid, are considered to act as sets of resonators, and the total energy of a system and its dissipation are evaluated in terms of these resonators, or modal energies. As yet only lightly damped and lightly coupled systems have been treated by this approach.

16.7.1 *Response of a simple resonator to wide band random excitation*

It is shown in chapter 14 that the mean square displacement of a lightly damped resonator responding to a wide band force is given by

$$\overline{w^2(t)} = \frac{\pi S_f(\omega_r)}{kc} = \frac{\pi}{2} \frac{G_f(\omega_r)}{\omega_r^2 mc}$$

where w is the oscillator displacement, $G_f(\omega_r)$ the spectral density of the exciting force at the resonant frequency ω_r, m the oscillator mass and c the oscillator damping coefficient. Since we are primarily interested in the kinetic energy of the oscillator we will write the mean square velocity as

$$\overline{v^2} = \overline{\dot{w}^2} \approx \omega_r^2 \overline{w^2} = \frac{\pi G_f(\omega_r)}{2mc}.$$

The half power frequency bandwidth is given by $\varDelta_{1/2} = \dfrac{c}{m}.$

The time averaged energy functions for the oscillator are given by:

Kinetic energy $\overline{T} \quad = \frac{1}{2}m\,\overline{v^2}$

Potential energy $\overline{U} \quad = \frac{1}{2}k\,\overline{w^2} = \frac{1}{2}m\omega_r^2\,\overline{w^2} = \overline{T}$

Total energy $\overline{E} \qquad = \overline{T} + \overline{U} = m\,\overline{v^2}$

Power dissipated $\overline{\varPi} \quad = c\,\overline{v^2} = \omega\eta\overline{E},$

where η is the loss factor and is defined by equation 18.1.

16.7.2 *Vibration energy of an extended structure*

Since acoustic excitation is the main source of high frequency wide band structural vibration, extended plate or shell type structures are often the subjects of high frequency vibration analysis. It is interesting to see how the vibration energy of such 'two-dimensional' structures can be expressed in terms of the energies of the normal modes of vibration.

We express the total instantaneous velocity of a point in terms of the contributions from each normal mode,

$$v(\bar{x}, t) = \sum_r f_r(\bar{x})\dot{q}_r(t)$$

where $f_r(\bar{x})$ represents a modal displacement function and $\dot{q}_r(t)$ represents a normal co-ordinate velocity.

The instantaneous kinetic energy of an element of surface area dA is given by

$$\delta T = \tfrac{1}{2}\mu(\bar{x})\, v^2(\bar{x}, t)\, dA$$

where $\mu(\bar{x})$ is the local mass per unit surface area. The total kinetic energy is given by

$$T = \int_{\text{surface}} \tfrac{1}{2}\mu(\bar{x})v^2(\bar{x}, t)\, dA.$$

Now

$$v^2(\bar{x}, t) = \sum_r \sum_s f_r(\bar{x})f_s(\bar{x})\dot{q}_r(t)\dot{q}_s(t).$$

The orthogonality condition for normal modes is

$$\int_{\text{surface}} \mu(\bar{x})f_r(\bar{x})f_s(\bar{x})\, dA = 0, \qquad r \neq s.$$

Hence

$$T = \tfrac{1}{2}\int_{\text{surface}} \sum_r \mu(\bar{x})f_r^2(x)\dot{q}_r^2(t)\, ds = \tfrac{1}{2}\sum_r M_r\dot{q}_r^2(t)$$

where M_r is the generalized mass of mode r. The time averaged kinetic energy is given by

$$\bar{T} = \tfrac{1}{2}\overline{\int_{\text{surface}} \mu(\bar{x})v^2(\bar{x}, t)\, ds} = \tfrac{1}{2}\sum_r M_r\,\overline{\dot{q}_r^2(t)}$$

which is equal to the sum of the modal kinetic energies. Also

$$\bar{E} = \bar{T} + \bar{U} = 2\bar{T} = \sum_r M_r\,\overline{\dot{q}_r^2(t)}.$$

16.7.3 *Response of a set of modes (resonators)*

Most structures have many normal modes and natural frequencies within the frequency bandwidth of a random excitation. If each mode is considered as a resonator the total response will have components from the response of each resonator. As a first approximation it is assumed that each resonator is independent of the others.

Random signals are often analysed in frequency bands (e.g. $\frac{1}{3}$ octave, 6% etc.) and it is convenient to consider the response of a group of resonators having natural frequencies in such a frequency band. The individual bandwidth energies can then be added to give the total system vibrational energy. The bandwidth is chosen to be of such a size that it does not blur significant changes in spectrum level with frequency, but that it contains a sufficient number of resonators to justify meaningful averaging over the set. In other words a slight shift of bandwidth centre frequency should make little difference to the quantities averaged over that band. Furthermore, the bandwidth should be large compared with the average effective resonator bandwidth.

It is assumed that if a resonator's resonant frequency lies within the bandwidth of interest then its entire energy, calculated as though responding to an *infinitely* wide flat spectrum, is included in the band's energy. The energies of resonators having resonant frequencies outside the band are excluded. In general the errors arising from these two assumptions cancel each other out, but it is not inconceivable that the non-resonant response of resonators external to the bandwidth, but very efficiently coupled by the excitation, could contribute substantially to the band's energy.

On this basis it is possible to calculate the average energy of a resonator in the band and hence, knowing the average number of resonators, the average energy in the band.

The mean square velocity of a typical resonator excited by a homogeneous pressure field is given by

$$\overline{v_r^2} = \frac{\frac{1}{2}\pi G_{f_r}(\omega_r)}{C_r M_r}$$

$$= \frac{\frac{1}{2}\pi G_p(\omega_r) A^2 j_{rr}^2}{C_r M_r}$$

where G_p is the spectral density of the pressure field, A is the area of the structure and j_{rr} is the joint acceptance of the structure and pressure field as defined in chapter 14. The total energy of vibration in a frequency range of bandwidth W is given by the sum of the energies of the N resonators with their natural frequencies in the frequency range, i.e.

$$\overline{E}_w = \sum_1^N \overline{E}_r = \sum_1^N M_r \overline{v_r^2}$$

$$= \frac{1}{2}\pi \overline{p_W^2} A^2 \frac{N}{W}\left[\frac{\sum(j_{rr}^2/C_r)}{N}\right]$$

where $\overline{p_W^2}$ is the total mean square pressure in W. The term in brackets is the average value of j_{rr}^2/C_r over the resonators in the band, i.e.

$$\frac{\sum(j_{rr}^2/C_r)}{N} = \langle j_{rr}^2/C_r \rangle.$$

N/W is the average number of resonant frequencies per unit bandwidth for the bandwidth of concern. It is known as the modal density, $n(\omega_c, W)$. Thus

$$\overline{E}_W = \tfrac{1}{2}\pi \, \overline{p_W^2} \, A^2 n(\omega_c, W) \langle j_{rr}^2/C_r \rangle$$

and

$$\overline{\Pi}_W = \tfrac{1}{2}\pi \, \overline{p_W^2} \, A^2 n(\omega_c, W) \langle j_{rr}^2/M_r \rangle$$

Therefore with a knowledge of the modal density of a structure and the average joint acceptances and loss factors in frequency bands we can calculate the average response energy to an applied pressure field.

16.7.4 *Modal density*

When a complex structure is excited by random forces having a wide frequency spectrum a very large number of modes contributes to the total vibration. It is generally impossible accurately to compute the normal modes and frequencies of such a structure over the entire frequency range. However it is sometimes possible, and more realistic, to estimate the average number of frequencies per unit bandwidth by considering the complex structure to be made up of simpler components (e.g. plates, beams) for which analytical expressions for the modal density have been derived. It is found that when the modal density is sufficiently large, its value is not significantly changed by small changes in the exact geometry of the idealized component. In consequence the modal density of a flat plate of irregular plan form can be assumed to be the same as that of a rectangular plate of the same area, thickness, density and elastic modulus. As a first approximation it can be assumed that the modal density of the whole structure will be given by the sum of the modal densities of the idealized component parts.

Modal densities for a few uniform systems are presented in Table 16.4.

16.7.5 *Joint acceptances and energy loss factors*

It is not possible to generalize about the calculation of bandwidth averaged joint acceptances. One approach is to try to classify the types of mode likely to occur in the bandwidth of interest in terms of characteristics of the input force distribution (e.g. ratio of structural wave speed to pressure field wave speed). It may then be possible to obtain estimates of the average response of a particular class of modes and then the total

Table 16.4

System	Modal density $n(\omega)$ (natural frequencies/radian/sec.)	Symbols
Stretched string	$\dfrac{L}{\pi}\sqrt{\dfrac{\rho A}{T}}$	L = length ρ = density A = cross-sectional area T = tension
Beam (Bending)	$\dfrac{L}{2\pi}\left(\omega\sqrt{\dfrac{EI}{\rho A}}\right)^{-1/2}$	E = elastic modulus I = second moment of area
Plate (Bending)	$\dfrac{A_s}{4\pi}\sqrt{\dfrac{Eh^2}{12\rho(1-\nu^2)}}$	A_s = surface area ν = Poisson's ratio h = plate thickness
Cylindrical shell (Bending)	As plate, for $\omega > \sqrt{\dfrac{E}{\rho r^2}}$	r = cylinder radius
	As plate $\times\ 0.7\left(\dfrac{\omega^2\rho r^2}{E}\right)^{1/4}$, for $\omega < \sqrt{\dfrac{E}{\rho r^2}}$	

response will depend upon the relative modal densities of the different classes of wave.

In dealing with the response of structures to acoustic fields it is convenient to use the following principle. The coupling to a structural mode of the field generated by a particular sound source is related to the acoustic pressures generated at the source position by the modal vibration. The generalized acoustic force corresponding to a particular mode is given by

$$L_r = \int \{p_{bl}(\bar{x}, t) + p_{\text{rad}}(\bar{x}, t)\} f_r(\bar{x})\, dA = L_{r_{bl}} + L_{r_{\text{rad}}}.$$

p_{bl} is the pressure produced by the source on the unmoving (rigid) structure and p_{rad} is the pressure produced by the modal vibration in the absence of the source. The blocked joint acceptance $j_{r,bl}$ is related to radiation properties of the mode by the relationship

$$(j_{r,bl}(\Omega))^2 = \frac{4\pi a D(\Omega)}{\rho_0 A^2 \omega_r^2} (R_{r_{\text{rad}}}).$$

$D(\Omega)$ is a directivity function relating the blocked generalized force from a source whose direction from the structure is given by Ω to the blocked generalized force averaged over all source directions

$$D(\Omega) = \frac{(j_{r,bl}(\Omega))^2}{\langle j_{r,\,bl}^2\rangle_\Omega}.$$

R_{rad} is called the radiation resistance of the mode and is a measure of the efficiency with which it radiates acoustic energy

$$\overline{\varPi}_{\mathrm{rad}} = R_{\mathrm{rad}}\,\overline{E}_r/M_r.$$

Hence the total joint acceptance can be expressed in terms of the radiated power. The use of these relations leads to the following expression for the bandwidth energy of vibration of a structure exposed to a diffuse acoustic field of bandwidth mean square pressure p_W^2,

$$\frac{\overline{E}_W}{p_W^2} = \frac{2\pi^2 an(\omega_c)}{\rho_0\omega_c^2}\,\frac{\langle R_{\mathrm{rad}}\rangle_W}{\langle R_{\mathrm{rad}}\rangle_W + \langle R_{\mathrm{mech}}\rangle_W}.$$

The radiation resistance $\langle R_{\mathrm{rad}}\rangle_W$ is averaged over the modes resonating in the band W. Likewise we define $\langle R_{\mathrm{mech}}\rangle_W$. Estimates of modal radiation resistance can be made either by the methods of chapter 18 or by considering volume source cancellation between internodal regions of a vibrating surface. R_{mech} is a measure of the power dissipated mechanically and must be estimated from experimental data as in the classical method. An interesting conclusion from the above relationship is that when $R_{\mathrm{rad}}\gg R_{\mathrm{mech}}$, as for instance with welded metal structures immersed in water, the response energy is very simply determined and is independent of structural detail, or mechanical damping.

16.7.6 *Vibration of lightly coupled systems*

One of the most important results of the statistical method of vibration analysis is that when two vibrating systems are lightly and conservatively coupled (i.e. the coupling forces are very much less than the individual internal system forces and no energy is dissipated by the coupling forces), the rate of energy flow from one system to the other is proportional to the difference between the average modal energies of the two systems. Such a consideration leads again to the expression for response energy quoted in 16.7.5. Vibration analysis in terms of energy flow from one vibrating system to another appears to have great potential in the very difficult study of the transmission of high frequency energy in solid structures (e.g. machinery induced building vibrations).

16.7.7 *Calculations of stresses from vibration energies*

The product of a theoretical analysis using the energy method is usually a space and time averaged vibration amplitude, velocity or acceleration. For the purpose of fatigue calculations the location and direction of the maximum stress in a structure are required. Very little is known about stress distributions in randomly vibrating structures and consequently it is difficult to generalize about the relationships between the vibration energy, the space averaged mean square stress and the maximum stress. However two observations can be made. First, the assumption of a diffuse vibration wave field in a uniform two dimensional plate structure leads to a relationship between the mean square bending stress and the mean square vibration velocity of the form $\overline{\sigma^2} = K(E\mu/h)\overline{v^2}$. In this E is the elastic modulus, μ the mass per unit area and h the material thickness. K is a factor which is approximately equal to unity. Just how diffuse are real fields remains to be seen. Secondly, dynamic stress concentrations occur at boundaries and at changes of material cross section. In particular, the r.m.s. stress at the clamped edge of a plate with a diffuse vibration field in it is between 2 and 3 times the r.m.s. stress in the plate far from the boundary.

We see therefore that much work remains to be done before the results of the statistical energy method can confidently be applied to the calculation of fatigue lives under random loading.

References

1. B. L. Clarkson and R. D. Ford. The response of a typical aircraft structure to jet noise, *J. Roy. Aeron. Soc.*, **66**, 31 (1962).
2. B. L. Clarkson and R. D. Ford. Random excitation of a tailplane section by jet noise, *Univ. Southampton Rep.*, A.A.S.U. 171 (1961).
3. P. M. Belcher, J. D. Van Dyke and A. L. Eshleman. Development of aircraft structure to withstand acoustic loads, *Aero/Space Eng.*, **18**, 24 (1959).
4. Y. K. Lin. Free vibrations of continuous skin-stringer panels, *J. appl. Mech.*, **27**, 669 (1960).
5. B. L. Clarkson and R. D. Ford. The response of a model structure to noise—Part II: Curved panel, *ASD Tech. Rep.*, TDR-62-706 (1962).
6. C. C. Kennedy and C. D. P. Pancu. Use of vectors in vibration measurement and analysis, *J. Roy. Aeron. Soc.*, **14**, 603 (1947).
7. D. J. Mead. The internal damping due to structural joints and techniques for general damping measurement, Aero. Res. Council (London), Current Paper No. CP 452 (1959).
8. B. L. Clarkson and C. A. Mercer. Use of cross correlation in studying the response of lightly damped structures to random forces, *AIAA Journal*, **3**, 2287 (1965).
9. Royal Aeronautical Society. *Data Sheets on Acoustic Fatigue*, Autumn 1966.
10. P. W. Smith, Jr. and R. H. Lyon. Sound and structural vibration, *N.A.S.A.*, CR-160 (1965).
11. S. H. Crandall (ed.). *Random vibration*, Vol. 2, M.I.T. Press, Cambridge, Mass. (1963).
12. R. H. Lyon and G. Maidanik. Statistical methods in vibration analysis, *AIAA Journal*, **2**, 1015 (1964).

CHAPTER 17

Design of Fatigue Resistant Structures

17.1 Introduction

It is clear from the results presented in chapter 16 and elsewhere that the response and subsequent fatigue life of an aircraft structure which is being subjected to acoustic loads depends on a number of parameters, many of which are only qualitatively understood. The stresses are known to be low in amplitude compared with normal fatigue standards (in the range 1,000–4,000 lb/in²) and in the frequency range 100 c/s to 1,000 c/s. It is not possible to draw any precise general conclusions from such measurements as have been made, although it does appear that in many cases of widely spaced frames or ribs (of the order of 20 in) the skin strain response may be predominantly uni-modal in form. In other cases of closely spaced ribs or frames the skin strain response is multi-modal in form. The majority of measurements of stringer, frame and rib flange strains have shown a multi-modal response. In these circumstances it is not possible to formulate rigid design rules which will ensure the integrity of any particular type of structure. As is discussed briefly in section 17.4 attempts to produce design charts for skin and rib thicknesses and pitches have been made by Douglas but these are based on one particular aircraft and have not yet been justified for aircraft structure in general.

For basic skin subjected to high-intensity noise pressures, tests have shown, and service experience has now confirmed, that honeycomb sandwich structures are able to withstand higher loads than an equivalent single metal skin. The overall noise level at which it becomes more economical in weight to use honeycomb instead of sheet skin is not precisely defined but is likely to be about 160 dB.

The majority of failures which have troubled aircraft manufacturers have been in detail attachments. Companies with experience of this problem have generally evolved design rules to eliminate or at least reduce failures of these points. One example of such a detail is the use of a symmetrical attachment for the rib-skin joint.

In addition to these general statements it is possible to go a little further with the design of a stiffened skin or control surface as a result of the investigations described in chapter 16. These designs are now discussed in more detail.

17.2 Stiffened skin design

17.2.1 *Single panel representation*

As a single panel represents one of the simplest structural models, much experimental and theoretical work has been done on the response and subsequent fatigue of panels in noise environments.[1, 2] Such studies cannot be applied directly to the majority of aircraft structures but nevertheless have served to illustrate some of the features of the more general problem. Panels have usually been fully fixed at the edges and fatigue failures have generally been initiated at the stress concentrations at the supports. Work by Dyer, Smith, Malme and George[3] gives a comprehensive study of the excitation of single panels by variable incidence, discrete frequency, sound.

As failures generally occur at the panel supports these areas should be stiffened up to get equal fatigue life for the plain centre and the stiffened edges. This can be done by bonding a strip to the skin along the rivet lines. Additional improvement can be achieved if a layer of visco-elastic material is sandwiched between the panel and the reinforcing plate. This will increase the damping and provide a general reduction in stress; but careful design is required to get the optimum width of strip and thickness of layer. Theoretical and experimental studies of this type of joint have been made by Mead and Eaton.[4]

17.2.2 *Coupled panels*

A more realistic representation of the skin of a fuselage, tailplane or control surface is that of a series of panels coupled by flexible supports at stringers and frames or ribs. In this way a nearer approximation to the types of responses found in practical structures can be achieved. The theoretical analysis due to Lin has formed the basis of the subsequent study of panel groups. The mathematical model is now a row of skin panels (theoretically infinite in number) between two frames or ribs and coupled by stringers which can bend and twist. The analyses described in chapter 16 show that this is a valid model for skin vibrations on the types of structure investigated. Practical difficulties arise due to any non-uniform stringer spacing, but even in these conditions coupling has been found to take place across two or three bays.

The theoretical analysis gives two bounding frequencies: the lower bound corresponds to the case where the stringers twist and adjacent panels vibrate out of phase; and the upper bound corresponds to the condition in which the stringers bend and adjacent panels vibrate in phase. In between the upper and lower bounds of frequency there is a large number of intermediate modes in which the stringers both twist and bend. Some of these have been measured on a curved model structure having uniform stringer spacing of $4\frac{1}{2}$ in and a frame pitch of 9 in.[5] A few typical measured intermediate mode shapes are shown in Fig. 17.1. There are so

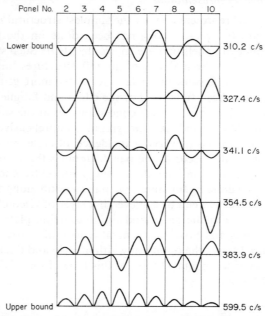

Fig. 17.1   Mode shapes of a curved stiffened panel.

many of these intermediate modes that their responses overlap to form a continuum in frequency. This gives the lowest group of frequencies and is generally the group of major significance in response to jet noise. The next group corresponds to overtones of the first and in the cases investigated had frequencies well above that of the maximum noise pressure.

As this form of vibration is considered to be the most important feature of the skin vibration of practical aircraft structures, it is worthwhile to study in more detail the effect of the structural parameters on the frequencies of the bounding modes and the separation between them for the lowest group of modes. For a given material the main structural parameters are: frame pitch, stringer pitch, torsional stiffness, flexural stiffness, skin thickness, and curvature. Owing to the wide difference in torsional stiffnesses of open and closed sections it is convenient to divide the curves into those relating to open section stringers and those relating to closed sections. In normal stiffened skin design when the stringer pitch is altered the stringer stiffness will also be changed. Thus, to present a realistic picture of the effect of change in stringer pitch the ratio flexural stiffness to torsional stiffness was kept constant and proportional to $(bh)^2$, which fitted reasonably well the dimensions used by Boeing[6] and Sud-Aviation[7] for open sections. The ratio of frame pitch to stringer pitch was also kept constant at two. On this basis the variation of the frequency of the two bounding modes with b and h is shown in Figs. 17.2 (a) and 17.2 (b) for open

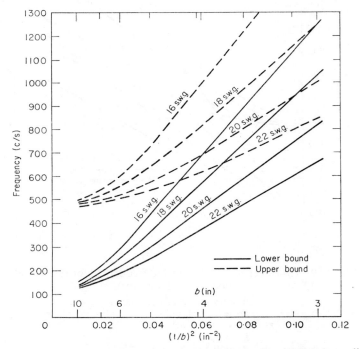

Fig. 17.2 (a) Effect of b and h on the frequencies of the bounding modes (open sections).

Rib pitch = l
Stringer pitch = b
Skin thickness = h
Stringer cross-sectional area = A_s
Stringer torsional stiffness = C_w (open), C (closed)
Stringer flexural stiffness = I
Radius of curvature = 60 in
l/b = 2 (inch units)
C = 0.085 $(bh)^2$
I = 1.08 $(bh)^2$
A = 0.69 $(bh)^2$

and closed section stringers. It can be seen that in general the frequency is approximately proportional to $1/b^2$, except for higher values of b, and also that the bounds converge as b decreases. The effect of an increase in panel aspect ratio from two to four is to decrease the slope of the curves.

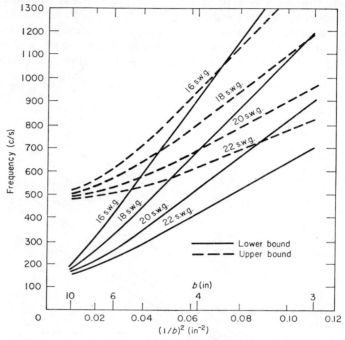

Fig. 17.2   (b) Effect of b and h on the frequencies of the bounding modes (closed sections).

Radius of curvature $= 60$ in
$l/b = 2$ (inch units)
$C = 0.36 (bh)^2$
$I = 0.36 (bh)^2$
$A_s = 0.52 (bh)$

As would be expected, stringer torsional stiffness has a marked effect on the separation between the upper and lower frequency bounds. Fig. 17.3 shows the effect of increase in torsional stiffness for constant flexural stiffness. The upper limit (stringers bending) remains constant while the lower limit rises with increase in torsional stiffness to give the minimum separation for closed section stringers. The second pair of curves on the figure also shows that the range of change of separation with torsional stiffness is greater for shorter stringer pitches.

In a particular design it will not be possible to change the curvature of

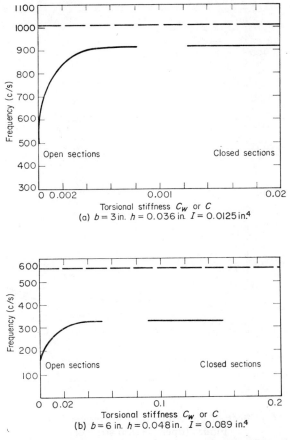

Fig. 17.3   (a) and (b) Effect of stringer torsional stiffness on the frequencies of the bounding modes.

a structural element, but nevertheless it is interesting to see the effect on the frequencies of skin curvature. Figs. 17.4 (a) and 17.4 (b) show that the main effect of increase in curvature is to increase the separation of the two bounding modes by making relatively small changes in the lower bound (stringers twisting), but making a very considerable change in the upper bound. Flat panels show the least difference between upper and lower limits.

Consider now the excitation of such a stiffened skin by acoustic pressures. As interframe panel rows are usually situated approximately at right angles to the jet centre line, the pressure correlation over the group is

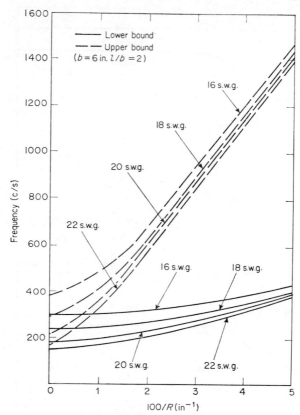

Fig. 17.4   (a) Effect of curvature on the frequencies of the bounding modes (open section stringers).

likely to be high and positive over distances of up to two feet for frequencies in the range 400–1,000 c/s. Thus, with stringer pitches of the order of 6 in or less, at least four panels will have pressures in phase over them. In this condition the lower bounding mode, having adjacent panels vibrating out of phase, should not be excited appreciably, but the upper bound and some of the higher frequency intermediate modes will respond. Use can now be made of the shape of the pressure spectrum to ensure that the frequency of the upper bound is well above the frequency of maximum sound energy.

A secondary consideration is the interaction of the intermediate modes situated between the upper and lower limit. Lin has attempted to allow for this effect using a statistical argument[8] and has achieved fair agreement between measured and computed stresses. From a physical point of view, however, it can be seen that when the bounds are close together the inter-

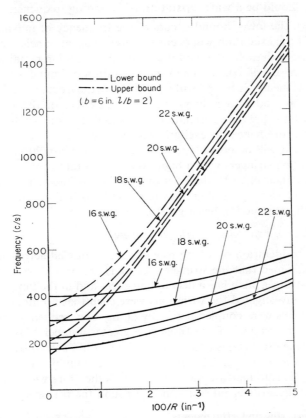

Fig. 17.4 (b) Effect of curvature on the frequencies of the bounding modes (closed section stringers).

mediate modes overlap and have coupled responses due to being excited by the same pressure field. It is difficult to see exactly what happens in general terms, but it seems likely that the overall r.m.s. stress in the skin will increase as the modes move closer together. A theoretical analysis adding further justification to this assumption has been published by Robson.[9] Lin's simplified analysis showed that the overall stress should vary as $1/\sqrt{d_n}$ where d_n is the difference in frequency between the upper and lower bounding modes.

With these effects in mind it is now possible to write down conditions which should minimize the response:

(1) The frequency of the lower bound should be equal to or higher than the frequency of maximum noise pressures.

(2) There should be a wide separation of bounding frequencies.

Although the lower bound occurs at the frequency of maximum noise pressures, little excitation will occur because adjacent panels are vibrating out of phase with each other and the pressure correlation remains high and positive over several panels. The wide separation of bounding frequencies ensures that the upper bound will be well above the frequency of maximum sound energy and also reduces the coupling effect of the intermediate modes. Also the pressure correlation at this higher frequency has a smaller spatial scale and hence the excitation is reduced.

Generally it will be possible to achieve these two conditions by the use of open section stringers with a high ratio of flexural to torsional stiffness.

For flat panels close to the jet it may be impossible to meet the above requirements with typical skin thicknesses and it is in these conditions that honeycomb construction has a great advantage.

17.2.3 *Effect of non-uniform stringer spacing*

In the consideration of a long row of panels coupled across stringers it has been assumed that the stringer spacing is uniform. A relatively small change in stringer pitch has a decoupling effect and in the tapering section of fuselage investigated the variations were such that no more than three adjacent panels were coupled markedly at any one frequency. In the constant diameter section of the fuselage a greater number of panels would have the same dimensions and thus coupling would extend over a greater circumferential distance. The effect of this decoupling in the stringer twisting mode is to increase the response. It is therefore very important to ensure that the stringer pitch is constant round the periphery.

17.3 Frames, ribs and attachments

Surveys have shown that the majority of the fatigue damage on aircraft has been confined to skins and attachments of skin stringer sheets to frames or ribs. Only where skins are very thick, as perhaps in a tailplane, have the ribs failed before the skin. Attachment loads are very difficult to estimate and the main point is that there should be good detail design to allow for the direct inertia loads from the skin in addition to the more usual structural shear loads. It is considered good design to make attachments symmetrical with back-to-back cleats or T-section extrusions but in some instances this has not guaranteed a satisfactory performance. The development of a skin–rib attachment by Boeing[10] is shown in Fig. 17.5. In some cases where this has not proved successful the cause of failure may be different and this has prompted the investigation of coupled skin–rib vibration discussed later in this section.

The end attachments of stringers need careful design. Generally the stringers will be continuous through ribs or frames but at their ends great care should be taken in the design of the stringer cap. The stringer, vibrating with the skin, has an inertia loading which is reacted at the ribs or frames. Examples of end attachments have also been given by Hubbard.[10]

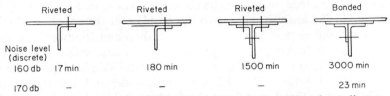

Fig. 17.5   Life of various skin–rib attachments subjected to discrete frequency siren testing.[1]

In the case of the Caravelle rear fuselage the predominant type of frame vibration has been found to be that associated with overall modes of the fuselage and not predominantly due to inertia forces from the skin and

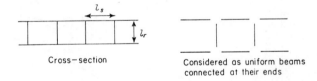

Fig. 17.6   (a) Two-dimensional model for coupled rib–skin vibrations.

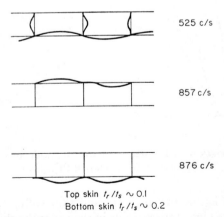

Fig. 17.6   (b) Predominant mode shapes measured on an actual tail-plane.

stringers. Stresses are generally very low and failure should not occur except possibly around any badly designed attachment points for skin cleats.

The coupling of a stiffened flat skin with a rib is of considerable interest in the case of tailplane and control surface vibration. To study this case initially a simplified two-dimensional model shown in Fig. 17.6 (a) has been used. The analysis reduces to the problem of the coupling of three beams and the method of receptances of Bishop and Johnson[11] has been used to get the characteristic equation and resultant mode shapes. The variation in the ratio of rib thickness to equivalent skin thickness is shown in Fig. 17.7 for ratios of 1 and 2 of rib pitch to depth. When the rib stiffness is equal to the skin stiffness the fundamental mode is one in which the skin and rib vibrate in a simply supported type of mode. For lower rib stiffnesses the lowest mode is one in which the rib vibrates in a fully fixed mode, whereas for higher stiffnesses the mode approaches that of a fully fixed skin mode of vibration. Thus if it is considered that the attachment of the rib to the skin is the weakest point from a fatigue point of view the

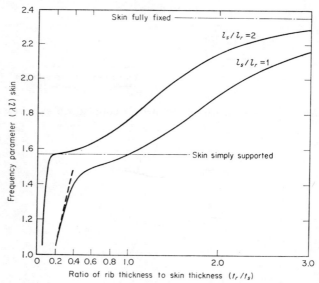

Fig. 17.7   Variation in frequency of lowest mode of coupled rib–skin vibrations with relative thickness.

optimum design is one in which rib and skin stiffnesses are equal. For lower rib stiffness high stresses occur in the rib at the skin attachment line whereas for higher rib stiffnesses higher stresses will occur in the skin along the rib attachment line. The condition for equal rib and skin stiffnesses is given by

$$m\sqrt{n} = 1 \quad \text{where} \quad m = \frac{l_r}{l_s}, \quad n = \frac{t_s}{t_r}.$$

In practice the skin is usually stiffened with open or closed section stringers. The above simplified theory can be modified to allow for this but it now becomes difficult to present the results in a general form. An interesting example of this form of coupling has recently been studied on a full-scale tailplane. In this case the top and bottom skins have different stiffnesses, the ratio of rib pitch to depth is of the order of 2, and the ratio of equivalent rib thickness to skin thickness is of the order of 0.2 for the bottom skin and 0.1 for the top skin. The predominant types of modes are shown diagrammatically in Fig. 17.6 (b). It can be seen that the lowest mode is a combination of an almost fully fixed rib mode and a simply supported bottom skin. At a higher frequency the bottom skin vibrates in a fully fixed type of mode and in this region also the stiffer top skin lowest mode appears.

17.4 The use of simple design charts

Many attempts have been made to produce design charts from which it is possible to read off skin thickness, rib thickness and pitch, etc., for a required noise level. The most comprehensive of these charts is that due to Belcher and his coworkers at Douglas,[12] and is shown in Fig. 17.8. Considerable attempts have been made to allow for variation in structural

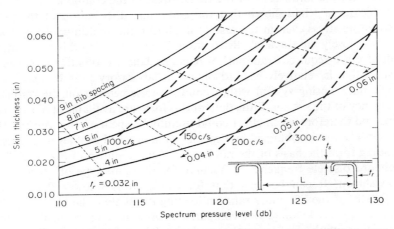

Fig. 17.8 Tentative design chart for plain skin on bent flange ribs (material 2024) for 1,000 hours.

design and at first sight it appears to offer the solution to all our problems. The major difficulty is that the chart is primarily based on tests of a limited number of components for the DC8 airliner. Extrapolation to cover almost the complete range of design is made from these few test results on the basis of the single degree of freedom analysis. The experimental analyses discussed in the previous chapter show that in only a very limited number of cases does the structure respond in only one mode and hence the single degree of freedom analysis does not apply. It can in fact be very misleading.

This type of approach could usefully form the starting point in the design but it is strongly suggested that the factors discussed in the previous section should be considered carefully before the details of the design are fixed.

17.5 Fatigue life

From the foregoing brief study of the pressure field and the types of response of actual structures it is clear that it is not possible to simplify the overall problems to the extent of being able to rely solely on design charts. In general it will not be possible to make an accurate estimate of stress and therefore the best that can be expected is an estimate showing an order of magnitude agreement with experiment. In these conditions it is not worthwhile to carry out a cumulative damage type of estimate of the fatigue life. We are faced nevertheless with the problem of making some estimates however inaccurate they may be. The most reasonable and meaningful approach is likely to be the use of a 'random' $S-N$ curve. If such a curve is available for the material being considered then the probable life is given directly from a knowledge of the r.m.s. stress in the component.

Schjelderup[13] has shown that the amplitude distribution of the peak alternating stress even in the case of nonlinear and multimodal responses follows closely the Rayleigh distribution. Thus tests using this distribution will give the required 'random' $S-N$ curve data. Fortunately a Rayleigh distribution is relatively easy to obtain in laboratory tests by exciting a single mass–spring system with a band of noise centred on the resonant frequency of the system. Little data of this form are available at the present time and therefore much effort could usefully be spent on this work in the future. The results obtained by Fralich[14] on a notched cantilever in reverse bending form the basis of the random $S-N$ curve shown in Fig. 17.9. The standard discrete frequency $S-N$ curve for the same type of specimen is also shown. It can be seen that for r.m.s. stresses lower than about 10,000 lb/in^2 the life under random loading is less than the life under discrete frequency loading. For comparison some discrete frequency results for higher numbers of reversals are shown. These data, due to Townsend and Corke,[15] relate to skin specimens in reverse bending along the rivet line. The life in these tests is higher than in the random tests due to the discrete frequency method of testing and also to the higher frequency of the test. With these factors in mind Fralich's curve is extended to produce a tentative random *S–N* curve for use in order of magnitude estimations. In the light of Townsend's results and the unknown effects of long aircraft

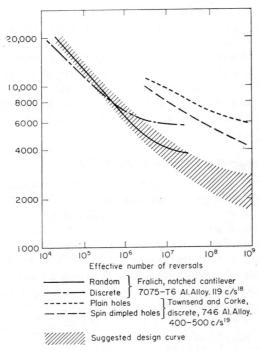

Fig. 17.9   Random *S–N* curve for aluminium alloy notched specimens.

life implied by the high number of reversals it is considered prudent to neglect the flattening of Fralich's curve. The suggested design curve is drawn as a band such that results for a high frequency, of the order of 1,000 c/s, are assumed to lie at the longer life edge of the band with 100 c/s lying at the lower life edge of the band. Recent work by Townsend using narrow-band random loading gives results which lie within the shaded area at the lower stress levels. These new results have formed the basis of a series of random *S–N* curves issued by the Royal Aeronautical Society as part of the *Data Sheets on Acoustic Fatigue*.[16]

The effective number of reversals to be used in calculations of life under random loads can be assumed to be given by half the number of zero crossings which the strain–time curve makes. Rice[17] has shown that the number of zero crossings per second of a random function is given by

$$2\left[\frac{\int_0^\infty f^2 G(f)\,df}{\overline{\sigma^2}}\right]^{1/2}$$

where $\overline{\sigma^2}$ is the mean square stress, $G(f)$ is the power spectral density of stress at frequency f (c/s). Thus the effective number of reversals per second will be *half* this value.

This approach has much to commend itself in estimating the improvement in life which is likely to result from structural changes whether the stresses have been computed or measured on a prototype. It is also possible to see what order of r.m.s. stress is required to ensure a given life and then if the prototype stresses are higher than this modifications can be put in hand before failure takes place.

17.6 Crack propagation under noise excitation

An important factor in the design of fail safe structure is the rate at which a fatigue crack propagates during service loads. Much work has been done on the propagation due to turbulence and pressurization loading but there could be an additional effect due to local vibration of the skin in the region of the crack due to high frequency pressure fluctuations from a jet efflux or turbulent boundary layer. Consider for example the case of the effect of pressurization load cycles on a cracked fuselage. For crack lengths less than the critical length each successive load produces a small increase in crack length. If during one flight, however, there is a considerable random pressure fluctuation on the side of the fuselage the crack may spread an additional amount due to this cause. Thus the number of flights which can be made before the crack reaches the critical length is less than that predicted from a consideration of the effect of pressurization alone.

Preliminary work on this secondary effect has been done at Southampton.[18] A standard flat specimen 10 in $\times$ 24 in $\times$ 0.064 in. is loaded in tension by the hydraulic jack to represent steady pressurization loads and an initial crack of the order of $3\frac{1}{2}$ in is made in the centre of the plate at right angles to the principal tensile stress and it is then subjected to high intensity acoustic pressure excitation from the random noise siren. The initial tests have been carried out at a noise level of 140 dB.

Fig. 17.10 shows the results of one of the tests. It is interesting to see that the behaviour of the system is closely associated with the behaviour of the two semi-elliptic areas of plate which have the crack as their minor axis. The predominant form of vibration is one in which these areas vibrate in their lowest mode. The overall tensile loading on the specimen produces compressive stresses in these semi-elliptic areas at right angles to the main tensile force. In any one test the overall tensile load on the specimen is kept constant. For short crack lengths the compressive stress is below the buckling stress of the plate. As the crack length increases a condition is reached when buckling occurs, and with further increase in crack length

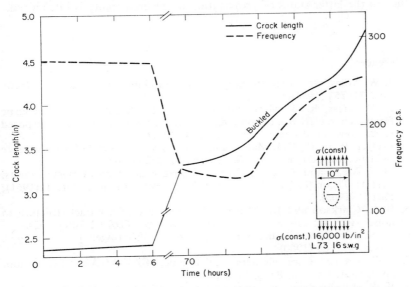

Fig. 17.10   Variation in crack length and frequency with time in noise crack propagation tests.

the area buckles out appreciably. In the prebuckled state the rate of crack propagation is very slow and increase in length (and hence compressive stress) reduces the frequency of vibration. When buckling occurs the semi-elliptic areas are unstable and move from one extreme position to the other. This damaging form of movement produces a higher rate of crack propagation and the frequency is at a minimum. For longer crack lengths the area becomes well buckled and stays in one extreme position vibrating as a curved plate. The rate of crack propagation slows down again and increase in crack length causes increase in buckling and hence the frequency of vibration increases. In the final stage the alternating stresses become high and fast crack propagation precedes the final failure.

Detailed examination of the cracked surface shows that the slow and fast propagation regions exhibit different metallurgical phenomena. Due to the manner in which the specimen is loaded by the acoustic excitation the stress distribution around the tip of the crack differs from that in the alternating tensile load testing of this type of specimen.

In practical cases where the tensile stress due to pressurization is likely to be of the order of 10,000 lb/in^2 tests indicate that crack propagation due to noise will be negligible until the crack reaches about 70% of the critical length. Thus providing that some allowance is made for this effect in the latter states of propagation no severe penalty is introduced.

References

1. J. W. Miles. On structural fatigue under random loading, *J. Roy. Aeron. Soc.*, **21**, 753 (1954).
2. R. W. Hess, R. W. Herr and W. H. Mayes. A study of the acoustic fatigue characteristics of some flat and curved aluminium panels exposed to random and discrete noise, *N.A.S.A. TN.DI.*
3. I. Dyer, P. Smith, C. I. Malme and C. M. George. Sonic fatigue resistance of structural designs, *ASD Tech. Rep.*, 61–262 (1961).
4. D. J. Mead and D. C. G. Eaton. Interface damping at riveted joints, Part I: Theoretical analysis. *ASD Tech. Rep.*, 61–467, Part I (1961); Part II (1965).
5. B. L. Clarkson and R. D. Ford. The response of a model structure to noise—Part II: Curved panel. *ASD Tech. Rep.*, TDR-62–706 (1962).
6. Y. K. Lin. Free vibrations of continuous skin stringer panels, *J. appl. Mech.*, **27**, 669 (1960).
7. J. G. Wagner. Caravelle acoustical fatigue, ICAF-AGARD Symposium on Aircraft Fatigue, Paris, May 1961.
8. Y. K. Lin. Stresses in continuous skin stiffener panels under random loading, *J. Roy. Aeron. Soc.*, **29**, 67 (1962).
9. J. D. Robson. The random vibration response of a system having many degrees of freedom, *Aero. Quart.*, **18**, 21 (1966).
10. H. H. Hubbard, P. M. Edge and C. T. Modlin. Design considerations for minimising acoustic fatigue, *WADC TR.*, 59–676.
11. R. E. D. Bishop and D. C. Johnson. *The Mechanics of Vibration*, Cambridge University Press (1960).
12. P. M. Belcher, J. D. Van Dyke and A. L. Eshleman. Development of aircraft structure to withstand acoustic loads, *Aero/Space Eng.*, **18**, 24 (1959).
13. H. C. Schjelderup and A. E. Galef. Aspects of the response of structures subject to sonic fatigue, *WADC TR.*, 61–187.
14. R. W. Fralich. Experimental investigation of effects of random loading on the fatigue life of notched cantilever-beam specimens of 7075-T6 aluminium alloy, *N.A.S.A. Memo.*, 4-12-59L.
15. N. A. Townsend and D. M. Corke. The fatigue strength between 10^6 and 10^{10} cycles to failure of light alloy skin–rib joints under fully reversed bending loads, de Havilland Aircraft Co. Note.
16. The Royal Aeronautical Society. *Data Sheets on Fatigue*, Sheets No. 66012 to 66022. Autumn 1966.

17. S. O. Rice. Mathematical analysis of random noise, *Bell System Tech. J.*, **24**, 46 (1945).
18. B. L. Clarkson. The propagation of fatigue cracks in a tensioned plate subjected to acoustic loads. Chap. 18 in *Acoustical Fatigue in Aerospace Structures* (Eds. W. J. Trapp and D. M. Forney), Syracuse University Press (1965).

CHAPTER 18

The Damping of Jet-Excited Structures

18.1 Introduction

The damping of a structure is one of its most important properties when it undergoes vibration of a resonant character. When a simple linear system is excited harmonically at its resonant frequency the damping is the only system characteristic which controls the response. When the same system is excited randomly, the mean square value of the displacement is inversely proportional to the product of the damping and stiffness, as shown by equation 14.10. Random response of this type exhibits a 'resonant peak' in its power spectrum and may be described as having a 'random-resonant' character. Chapter 16 gives conclusive evidence that jet-excited vibrations are of this type. Their amplitudes are therefore limited by the damping of the structure. For this reason any increase of the damping of the significant modes of vibration of the structure offers a means of reducing the vibration amplitudes and associated stresses.

In this chapter we shall consider the sources of damping in a conventional aeroplane structure and we shall quote some typical values. We shall then proceed to discuss ways whereby the damping can be increased by the addition of visco-elastic damping treatments. The likely stress reductions obtainable with certain treatments are given.

18.2 The representation of the damping in the equation of motion

The generalized damping force and coefficient corresponding to a particular mode of vibration have been discussed in chapter 13. The equation of forced harmonic motion corresponding to one of the modes is of the form:

$$M \frac{d^2q}{dt^2} + C \frac{dq}{dt} + Kq = P\,e^{i\omega t}$$

when the damping is viscous.

The critical damping coefficient is defined as that value of the damping coefficient which causes a free motion to be just aperiodic, and its value is readily found to be $2\sqrt{MK}$. We then define the *damping ratio* or *damping factor* as the ratio of the actual damping to the critical value, thus:

$$\text{Damping ratio, } \zeta = C/C_{\text{crit}} = C/2M\omega_r.$$

In this ω_r is the undamped natural frequency of the mode, $\sqrt{K/M}$. When the mode executes one complete cycle of harmonic motion at its natural frequency, it can be shown that

$$4\pi\zeta = \frac{\text{Energy dissipated during the cycle}}{\text{Maximum amount of potential energy stored during the cycle}}.$$

When the complex stiffness or hysteretic form of damping is used instead of viscous damping (as in section 13.1) we may define a hysteretic damping factor using the above energy definition. The energy dissipated during a cycle is now independent of the frequency of vibration, so that the frequency of the harmonic cycle does not have to be specified. In this way we may show that if the generalized hysteretic damping coefficient is H, then the corresponding hysteretic damping ratio, ζ_H, is $H/2K$. This is also shown by equation 13.19.

Another non-dimensional measure of the damping which is frequently used is the *loss factor*, η, which may be defined by the energy equation:

$$2\pi\eta = \frac{\text{Energy dissipated per harmonic cycle}}{\text{Maximum energy stored during the cycle}}. \tag{18.1}$$

From this it is evident that

$$\eta = 2\zeta_H.$$

18.3 The initial damping of aeroplane structures

By 'initial' we imply the damping of the structural modes before any deliberate attempt is made to increase it. The sources of the damping are threefold:

(a) From the structural hysteresis itself, deriving principally from the numerous joints in the structure. The damping of the material of the structure is very small in comparison.

(b) From the acoustic radiation away from the structure, energy being dissipated into the adjacent air by virtue of the velocity of vibration.

(c) From sundry sources such as frictional effects from stowed equipment, etc.

The structural and acoustic damping will now be discussed in some detail.

18.3.1 *The structural damping*

A possible mechanism for the damping arising from the joints in the structure has been suggested by Mead.[1] Actual slipping of one joint face

13

against the other (as the rivet or bolt deflects) is considered to be the last phase of the mechanism. The damping arising from joints has been found from experiment and theory to be a non-linear function of amplitude. Early work at Southampton[1] showed that the energy dissipated per cycle may be proportional to the load amplitude raised to the power of 2.2 or even 2.6. For complex stiffness damping this power is 2.0. Later work[2] has shown that over the low load range the power is sensibly constant at 2, but the non-linearity (change to a higher power) occurs at a certain critical load level. At loads higher than this, the damping changes with duration of load excitation, as a fretting–eroding mechanism proceeds. The damping actually drops in this period, until it is equivalent to the low load value. At the same time, the critical load for the onset of non-linearity *increases*. The overall effect is that the curve of energy dissipation against load amplitude is shifted to the right.

When a fuselage structure vibrates in the overall or local modes, described in chapter 15, each joint in the structure is subjected to oscillating loads. The magnitude of these loads depends on the inertia loading of the mode. It follows that some modes will cause certain joints to be highly loaded, while other modes cause them to be lightly loaded. The contribution to the damping of a mode from any one joint will, therefore, vary from mode to mode. It appears that stringer-skin joints under shear contribute little to the damping of stringer bending modes or of the overall modes. Skin to skin joints probably contribute considerably to the damping of the overall modes, and stringer-skin joints under tension loading to the stringer bending modes.

Tests on a large riveted fuselage-type cylinder have shown structural damping ratios of the order of 0.0007 to 0.003 for the overall modes. The modal wavelengths were relatively large and the frequencies below those usually encountered in jet efflux problems.

Tests on model reinforced panels (with stringers, frames, rivets, etc. of the same size as full-scale components) have yielded damping ratios varying from 0.002 to 0.012. It appears to be generally true that stringer bending modes have a higher damping ratio than stringer torsion modes. The intermediate modes tend to have damping ratios in between these two extremes. On a large model structure with closely spaced stringers and frames the damping of the two bounding modes was found to be 0.005 and 0.008 respectively. On another smaller model with more widely spaced reinforcing these figures have varied from 0.005 and 0.006 to 0.007 and 0.013. The last pair were measured on a model with a 16 S.W.G. skin, whereas the first pair came from a model with 20 S.W.G. skin. This suggests that the damping ratio increases with skin thickness, but it was also found that reducing the skin thickness below 20 S.W.G. also increased the damping. *For this particular model* the 20 S.W.G. skin gave the minimum damping. For other models having different stringer and frame configurations, the minimum will occur at other skin thicknesses. For a fuller discussion of the effect of skin thickness and mode of vibration on the damping see reference 3.

From the results of tests on the Caravelle aircraft, damping ratios of 0.016–0.018 have been measured for stringer torsion modes. These were deduced from measurements of the random vibration of the structure excited by jet noise and may have been subject to certain errors which could make the apparent damping greater than the true damping.

Some damping measurements have also been made at Southampton on aluminium honeycomb sandwich structures.[4] In simple beams of honeycomb sandwich (light alloy skins and cores, Redux bonded) the damping due to the adhesive bond was extremely small. On actual honeycomb structures the structural damping must therefore derive principally from the interpanel joints and not from shearing of the bond.

18.3.2 *The acoustic damping*

This has been discussed in reference 5, consideration being given to both single panel modes and to overall modes. A simple theory may be derived for the damping of panel modes by splitting up the panel into a large number of elementary 'pistons', each vibrating with the local amplitude of the panel. If the panel is considered to be set in an infinite plane and rigid wall, then the acoustic pressure distribution from a simple small piston is readily obtainable from classical theory. The pressure has components in phase with and in quadrature with the piston velocity. The in-phase component opposes the piston velocity at the piston, but at a quarter wavelength away from the piston its phase changes by 180°. The pressure due to one piston will therefore oppose the velocity of all the other in-phase pistons in the region within a quarter wavelength, and will act *with* the velocity of those pistons 180° out of phase with it. If the velocity and pressure are opposed, sound energy is radiated away from the panel. Where they act together, sound energy radiated from another part of the panel is being put back into the panel.

Now the damping of the whole panel is proportional to the product of velocity and total in-phase pressure, integrated all over the panel. Reference 5 gives the result of such integration for a rectangular aluminium panel of any length:breadth ratio, on the assumption that the wavelength of the radiated sound is greater than about two or three times the panel length. A simply supported panel has a slightly lower acoustic damping ratio than a fully-fixed panel, as shown in Fig. 18.1.

If two adjacent (side by side) panels are vibrating in anti-phase, the acoustic damping pressure from one will annul the damping pressure from the other, whereas if they vibrate in-phase the damping is doubled.

For aluminium panels of length:breadth ratio (n) of about 3, the acoustic damping ratio is approximately 0.004. If $n = 1$, the acoustic damping ratio has a minimum value of 0.003 for a fully-fixed panel. It should be pointed out that the acoustic damping coefficient in this case is proportional to the frequency squared, and this must be borne in mind if the frequency of the panel is changed by the addition of damping material, etc.

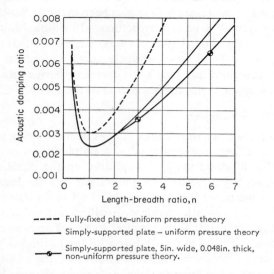

---- Fully-fixed plate–uniform pressure theory

—— Simply-supported plate – uniform pressure theory

—o— Simply-supported plate, 5in. wide, 0.048in. thick, non-uniform pressure theory.

Fig. 18.1   Acoustic damping ratio of rectangular plates vibrating in fundamental modes.

The acoustic radiation from an infinite cylinder vibrating in overall modes has been investigated by Junger.[6] The damping coefficients of these modes vary with the number of waves around the circumference, with the ratio of radiated sound wavelength to cylinder diameter, and with the ratio of sound wavelength to wavelength of the mode *along* the cylinder. The latter ratio varies inversely with frequency, and from Junger's work we may deduce that the acoustic damping coefficient varies with frequency in the manner illustrated in Fig. 18.2. At a certain value of the frequency for each mode the acoustic damping drops to zero. Below this frequency, it remains at zero. This 'cut-off' frequency corresponds to a sound wavelength equal to the longitudinal modal wavelength. As the frequency rises above the cut-off frequency, the damping increases rapidly but then reaches a maximum after which it falls off and approaches an asymptotic value. This asymptote corresponds to the condition of the local damping pressure everywhere being equal to $\rho a \times$ the local harmonic velocity. (ρa = characteristic impedance of the air surrounding the cylinder.)

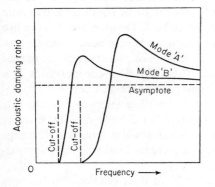

Fig. 18.2 Variation with frequency of the acoustic damping ratio of two typical overall modes.

Below the cut-off frequency, the air has only a 'reactive' effect upon the vibrating cylinder, i.e. its effect can be regarded solely as a change of the mass or stiffness of the cylinder.

Typical values for the acoustic damping ratio have been quoted[5] for a 10 ft diameter stiffened fuselage structure vibrating at frequency f and having n full-waves around the circumference and a longitudinal wavelength λ. They are reproduced in Table 18.1.

Table 18.1 Values of the acoustic damping ratio of a stiffened cylindrical fuselage (overall modes)

f (c/s)	n	λ (ft)	Acoustic damping ratio
400	2	6	0.0208
400	3	3.75	0.0306
400	7	5	0.0299
200	4	10	0.0306

18.4 Methods of increasing the damping

It seems unlikely that an increase in the friction damping of structural joints can be obtained without a serious increase in fretting fatigue troubles. By careful structural design, it might be possible to optimize the acoustic damping by ensuring that the natural frequencies of the important modes occur in the region of the peaks of the curves of Fig. 18.2, but when this happens the modes will be found to be good acceptors of acoustic energy, as well as good radiators. Altogether, then, little or nothing may be gained.

The most promising, and so far *only* method of increasing the damping is to add to the structure certain anti-vibration materials. These are usually high-polymers and have very high natural damping properties. One such material ('Aquaplas') has a loss factor of 0.34 and a dynamic Young's modulus of about 10^6 lb/in^2 at 200 c/s. Greater loss factors are obtainable in the basic polymer, but its stiffness is very much lower.

Since these materials come under the category of plastics their properties are markedly temperature dependent. Their chemical compositions are, however, usually arranged so that the damping properties have their maximum values in the region of the normal operating temperature. Outside this temperature range the properties may fall off quite rapidly. Further developments of the materials have provided a wider temperature range over which satisfactory damping properties are maintained, but the maximum values are not usually as high.

It has been well-established that there is an inter-relationship between the effect of temperature and frequency on these damping properties, i.e. a change of temperature can have the same effect as a change of frequency. Although the damping properties do have this frequency dependence, the stiffnesses and loss factors are commonly represented by a complex Young's modulus, which is directly analogous to the complex stiffness of chapter 13. It should be remembered, however, that both parts of the complex modulus are frequency dependent. Fig. 18.3 shows how the complex shear modulus $G(1 + i\beta)$ of a soft damping material varies with both temperature and frequency.

When damping materials are added to plate-like structures they may be added in the form of unconstrained or constrained layers. An unconstrained layer has one surface which is perfectly free and is obtained when the treatment is sprayed or trowelled on to the plate surface where it dries and hardens. Aquaplas is added in this way. Alternatively a pre-formed layer of the material may be stuck straight on to the surface to be damped (e.g. LD 400, made by the Lord Mfg. Co.). When the composite plate undergoes flexural vibration there is a linear variation of direct bending strain across the section of the plate. Energy is dissipated as the damping material undergoes this oscillating direct strain (see Fig. 18.4a).

An unconstrained layer, on the other hand, has no free surface but is sandwiched between two stiff layers, one of which is the basic plate, to be damped. The other may be a thin metallic foil, which together with the damping material constitutes 'damping tape'. Both the foil and the damping layer may be very thin (0.002 in and 0.005 in respectively) for quite good damping properties. The principal damping mechanism here is

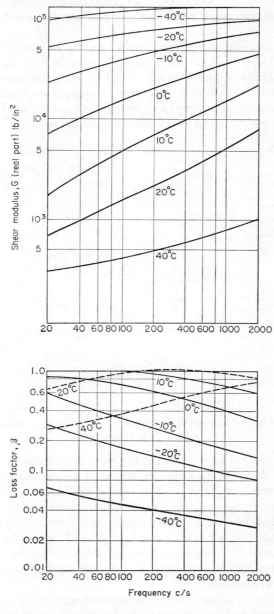

Fig. 18.3 Variation of the complex shear modulus, $G(1 + i\beta)$ of a high damping material.

the shearing of the damping layer which dissipates energy by virtue of the shear strain it undergoes when the composite plate vibrates (see Fig. 18.4b).

The same shearing mechanism applies to the double-skin sandwich configuration (see Fig. 18.4c). Here the structure must be designed *ab initio* with double skins of equal thickness. The two skins are bonded together with a visco-elastic material having high damping properties. The thickness of the damping layer may be as little as 10% of the thickness of one of the outer skins or as great as five times the thickness.

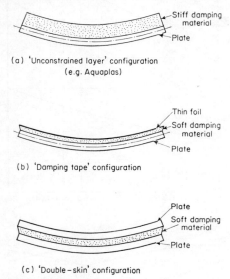

(a) 'Unconstrained layer' configuration
(e.g. Aquaplas)

(b) 'Damping tape' configuration

(c) 'Double-skin' configuration

Fig. 18.4   Energy dissipating mechanisms of constrained and unconstrained layer treatments.

If the conventional type of skin-stringer construction is to be retained, the structural damping may also be increased by including visco-elastic layers (or 'inserts') at the interfaces of the riveted joints in the structure. The layer helps to transmit the load across the joint and so relieves the rivets of some of the load and, furthermore, prevents fretting between the joint plates around the rivets. It has been found[8] that when the joint is correctly designed for maximum damping, the load on the rivet is only about one half of the total load on the joint and a significant improvement in fatigue life is obtained.

The latter method suggests that structural adhesives with good damping properties should be very beneficial. Rivets and associated stress raisers would then be eliminated. Unfortunately, such materials are not yet available and the properties of high strength and high damping capacity seem to be incompatible.

18.5 The effect of visco-elastic layers on the damping and stiffness of plates

18.5.1 *The unconstrained damping layer*

When an unconstrained damping layer is added to a plate, both the flexural stiffness and the damping ratio of the plate are increased. Oberst[7] first derived expressions for these quantities in terms of the material properties and the ratio of the thicknesses of the layer and plate. If the flexural rigidity of the undamped plate is D_u and that of the damped plate is D_d (it being assumed that the plate and damping layer have a uniform thickness throughout) then the 'stiffness ratio', κ, is given by

$$\kappa = \frac{D_u}{D_d} = 1 + \frac{3e\tau_3(1 + \tau_3)^2}{(1 + e\tau_3)^2} + e\tau_3^3 \tag{18.2}$$

and the damping ratio, ζ, of the plate when vibrating in *any* flexural mode is given by

$$\zeta = \frac{\eta_d}{2}\left[e\tau_3^3 + \frac{3e\tau_3(1 + \tau_3)^2}{(1 + e\tau_3)^2}\right]\kappa^{-1}. \tag{18.3}$$

In these expressions, $e = E_d/E_m$; E_d is the real part of the complex Young's modulus of the damping material and E_m is the Young's modulus of the plate material. η_d is the loss factor of the damping material. τ_3 is the thickness ratio defined by

$$\tau_3 = \frac{\text{Thickness of damping layer}}{\text{Thickness of basic plate}}.$$

Fig. 18.5 shows how ζ and κ vary with τ_3 when a grade of Aquaplas is applied to an aluminium plate. In this case $E_d = 1.08 \times 10^6$ lb.in^{-2}; $E_m = 10 \times 10^6$ lb.in^{-2}; $\eta_d = 0.33$.

If the damping layer is very thin compared with the plate thickness, then equation 18.3 reduces to

$$\zeta \approx \frac{\eta_d}{2} \cdot 3e\tau_3 = \frac{3}{2}\frac{E_d\eta_d\tau_3}{E_m} \tag{18.4}$$

and equation 18.2 becomes

$$\kappa \approx 1. \tag{18.5}$$

Equation 18.4 shows clearly that the important physical property of an unconstrained layer damping material is the term $E_d\eta_d$, and not just η_d on its own. The problem facing manufacturers of damping materials is to make this property as large as possible over a wide range of temperature and frequency.

The above expressions apply only to plates which are vibrating in flexural modes and which are not influenced by stiffeners or constraints which contribute to the energy of vibration. If stiffeners on the plate do contribute to the energy then the mode of vibration affects the damping and stiffness ratios.[3]

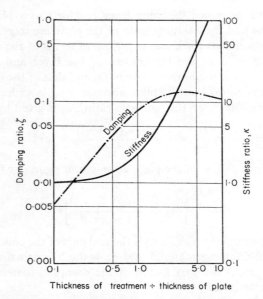

Fig. 18.5   Damping and stiffness ratios for an aluminium plate with a uniform unconstrained layer of damping treatment.

18.5.2 *The constrained damping layer*

The constrained damping layer configuration depends for its action upon the shearing of the damping material. When such shearing deformation plays an important part in the bending of beams or plates, the effective flexural rigidity depends upon the mode of bending of the plate. On account of this, the stiffness and damping ratios of plates with constrained layer treatments are dependent upon the mode of vibration. It has been usual in the analyses of these plates to assume that they bend in sinusoidal modes. The mode can then be uniquely defined and identified by its wavelength.

It has been shown[9] that the damping and stiffness ratios of a plate with a thin damping tape on one surface are given by

$$\zeta = \frac{\beta}{2} \left[\frac{3\Phi\tau_2}{(\Phi + 1)^2 + \beta^2} \right] \qquad (18.6)$$

and

$$\kappa = 1 + 3\tau_2 \frac{(\Phi + 1 + \beta^2)}{((\Phi + 1)^2 + \beta^2)} \tag{18.7}$$

in which the symbols used have the following definitions:

$$\tau_2 = \frac{\text{Thickness of the tape foil}}{\text{Thickness of the basic plate}}$$

$$\Phi = \pi^2 \frac{E}{G} \left(\frac{h}{\lambda}\right)^2 \tau_2 \tau_3$$

$$\tau_3 = \frac{\text{Thickness of the damping layer}}{\text{Thickness of the basic plate}}$$

E = Young's modulus of the plate and foil material (assumed to be the same)

h = Thickness of the basic plate

λ = Semi-wavelength of the mode.

G and β together constitute the complex shear modulus of the damping material, $G(1 + i\beta)$.

The above expressions only apply when the foil is very much thinner than the basic plate, i.e. when $\tau_2 \ll 1$.

$1/\Phi$ has been called the *shear parameter*.[9] For a plate and tape of given geometry, E and G, it is proportional to λ^2 and hence is approximately

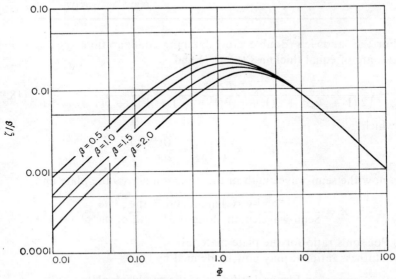

Fig. 18.6 Variation with Φ and β of ζ/β for a plate with a damping tape attached. $\tau_2 = 0.07$.

inversely proportional to frequency. Alternatively, for a given wavelength of a given plate (λ, E and h given) it is proportional only to the tape characteristics G, $1/\tau_2$ and $1/\tau_3$.

Fig. 18.6 shows how ζ/β varies with Φ when $\tau_2 = 0.07$ and β takes different values. ζ has a characteristic maximum which is obviously the optimum operating value for the particular configuration. As Φ varies from large to small values, κ varies from a value which is only just greater than 1 to a slightly greater value. For practical purposes it can be taken to be equal to 1.

Some values which are typical of the damping ratios obtainable with damping tapes are given in Table 18.2. These apply to certain commercial tapes applied to an aluminium plate, 0.04 in thick.

Table 18.2   Values of the damping ratio of an aluminium plate, 0.04 in thick, with damping tape on one side.

Foil thickness (in)	Damping layer thickness (in)	Frequency (c/s)	ζ
0.0055	0.0025	100	0.0125
		1,000	0.03
0.008	0.005	100	0.032
		1,000	0.055
0.012	0.005	100	0.047
		1,000	0.07

For the sandwich (double skin) damping configuration, the two outer plates are of equal thickness ($\tau_2 = 1$) and

$$\zeta = \left\{ \frac{\beta}{2} \frac{\Phi_n 3(1 + \tau_3)^2}{(1 + \Phi_n)^2 + 3(1 + \tau_3)^2(1 + \Phi_n) + \beta^2[1 + 3(1 + \tau_3)^2]} \right\} \quad (18.8)$$

in which

$$\Phi_n = \pi^2 \frac{E}{G} \left(\frac{r}{\lambda}\right)^2 \tau_3 \frac{(n^2 + 1)}{2(1 - \nu^2)}.$$

λ is now the semi-wavelength in the direction across the plate and

$$n = \frac{\text{Semi-wavelength across the plate}}{\text{Semi-wavelength along the length of the plate}}.$$

ν is Poisson's ratio for the plate material.
The stiffness ratio, κ, may now be defined by

$$\kappa = \frac{\text{Flexural stiffness of the sandwich plate}}{\text{Flexural stiffness of a solid plate of thickness } 2h}.$$

This has been found[10] to be given by

$$\kappa = \frac{1}{4}\left[1 + \frac{3(1 + \tau_3)^2(1 + \Phi_n + \beta^2)}{(1 + \Phi_n)^2 + \beta^2}\right]. \tag{18.9}$$

Fig. 18.7 shows that ζ varies with $1/\Phi_n$ in the same general way as ζ for the damping tape. Likewise κ varies in the same way as for the damping tape but it always starts at $\frac{1}{4}$ and rises to a value greater than 1. If the damping layer is very thick, κ can rise to values as high as 5, or more.

In general, relatively soft materials can be used with great effectiveness in sandwich configurations. As soft materials can usually be produced with higher loss factors, β, than stiff materials, it is found that damped sandwich plates can be made with damping ratios higher than those of plates with unconstrained layers.

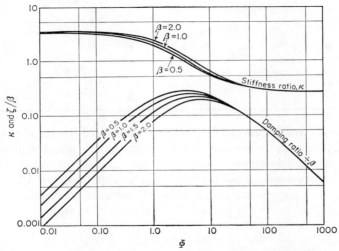

Fig. 18.7 Variation with Φ and β of the damping and stiffness ratios of a damped sandwich plate: $\tau_3 = 1.0$.

As examples of the damping ratios and stiffness ratios obtainable with damped sandwich plates, the values for three commercial products are given in Table 18.3. As these were all of different weights, direct comparison between them should not be made.

18.6 Criteria for comparing damping treatments

In order to compare one damping treatment with another, it is not generally sufficient to compare the magnitudes of the damping ratios they produce. The treatment giving the highest damping ratio is not necessarily the most effective in reducing the vibration response. The effect of the treatment on the stiffness of the system must also be taken into account. It is possible to devise a damping treatment which gives a very high damping ratio at the expense of the stiffness of the system. This treatment may be much less effective than another which gives a smaller damping ratio but which adds to the stiffness of the system.

Table 18.3 Damping ratios and stiffness ratios of some damped sandwich plates

(Specimen A—Hycadamp; Specimen B—Dynadamp; Specimen C—product of Grünzweig and Hartmann, Germany).

Specimen identification	Face-plate thickness (in)	τ_3	Low frequency properties (at 100 c/s)	High frequency properties (at 1,000 c/s)
A	0.0188	1.72	$\zeta_s = 0.06$ $\kappa = 4.1$	$\zeta_s = 0.29$ $\kappa = 1.3$
B	0.0388	0.204	$\zeta_s = 0.023$ $\kappa = 1.6$	$\zeta_s = 0.029$ $\kappa = 0.70$
C	0·025	2.90	$\zeta_s = 0.35$ $\kappa = 2·2$	$\zeta_s = 0.38$ $\kappa = 0.5$

The importance of considering both the damping ratio and the stiffness ratio may be seen by considering a simple system whose equation of motion can be reduced to the form of equation 13.29 (or to its 'complex stiffness' equivalent). A plate vibrating in a single mode of vibration is such a system. Suppose the system is initially without a damping treatment and is being harmonically excited. The equation of motion is

$$M_i \frac{d^2q}{dt^2} + K_i(1 + i2\zeta_i)q = P\,e^{i\omega t}.$$

The subscript i refers to the initial, untreated condition. The initial damping of the system has been included in the complex stiffness, in the way outlined in section 18.2.

At the resonant frequency of the system ($\omega_r = (K_i/M_i)^{1/2}$) the amplitude of q reaches its maximum resonant value given by

$$q_{max} = \frac{P}{2K_i\zeta_i}. \tag{18.10}$$

Now suppose a damping treatment is added to the system so that the damping ratio increases to ζ_t and the stiffness to $K_t = K_i\kappa$. κ is the stiffness ratio mentioned in section 18.5. For most practical purposes ζ_t can be taken to be the damping ratio increment provided by the treatment since

ζ_i is usually very small compared with this. The amplitude of the resonant response of the treated system at the new resonant frequency is evidently

$$q_{\max, t} = \frac{P}{2K_t \zeta_t} = \frac{P}{2K_i \kappa \zeta_t}. \tag{18.11}$$

If we now have several possible damping treatments, each of which gives a different value of the product $\kappa \zeta_t$, then the treatment which is the most effective is that which provides the highest value of $\kappa \zeta_t$ since it yields the lowest resonant response. The product $\kappa \zeta_t$ is therefore the criterion by which to compare the effectiveness of different treatments on resonant harmonic response. The larger the value of $\kappa \zeta_t$, the smaller is the response.

If the same system is randomly excited, equation 14.10 may be used to estimate the mean square value of the random response of the untreated system, $\overline{q_i^2(t)}$. i.e.

$$\overline{q_i^2(t)} = \frac{\pi}{2} S_f(\omega_r) \frac{\omega_r}{K_i^2 \zeta_i}$$

$$= \frac{\pi}{2} S_f(\omega_r) \frac{1}{M_i^{1/2} K_i^{3/2} \zeta_i} \tag{18.12}$$

When the damping treatment is added to the system, increasing M_i to θM_i, the mean square value of the response becomes

$$\overline{q_t^2(t)} = \frac{\pi}{2} S_f(\omega_r) \frac{1}{M_i^{1/2} K_i^{3/2} \theta^{1/2} \kappa^{3/2} \zeta_t}. \tag{18.13}$$

The r.m.s. value is

$$q_{\mathrm{rms}, t} = \left(\frac{\pi S_f(\omega_r)}{2 M_i^{1/2} K_i^{3/2}} \right)^{1/2} \frac{1}{\theta^{1/4} \kappa^{3/4} \zeta_t^{1/2}}. \tag{18.14}$$

The effects of different damping treatments on the r.m.s. displacement response of the system may therefore be judged by comparing the values of the product $\theta^{1/4} \kappa^{3/4} \zeta^{1/2}$ applicable to the different treatments. The higher the value of this product, the lower will be the displacement response. We call this product the *random displacement criterion*.

In a similar way we may derive criteria applicable to the bending stresses in the randomly vibrating plate, or to the forces which the plate is exerting on its boundaries. The latter can be regarded as those forces which load the rivets which attach a skin-plate to the reinforcing sub-structure. The r.m.s. values of these are proportional to the modulus of the complex stiffness, $K_i(1 + 4\zeta_t^2)^{1/2}$, times the r.m.s. displacement; i.e. to $1/\theta_i^{1/4} \kappa^{-1/4} \zeta_t^{1/2}(1 + 4\zeta_t^2)^{-1/2}$. $\theta^{1/4} \kappa^{-1/4} \zeta_t^{1/2}(1 + 4\zeta_t^2)^{-1/2}$, then, is the *random boundary force* criterion. The higher its value, the smaller are the forces on the plate boundaries.

These criteria, and others appropriate to other response quantities, have been used to compare different grades of Aquaplas.[11] They may also be used to optimize the design of sandwich plates (chapter 5 of reference 10). Some of the results of this work are summarized in the next section.

18.7 The optimum use of damping treatments

18.7.1 *The unconstrained layer*

Since unconstrained layers dissipate energy by virtue of the bending strains they undergo, it follows that they act most efficiently on vibrating plates when they are placed in the regions of highest bending strain, viz. in the regions of vibration anti-nodes, and as far as possible from the neutral surface of bending. Near the nodes, the material undergoes very little strain and dissipates very little energy. If no material is placed in these regions but is concentrated instead near the anti-nodes, more of it is subjected to higher strain and the overall damping effect of the damping treatment is improved. This technique is known as the *anti-nodal treatment*.

There is a limit, however, to the degree of concentration that is allowable. If the material is packed into a very small region, it stiffens that region so much that only very small bending strains can occur there. Most of the bending of the plate occurs outside the region, so that the arrangement is again inefficient from the damping viewpoint. There is, in fact, an optimum proportion of the plate that should be covered to give the highest damping, as shown by Fig. 18.8. This relates to a beam treated with a given amount of an unconstrained layer which covers the centre $p\%$ of the beam with a uniform thickness. The beam was taken to be simply supported at each end and was allowed to vibrate only in its fundamental mode. Fig. 18.8 shows how the damping ratio, the random boundary force criterion and the harmonic resonant criterion $\kappa\zeta$ vary as p is reduced, the thickness of the layer being increased at the same time. These curves were calculated for the special conditions: $E_d = 0.70 \times 10^6$ lb.in^{-2}; $E_m = 10 \times 10^6$ lb.in^{-2}; $\eta_d = 0.60$, and, when $p = 100\%$, $\tau = 0.8$. Similar curves are obtained with different values of these characteristics, but the maxima occur at different values of p. Evidently, with the above values the maximum efficiency of damping is obtained when $p \approx 60\%$, as judged by any of the three criteria illustrated.

In a more exhaustive analysis of this technique[12] it has been shown that smaller amounts of material, and less stiff materials should be more concentrated (i.e. lower value of optimum coverage, p) than large amounts of stiffer materials. For very large amounts of material, uniform coverage is recommended. Uniform coverage is also recommended if more than one mode of the plate (or beam) is to be damped, for it is clearly impossible to place the material at the anti-nodes of all the modes unless the plate is uniformly covered.

If the vibrating plate is reinforced by stringers which have very low flexural stiffnesses or which are very closely spaced, it may be advantageous to apply unconstrained layer treatments to the crowns of the stringers rather than to the surface of the plate. It has been shown[13] that in this way the damping ratio obtainable may be up to 1,000 times that obtainable by adding the treatment to the skin surface.

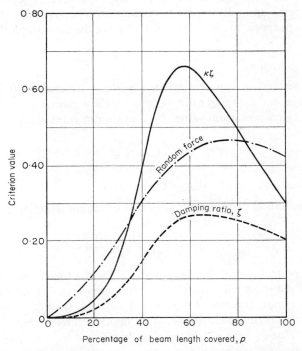

Fig. 18.8 Variation of ζ, $\kappa\zeta$ and the random boundary force criterion with proportion of the beam length covered with an unconstrained layer damping treatment.

When the mode of vibration involves several plate-like members (e.g. the top and bottom skins and ribs of a tailplane) an unconstrained layer treatment is most effective when it is added to the member (or members) with which is associated the greatest amount of the vibrational potential energy. This may or may not be the member which is most fatigue prone in the untreated state. It is conceivable that high stresses in an outer skin may be most effectively reduced by adding a damping treatment to the internal ribs! This, of course, is not a general rule, and each problem must be considered on its own merits.

Another method of increasing the efficiency of an unconstrained layer is to separate it from the plate surface by a shear-stiff spacer layer. The distance between the damping layer and the neutral surface of the composite section is then greatly increased and, in consequence, so also is the bending strain in the damping layer for a given plate deflection. It is important that the spacer layer should be stiff, otherwise its shear flexibility can reduce the bending strains in the damping layer.[14] Certain polyurethane foams and polystyrene mouldings have been produced by 3 M's Co. for this purpose, the foam materials having the advantage of lightness of weight.

18.7.2 *Damping tape*

Figure 18.6 has shown that when a plate is covered uniformly by a damping tape, the damping ratio varies with the shear parameter $1/\Phi$ and passes through a maximum value. This maximum value is given by

$$\zeta_{max} = \frac{3}{4} \tau_2 \frac{\beta}{1 + (1 + \beta^2)^{1/2}}$$

and occurs when

$$\Phi_{opt} = (1 + \beta^2)^{1/2}.$$

From the definition of Φ, this implies that optimum damping is obtained when

$$\tau_{3,\,opt} = \frac{G}{E} \left(\frac{\lambda}{h}\right)^2 \frac{(1 + \beta^2)^{1/2}}{\pi^2 \tau_2}.$$

For a given plate, foil and damping material there is therefore an optimum thickness of damping material for maximum damping. The user of damping tapes does not, however, have control over this thickness as the tapes are usually marketed readymade. The manufacturers have previously chosen the value of τ_3 to give maximum damping over a certain temperature and frequency range when the tape is applied to a plate of specified thickness.

Damping tapes can be applied in multiple layers, one upon the other. Theoretical and experimental work[15] has shown that additional tapes provide a large increase in damping ratio at low frequencies (long wavelengths) but only a small increase at high frequencies. Multiple tapes and a simple tape incorporating an equivalent amount of metal in the foil provide nearly the same damping. In this work, no reference was made to the increase of stiffness obtained with multiple tapes. Neglect of this may have resulted in an underestimate of the effectiveness of multiple tapes at high frequencies.

The low frequency effectiveness of damping tapes can sometimes be improved by cutting the tape in a direction perpendicular to the shorter dimension of the vibrating plate. The cuts cause high local shear stresses to develop in their immediate vicinity, and this causes an increase in the local energy dissipation and overall damping capacity. The conditions under which the cuts are beneficial have been studied by Lambeth and Parfitt.[16]

Spacer layers may be used to increase the damping due to tapes in the same way as they are used for unconstrained layers. Large increases in the stiffness of the plate are also obtained in this way.

Thin aluminium foils are usually used for the constraining layers of damping tapes, but as an alternative it is possible to use thicker constraining layers of softer materials, e.g. a polyurethane foam with a vinyl covering. These can be attached to the plate with pressure sensitive adhesives which act as the damping layer. This is known as 'spacer foil' treatment. The damping capacity of the vinyl covering can also be utilized in this configuration at the same time as the vinyl provides an attractive internal decorative finish.

18.7.3 *The damped sandwich configuration*

With a given thickness of face-plate and damping layer, the damping ratio of the damped sandwich depends only upon the shear parameter and the loss factor β of the damping material. The damping ratio has the maximum value

$$\zeta_{max} = \frac{\beta}{2}\left[\frac{3(1 + \tau_3)^2}{(2 + 3(1 + \tau_3)^2) + 2(1 + 3(1 + \tau_3)^2)^{1/2}(1 + \beta^2)^{1/2}}\right] \quad (18.15)$$

which is achieved when

$$\Phi_n = (1 + \beta^2)^{1/2}(1 + 3(1 + \tau_3)^2)^{1/2}. \quad (18.16)$$

Fig. 18.9 shows the maximum values of ζ which can be achieved with different damping layer thicknesses and loss factors.

The criterion $\kappa\zeta$, which relates to resonant harmonic displacements, varies with Φ_n in the same general way as ζ but its maximum value for a given β occurs when

$$\Phi_n = (1 + \beta^2)^{1/2} \quad (18.17)$$

which is independent of τ_3.

The maximum value of $\kappa\zeta$ is given by

$$(\kappa\zeta)_{max} = \frac{3(1 + \tau_3)^2\,\beta}{16[1 + (1 + \beta^2)^{1/2}]}. \quad (18.18)$$

Notice that the optimum value of Φ for maximum ζ is $(1 + 3(1 + \tau_3)^2)^{1/2}$ times that for maximum $\kappa\zeta$. If $\tau_3 = 3$ then, the optimum values of Φ differ by a factor of 7. It is impossible therefore to design a sandwich plate such that all the criteria are maximized (and all the responses are minimized) at the same conditions of frequency, wavelength or temperature.

If the problem is to minimize random bending stresses in the plate or random boundary forces, still different optimum values of Φ are required.[10] To minimize the boundary forces, a low stiffness ratio is required. To minimize the random bending stresses a high stiffness ratio is required. The art of designing the ideal damped sandwich plate is therefore one of compromise.

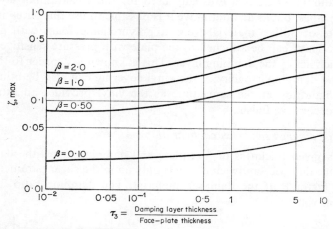

Fig. 18.9   Maximum possible values of the damping ratio of a damped sandwich plate.

The basic problem facing the designer of a damped structure which will resist acoustic fatigue can be stated in the following terms:

A certain weight of skin plating and damping treatment is allowed to cover an area of specified size and stringer spacing. How should this weight be distributed between the skin plating and the damping treatment in order to give the minimum response?

To answer this question, a 'constant weight' theory of the damped sandwich plate is required, in which the thickness of the face-plates is reduced as the thickness of the damping layer is increased. The total weight for any thickness ratio should be kept constant. Such a theory has been developed[10] from which it has been concluded that the damping layer should be about eight times as thick as the aluminium face-plate, if random bending stresses are to be kept to a minimum. The theory assumed a specific gravity for the damping material of 1. This would be an 'optimum' configuration, so that little loss in effectiveness would occur if the thickness ratio were reduced to, say, 3.

It is worth noting that at these high thickness ratios the plate response seems to be less sensitive to the damping layer properties (and hence to temperature) than it is at lower thickness ratios.

18.7.4 *Visco-elastic inserts at joints*

As with the other constrained layer damping mechanisms, the joint insert gives maximum damping with a certain optimum thickness of the insert. Provided the extensional stiffness of the overlapping part of the joint plate is high compared with the shear stiffness of the rivet, the optimum thickness of the insert is given by[8]

$$t_{opt} = \frac{2lbG(1 + \beta^2)^{1/2}}{k_r}.$$

In this, $2l$ is the total overlapping length of the joint plate, b is the rivet pitch, k_r is the shear stiffness of a single rivet. The above expression applies only to joints which are loaded longitudinally and which have a single line of rivets. If the extensional stiffness of the plates is low, the optimum layer thickness must be smaller than the value given by the above expression. It is not possible to state what the values of modal damping ratios will be when joint inserts are used. They will depend very much on the detail of the structure and of the particular mode concerned. To calculate the damping ratio, it is necessary to determine the total amount of energy dissipated per cycle by all the damped joints, and also the maximum energy stored when the structure vibrates in the particular mode. The damping ratio is then found by using the energy definition given in section 18.2. It is not likely that the damping ratios obtained by the use of inserts will be anywhere near as large as those obtained with the layered treatments applied to the plate surfaces.

18.8 The effect on stresses of increasing the damping

In this section we shall compare the stresses in a treated structure with the corresponding stresses in an untreated structure of the same type. Comparisons based on theoretical analyses have been made for Aquaplas[5] and for the double-skin sandwich.[10] Only a limited amount of experimental work has been conducted.[3]

Theoretical comparisons can be made quite easily by using the values of the criteria discussed in section 18.6. The maximum r.m.s. bending stress in a plate which vibrates randomly in a single mode is proportional to the r.m.s. bending deflection, q_{rms}, times the distance, $\bar{y}$, between the free plate surface and the plate neutral surface, i.e. $\sigma_{rms} \propto q_{rms}\bar{y}$.

Using the subscript i to denote quantities corresponding to the initial, untreated structure, and the subscript t to relate to the treated structure, we have

$$\sigma_{\mathrm{rms},\,i} = q_{\mathrm{rms},\,i}\,\bar{y}_i$$

and

$$\sigma_{\mathrm{rms},\,t} = q_{\mathrm{rms},\,t}\,\bar{y}_t.$$

Using these two equations and equations 18.12 and 18.14 we have

$$\frac{\sigma_{\mathrm{rms},\,t}}{\sigma_{\mathrm{rms},\,i}} = \frac{\zeta_i^{1/2}}{\theta^{1/4}\kappa^{3/4}\zeta_t^{1/2}\bar{y}_i/\bar{y}_t}$$

This assumes that the power spectral density of the generalized exciting forces are the same for the treated and untreated structures at their respective natural frequencies. The ratio of the stresses is therefore seen to be dependent on the initial damping ratio, ζ_i, and also on the product $\theta^{1/4}\kappa^{3/4}\zeta_t^{1/2}\bar{y}_i/\bar{y}_t$. This product has been called the 'random stress criterion', and is applicable as it stands to plates treated with unconstrained layers. If sandwich plates are under consideration, the term $\bar{y}_i/\bar{y}_t$ must be replaced by another more complicated function of the sandwich plate dimensions.[10]

In a similar way, it may be shown that the corresponding ratio of the r.m.s. boundary forces (and stresses) at the edges of randomly vibrating plates is given by

$$\frac{\zeta_i^{1/2}}{\theta^{1/4}\kappa^{-1/4}\zeta_t^{1/2}(1 + 4\zeta_t^2)^{-1/2}}$$

The denominator here is the random boundary force criterion of section 18.6.

Specific values of these ratios can only be quoted when the damping ratio, ζ_i, of the untreated structure is specified. In this section, we shall take this to be 0.02 which is a rather greater value than those measured by Clarkson and Ford (see chapter 16). However, by taking a high value for ζ_i, the effect of the damping treatment on the stresses is shown in a conservative light.

Fig. 18.10 shows the random stress and random boundary force ratios obtained as above for an aluminium plate coated uniformly with an unconstrained damping layer. The properties of the material are the same as those pertaining to Fig. 18.5, viz. $E_d = 1.08 \times 10^6$ lb.in^{-2}, $\zeta_d = 0.33$. It is evident from Fig. 18.10 that quite a large amount of damping material is required before the random stresses are reduced by 90% (to a stress ratio of 0.10). If the value of ζ_i had been lower, then less damping material would have been required.

The random boundary forces, it is seen, cannot be reduced by more than about 40%. If very large amounts of damping material are applied, then the effect of the added stiffness counteracts the effect of the additional damping, and no further gain is achieved.

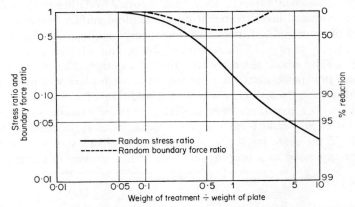

Fig. 18.10 Variation of the random stress ratio and random boundary force ratio with weight of damping treatment applied uniformly to an aluminium plate.

Fig. 18.11 shows the stress ratios and reductions obtained by using the ideal sandwich plate of constant weight, i.e. as the thickness of the damping layer increases (τ_3 increasing) the thickness of the face-plates is reduced to maintain the total weight at a constant value. Furthermore, as the thicknesses change, the shear modulus of the damping layer also changes to the optimum value for maximum effectiveness. The loss factor of the damping material has been taken to be constant throughout at 1.0. Random stress reductions of 80% are evidently obtainable with a thick damping layer. Slightly greater reductions of random boundary forces are obtainable.

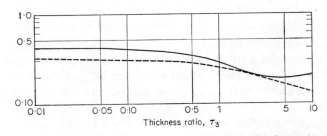

Fig. 18.11 Variation with damping layer thickness ratio of the random stress ratio and the random boundary force ratio for an ideal damped sandwich plate (constant weight as τ_3 changes).

When the damped plates form part of a structural surface and are reinforced by stringers, frames and ribs, the stress reductions will be rather smaller than suggested above. This is due to the effects of the reinforcing on the stiffness and damping ratios.[3]

Some experimental results have been reported[3] relating to the effect of an unconstrained layer treatment on the random surface-bending stresses in a set of structural test specimens. The specimens consisted of aluminium alloy plates, 48 in × 16 in, reinforced by a pair of frames and seven stringers. The whole assembly was riveted together. The plates were of different thicknesses, but the total weight of each plate was made up to that of a solid 16 S.W.G. plate by the addition of a uniform layer of Aquaplas. The structures were subjected to plane random sound waves from a siren at grazing incidence. The r.m.s. stresses at the centre and edge of one of the plate bays were measured. Fig. 18.12 shows these r.m.s. stresses expressed as a proportion of the r.m.s. stresses in the untreated

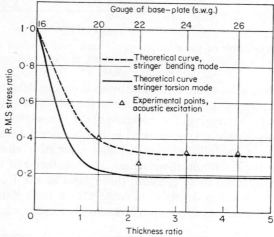

Fig. 18.12   Measured r.m.s. stress ratios compared with theoretical values.

16 S.W.G. specimen. The damping ratios of the significant modes of the 16 S.W.G. specimen were about 0.007. The treated specimens are seen to have measured r.m.s. stresses which are about 30% of the stresses in the untreated specimen, and this proportion agrees quite well with a theoretical prediction.

Moskal[17] has reported on a series of acoustic fatigue tests on damped sandwich plates reinforced with stringers. The measured r.m.s. stresses were about 37% of those in an equal-weight untreated structure when the noise level was 160 dB, but were 24% when the noise level was 154 dB. The acoustic fatigue life of the damped plate was 1.6 times that of the untreated plate at 160 dB, rising to 5.8 at 154 dB. The drop in effectiveness of the damping at higher noise levels and amplitudes is probably due to the effects of increasing membrane tension stresses in the plates as the amplitudes increase.

Much more experimental work needs to be done in determining the effects of damping on stress and fatigue life.

References
1. D. J. Mead. The internal damping due to structural joints and techniques for general damping measurement, Aero. Res. Council (London), Current Paper No. CP452 (1959).
2. D. J. Mead. The damping, stiffness and fatigue properties of joints and configurations representative of aircraft structures, *W.A.D.C. Tech. Rep.*, 59–676, pp. 236–261 (1961).
3. D. J. Mead. The damping of stiffened plate structures, Chapter 26 in *Acoustical Fatigue in Aerospace Structures* (ed. W. J. Trapp and D. M. Forney), Syracuse University Press (1965).
4. D. J. Mead and G. R. Froud. The damping of aluminium honeycomb beams, *A.S.D.-TDR-62-1096* (1963).
5. D. J. Mead. The effect of a damping compound on jet-efflux excited vibration: Part II—The reduction of vibration and stress level due to the compound, *Aircraft Eng.*, **32**, 106 (1960).
6. M. Junger. The physical interpretation of the expression for an outgoing wave in cylindrical co-ordinates, *J. acoust. Soc. Amer.*, **25**, 40 (1953).
7. H. Oberst and K. Frankenfeld. Über die Dämpfung der Biegeschwingungen dünner Bleche durch festhaftende Beläge, I: *Acustica*, **2**, *Akustische Beihefte* 4, 181 (1952).
8. D. J. Mead and D. C. G. Eaton. Interface damping at riveted joints. (Part II—Damping and fatigue measurements). *A.S.D.-TR-61-467*, Part II (1965).
9. E. M. Kerwin. Damping of flexural waves by a constrained visco-elastic layer, *J. acoust. Soc. Amer.* **31**, 952 (1959).
10. D. J. Mead. The effect of certain damping treatments on the response of idealized aeroplane structures excited by noise, *A.F.M.L.-TR-65-284* (1965).
11. D. J. Mead. Criteria for comparing the effectiveness of damping materials, *Noise Control*, **7**, No. 3, 27 (1961).
12. D. J. Mead and T. G. Pearce. The optimum use of unconstrained layer damping treatments, *ML-TDR-64-51* (1964).
13. D. J. Mead. The effect of a damping compound on jet-efflux excited vibration: Part I—The structural damping due to the compound, *Aircraft Eng.*, **32**, 64 (1960).
14. D. Ross, E. E. Ungar and E. M. Kerwin. Damping of plate flexural vibrations by means of visco-elastic laminae, Chapter 3 of *Structural Damping* (ed. J. E. Ruzika), Pergamon Press, New York (1960).
15. E. E. Ungar, D. Ross and E. M. Kerwin. Damping of flexural vibrations by alternate visco-elastic and elastic layers, *W.A.D.C. Tech. Rep.*, 59-509 (1959).
16. G. G. Parfitt and D. Lambeth. The damping of structural vibrations. Aero. Res. Council (London), Current Paper No. CP596 (1962).
17. B. J. Moskal. Investigation of the sonic fatigue characteristics of randomly excited aluminium visco-elastic panels at ambient temperatures, *N.A.S.A. CR-425* (1966).

CHAPTER 19

Acoustic Fatigue Test Facilities

19.1 Introduction

High intensity noise forms an environmental problem in aircraft and missiles where the jet or rocket engines provide sources of high acoustic energy. Any parts of relatively light structure close to the engine and internally mounted electronic and even hydraulic equipments are liable to failure or malfunction. The source of the acoustic environment is a very complicated region of high turbulence in the shear layer of the exhaust stream, together with rotating machinery within the engine. Usually the most severe noise levels occur at take-off or landing. It is not possible to calculate accurately the noise levels and spectra at any point of interest close to the jet, although the typical form of the spectra is known and an order of magnitude estimate of the overall level can be made. Even if the environment could be completely specified it would not be possible to calculate with any accuracy the life of components of a practical flight vehicle. Experience has shown that good detail design of conventional structures can provide a satisfactory vehicle for noise environments having external overall noise levels of around 145–150 dB. For conditions of higher noise levels it is generally necessary to take special precautions in design (e.g. the use of honeycomb structures).

The design, development and proving stages can be considered under two headings: (1) primary structure; (2) electronic and other internally stowed equipment. In the case of the structure the spatial distribution and phase relationship of the acoustic pressures are generally important and a relatively large section of test structure may be required. In the case of internal equipment relatively small items are involved and the sound field inside the equipment bay may be so diffuse that a reverberant test chamber may be a good representation. When the details of any such test are considered it is found that the environment cannot be reproduced accurately and the value of any particular test configuration must be assessed in the light of the effectiveness with which it reproduces the following parameters:

398

(1) The noise environment
   (a) overall intensity and distribution over the specimen;
   (b) pressure spectrum and distribution over the specimen;
   (c) pressure correlation over the test surface;
   (d) variation in r.m.s. level with time.
(2) The structural sample or equipment
   (a) size of specimen as affected by (1);
   (b) edge conditions or mountings;
   (c) damping;
   (d) acoustic properties of cavities and enclosures;
   (e) other stresses present during acoustic exposure.

Many full-scale proof tests of complete aircraft have been made in truly representative conditions but this is a very expensive and time-consuming process. Even when this apparently exact simulation is made it is necessary to make an empirical allowance for damage done at lower noise levels, such as occur during climb and cruise conditions. Generally much simpler tests, often at higher noise levels and hence accelerated testing rates, have been used in the design development phases of an aircraft project. The present policy of many of the companies is to use a small-scale test facility of the siren type for design and then check larger portions of the structure behind a jet engine.

A comprehensive survey of test facilities is given in references 1 and 2 and some discussion is given in reference 3. The various methods of producing a high intensity noise environment and the associated testing rigs are now discussed in some detail.

19.2 Noise sources

In discussing the various types of noise sources which can be used, the complete case of the actual engine on the full aeroplane is not considered. Here it only remains to make some empirical allowance for the damage done at conditions other than take-off. A reasonably accurate allowance can only be made after flight test measurements have enabled a stress–time history for each type of flight plan to be obtained. The effect of different types of flight plan and a suggested design method to allow for these effects is given in reference 4. If it is desired to accelerate the testing this can only be done by carrying out the test in a much colder climate. Schjelderup[5] describes such an accelerated test carried out on a Douglas RB66 aircraft in Alaska. The noise levels were increased by about 3 dB.

Methods of simulating the noise environment artificially are now discussed.

19.2.1 *Jet engine on an open-air stand*

In the case of a jet engine on an open-air stand it is not always possible to use the actual engine designed for the aircraft but generally another similar engine is used. The specimen is then moved closer to the exhaust to get the higher intensities required for the new design. It will not be possible to reproduce the spectra exactly in this way but it should be possible to reproduce reasonably accurately the spectrum in the range from, say, 150 to 1000 c/s, the important structural region. The significance of the change of correlation pattern will depend on the types of modes being excited. At the Royal Aircraft Establishment two full-scale jet engines (one with after-burner) have been used for structural response and fatigue tests. Sections of the Blue Streak skirt structure were tested close to the engine with after-burner to simulate the levels produced by the rocket environment. A programme of proof tests on a series of tailplane test specimens was also carried out and stress levels and spectra were measured.

This R.A.E. facility has now been closed down but another has been made available at Spadeadam based on a jet engine with an after-burner. This facility is not used for extensive proof testing which would be very expensive in fuel alone.

This type of test facility is excellent when an engine of similar noise output to the one of the new design is available. In many cases, however, an engine of sufficient power may not be available and there is a limit to the increase in noise level which can be obtained by positioning the test specimen closer to the jet efflux. In such cases it will be necessary to go to some other noise source—such as a siren—for the initial design development testing.

19.2.2 *Jet engine in a test cell*

Much engine running is carried out in the development of a jet engine and therefore it was considered that the engine test cells might form a useful noise environment in which to test structures. Such test cells have been used to some extent in America as reported in references 1 and 5: but the main difficulty is that the noise spectrum is usually very distorted owing to the reverberant conditions which exist in most test cells. There is also the additional difficulty that in some designs the low-frequency sources downstream in the efflux may be enclosed in the exhaust ducting. In some cases attempts have been made to reduce standing waves by attaching sound-absorbent material to the walls of the test cell. Companies using this method generally strain gauge the structure on the aircraft, if possible, in order to measure actual operating strains. The specimen is then positioned in the test cell, usually with a sound-absorbent backing box, at a position which gives approximately the same r.m.s. strain level. In general it will only be possible to reproduce the r.m.s. level, whereas the strain spectrum will have changed. This limits testing to the time after the first flight of the prototype and cannot therefore be used with any confidence during the design development.

19.2.3 *Loudspeaker arrays*

Loudspeaker arrays have been used by some companies, notably North American, Columbus, to reproduce lower intensity environments for tests on electronic equipment.[6] Up to the economic limit of intensity of the order of 145 dB a reasonably acurate spectrum can be produced. The Columbus installation produces a wave of sound which passes over the test item from an exponential horn. This will not reproduce exactly the pressure correlation patterns present in an instrument compartment which is likely to be a reverberant chamber.

The intensity limitation of the order of 145 dB makes this type of noise source unsuitable for structural fatigue tests, but the ability to reproduce the noise spectrum makes the method very useful for proof tests on electronic components. Loudspeakers are also sometimes used in siren facilities to improve the high frequency content of the noise in the test section.

A large loudspeaker has been used at Southampton University to investigate the response of relatively large pieces of structure, e.g. a Sea Vixen tailplane, Comet tailplane section, and a 6 ft × 4 ft fuselage section. Noise levels of the order of 120–130 dB over these large areas produce measurable strains in the structure and the predominant modes of vibration can be studied. Current work is aimed at relating response under loudspeaker excitation to response to jet noise.

19.2.4 *Air jets*

Where a large compressed air supply is available, a cold air jet is a possible source of high intensity noise. At N.A.S.A., Langley, a 12-in diameter air jet is capable of producing an overall noise intensity of 157 dB over a test area of several square feet close to the nozzle. Four 90° bends in the pipe just upstream of the nozzle increase the low-frequency content of the jet noise. However, in this installation the noise spectrum changes appreciably in the region close to the jet. A constant spectrum cannot be produced over an area much greater than about 1 ft². Larger specimens cannot therefore be tested in this installation. It can, however, be used for comparative panel tests.

In more recent work the 12-ft diameter diffuser exit of the thermal structures tunnel has been used to provide high intensities over large areas.

19.2.5 *Sirens*

The most commonly used high intensity noise device is the siren. This seems to be the only practical way of achieving sound pressure levels of the order of 160–170 dB without using an actual jet engine. One advantage of the siren over the engine is that higher equivalent noise pressures can be achieved, thus making it possible to carry out accelerated fatigue tests. It may also be much cheaper to operate. The earlier sirens were of the pure tone or discrete frequency type as it did not seem feasible to produce a random noise generator which could create the pressure levels required. Recently such random noise generators have been developed but it may still be necessary to use occasionally the more efficient discrete frequency type to carry out accelerated tests. Typical efficiencies which can be achieved are of the order of 50% for the discrete siren and 10% for the random air modulator. A summary of some of the siren test facilities now in operation is given in Table 19.1.

Discrete frequency sirens. In the discrete frequency siren the air flows through a series of ports in the stator and is then modulated by a rotor. The ports in the rotor must be shaped to allow a sinusoidal flow of air as they pass over the stator ports. Poor shaping of the ports produces harmonics in the sound output. The design of a discrete frequency siren is discussed by Clark Jones.[8] For maximum efficiency the air supply pressure should only be about 2–5 lb/in². The sound intensity continues to increase with increase in plenum chamber pressure but the efficiency drops owing to turbulence losses. Increases in pressure above that required to give choked conditions at the ports produce relatively small increases in sound output. In this case shock wave losses cause a further decrease in efficiency. This has the advantage that, with a plenum chamber pressure higher than that required to choke the nozzles, small variations in plenum chamber pressure do not significantly change the sound pressure level in the working section.

As can be seen from Table 19.1, pressure levels of the order of 170 dB in the working section can be achieved with the discrete frequency sirens. The frequency is usually in the range 50–1000 c/s. The lower limit depends on the horn design and the upper limit, which can be extended relatively easily, is fixed by the structural design of the rotor and the power available to drive the rotor. Many of the sirens of this type operating in America are based on the designs of Leonard and Rudnick of the University of California at Los Angeles.

It can be concluded that the main factors affecting the operation and performance of discrete frequency sirens are now understood and satisfactory designs can be produced. As is discussed under section 19.4.3 it is essential to have a fine control of the frequency of the siren for fatigue testing and therefore an accurate speed control of the driving motor is required.

Random noise generators. In practice most structures or electronic components do not respond at just one natural frequency when excited by broad-band acoustic noise. If discrete frequency testing is used there is the difficult problem of interpreting the results in terms of flight vehicle life. Thus great interest has arisen recently in the use of random noise generators for test purposes. Currently there are four types in various stages of development as listed in Table 19.2.

Von Gierke design. The first type of random siren, by H. E. Von Gierke and his co-workers at the Aeromedical Laboratory, A.S.D., Dayton, Ohio, has a single nozzle and four overlapping rotors which each have non-uniformly spaced ports.[10] Adjacent rotors pass in opposite directions with slightly different speeds so as to interrupt the flow of air from the nozzle in a non-uniform way. The output spectrum is composed of a series of lines occurring at frequencies which are sums and differences of multiples of the rotational speeds of the rotors. By suitable shaping of the ports and adjustment of the relative velocity of the rotors a spectrum is attainable which has a relatively high density of line components in the required frequency range. The spectrum can be changed in this way but the region of maximum spectral density is usually in the range 200–300 c/s. The overall noise level produced in a typical installation is of the order of 168 dB.

Altec-Lansing design. In the second type of random siren, designed by Altec-Lansing, the area of the air ports is modulated electromagnetically. Air passes radially inwards through a slotted annular ring which is moved electromagnetically over a fixed slotted ring. The air flow is then deflected to pass axially out of the ring into a horn. In this case the device is relatively small and the noise intensity in a working section is lower (typically 155–160 dB) but the spectrum can be shaped within the operational bandwidth of the siren. The spectrum is limited at the low frequency end by the horn characteristics and at the upper end by the inertia of the vibrating mass. Thus if the rings were to be made larger to allow greater airflow the upper limiting frequency would be reduced owing to the increase in mass of the ring. A higher intensity device has now been developed in this way for simulation of the environment close to very large rocket motors where the frequency of the spectrum peak is very much lower (around 50 c/s). The spectrum of the standard design is shown in Fig. 19.1. Peaks in the spectrum are attributable to coupled horn duct resonances rather than to the transducer itself. Detailed design improvements are being made to raise the upper frequency limit to the region of 1500 c/s.

The advantage of these types of noise generator is that any type of

Table 19.1   Details of some siren test facilities

Facility	Sound level (dB)	Type	Frequency range (c/s)	Air supply		Test configuration	
				Pressure (lb/in²)	Flow (ft³/min)	Size	Angle of incidence
Douglas— Santa Monica	163	D	50–1000	5	4080	up to 4 ft × 5 ft	Grazing
Convair— San Diego	170	D	50–1000	2	1800	22 in × 44 in	Grazing
Convair— Fort Worth	170–180	D	100–2500	10	6000		Normal
Boeing— Seattle	170	D	100–950	50	1400	24 in × 24 in	Normal
Boeing— Seattle	160–165	D or RV	100–1000	40	4 × 200	6 in × 6 in up to 48 in × 48 in	Grazing
Boeing— Wichita	170	D	100–1000	40	1200	24 in × 24 in	Normal
Lockheed— Georgia	170	ML		40	7000		
North American† Columbus	168	ML	50–800	40	7000	up to 4 ft × 14 ft	Grazing
North American— Columbus	173	D	50–600	2–3		up to 4 ft × 14 ft	Grazing
North American— Los Angeles	170	D	50–10,000			3 ft × 5 ft	Grazing
A.S.D. Dayton— Aircraft Laboratory (under construction)	160	D or ML	50–10,000	30	300,000	70 ft × 50 ft × 42 ft 50 ft² floor area	Reverberant Grazing or normal
Aeromedical	165	ML	200–10,000†	40		Used for subjective tests in anechoic room	

British Aircraft Corporation—Filton	153?	D or RV	100–1000	40	300	9 ft × 6 ft	Grazing
Southampton University	159	D or RV	130–1000	30	200	2 ft × 3 ft	Grazing
Hawker-Blackburn—Brough	165?	ML	60–10,000†	15	3200	10 ft × 4 ft	Grazing

D—Discrete frequency; R—Random; V—Variable spectrum; ML—Multi-line spectrum.
† See Fig. 19.1 and under the heading 'Random noise generators'.

Table 19.2   Details of random noise generators

Typical facilities	Siren design and manufacture	Design	Spectrum		Level, dB
			Type	Range, c/s	
A.S.D. (Aeromedical), Dayton Lockheed, North American	Van Gierke Noise-Unlimited	Multi-rotor	Variable line spectrum	Predominantly 100–1000	165–170
Boeing, Seattle	Altec-Lansing	Electromagnetic air modulator	Variable	100–1000	160–165
Boeing, Seattle	Norair	Electromagnetic poppet valve	Fixed	100–1000	165
A.S.D. (Flight Dynamics Laboratory), Dayton	B.B.N.	Discrete frequency, rotor plus amplitude modulator	Single frequency amplitude modulated	50–1000	170

signal, discrete frequency, narrow-band or broad-band noise, can be fed
into the coil of the electromagnetic driver. Fig. 19.1 also shows a comparison between the spectra from a typical jet engine and that produced
by the air modulator in the Southampton University facility. If the full
bandwidth of the modulator is used the spectrum level is that appropriate
to an overall jet noise level of about 159 dB. If, however, the structure of
interest—say the ribs of a control surface—is responding in a relatively
narrow band of frequencies it is not necessary to use the full bandwidth of
the siren. To illustrate the increase in equivalent test level obtained by
restricting the bandwidth of the excitation consider the testing of a structural component whose response is in a relatively narrow band centred at
400 c/s. Table 19.3 shows the increase in spectrum level (and hence structural response) obtained by reducing the bandwidth of the excitation.

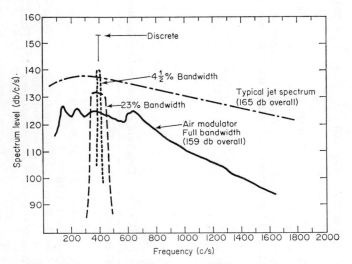

Fig. 19.1   Output spectra for random noise sources.

Thus if the bandwidth of the response is narrow the Altec–Lansing air
modulator can be used to simulate a considerably higher overall level of
jet noise. The appropriate bandwidth to use in any particular test must be
determined from the structural response spectrum. The overall levels can
be increased by using several air modulators. If four are used the overall
level is increased by 6 dB, which makes the figures in the last column of
Table 19.3 appear more useful for the testing of practical structures.
Further increase in the number of modulator units brings a relatively
small increase in noise level at a steeply rising cost.

Table 19.3
Use of the Altec–Lansing air modulator to simulate higher jet overall
levels

Overall level from modulator (dB)	Bandwidth centred on 400 c/s	Spectrum level at 400 c/s, dB/c/s	Equivalent jet overall (dB)	
			1 Siren	4 Sirens
159	Full, 100–800	130	159	165
159	⅓ octave, 23%	137	166	172
159	Narrow band, 4½%	145	174	180
159	Discrete	159	174–176†	180–182†

† Depends on the number of modes and their damping.

Norair design. The Norair siren is similar in principle to the Altec–Lansing design. A large vibration generator is used to operate a poppet valve and thus modulate the airflow. The advantage of the device is that the valve is designed to pass a large airflow and hence high noise levels can be achieved. The disadvantage is that because of the mass of the moving part of the valve the spectrum of the noise output shows a peak at around 100 c/s, and falls continuously with increase in frequency. The spectrum is fixed in shape. Details of the performance of the device in a typical facility have not yet been published, although Boeing, Seattle, have carried out an initial evaluation.

The overall intensity is of the same order as that produced by the Von Gierke design but the peak in the spectrum is at a lower frequency.

B.B.N. design. Bolt, Beranek and Newman Inc. have designed a group of modified discrete frequency sirens for use in the A.S.D. Sonic Test Facility at the Wright-Patterson Air Force Base, Dayton, Ohio.[11] The basic unit in the design is a discrete frequency siren which is modified by the addition of a flow modulator adjacent to the stator. The modulator has ports which alter the effective area of the stator ports when the modulator is oscillating. The modulator can be moved through a total angle slightly greater than that required to completely open and completely close the stator ports. Two sirens size are proposed—one 24 in in diameter having a frequency range from 50 to 2000 c/s and the other 12 in in diameter to work in the higher frequency range from 500 to 10,000 c/s. The modulator can produce a 20 dB modulation of the discrete frequency at frequencies up to about 40 c/s for the 24 in and 80 c/s for the 12 in siren.

The idea is now to replace the truly random noise sources such as jets by a series of sirens which will each give a discrete frequency amplitude modulated in the way described. This is not a true representation as the noise spectrum has a series of lines in it—one for each siren—whereas the

jet or rocket noise has a continuous spectrum. Thus the interpretation of
the test results will depend on the form of response of the structure. It is
difficult to see what advantage this system of modulation has over the
standard discrete frequency form of excitation.

19.3 Test configuration

As can be seen from Table 19.1 the location of the test specimen relative
to the sound wave from the siren remains an unresolved and controversial
question. The types of installation are shown diagrammatically in Fig.
19.2. Some workers feel that in practice a structure experiences primarily
grazing incidence excitation from a jet or rocket exhaust whilst others
maintain that reflections from the runway and adjacent flight vehicle
structure result in mixed or nearly normal impingement.

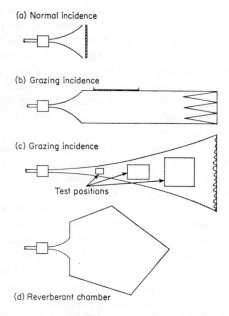

Fig. 19.2   Test configurations. (a) Normal incidence, (b) grazing incidence, (c) grazing incidence, (d) reverberant chamber.

19.3.1 *Normal incidence*

The early siren installations such as those at Boeing and Convair, Fort
Worth, were of the normal incidence type. In this configuration the test
panel is close to the mouth of the horn and reflection of the sound from

the test surface causes pressure reinforcement and therefore results in a higher intensity loading on the panel for a given siren condition. This, however, has the disadvantage that in some cases, when the initial frequency sweep is being made to determine the structural resonances, a coupled panel-horn resonance may be picked out instead of a pure panel resonance. The test may then be carried out inadvertently at this resonance rather than at the true structural resonance. It is also difficult to monitor the pressure level accurately as the pressure level changes rapidly with position in the mouth of the horn.

19.3.2 *Grazing incidence*

As a result of the practical difficulties associated with normal incidence testing the majority of the later and all current designs are based on the grazing incidence configuration. This removes the possibility of coupled panel-horn resonances but loses the pressure-doubling effect and hence the levels are 5–6 dB lower. The earlier types of grazing incidence layouts are of the form shown in Fig. 19.2(b) with an exponential horn connected to a parallel-sided duct which has wedge termination. In later improvements of this layout the whole section becomes an exponential horn, as shown in Fig. 19.2(c) with provision for specimens to be mounted in the side. The advantage of this is that a very good low frequency response can be obtained and test sections of varying sizes (and hence noise levels) can be made in the same test rig. In either design larger structural specimens can generally be incorporated than in the normal incidence configuration and it is often possible to test panel groups, not just single panels. In the grazing incidence case it is possible to excite modes having an even number of half wavelengths along the span as well as odd numbers, but in the normal incidence case it is only possible to excite modes having odd numbers of half wavelengths.

19.3.3 *Reverberant chambers*

The third possible configuration is one in which the sound impinges on the specimen at random incidence. This condition occurs in a reverberant room and may be more representative of the conditions which exist in instrument compartments. Thus more tests on electronic components have been carried out in reverberant rooms. One such large room with random noise sources has been constructed by the Bell Aero Systems Company.

19.3.4 *Comparison of normal and grazing incidence configurations*

The question as to which is the more realistic representation really rests on the problem of the pressure correlation over the test specimen. The spatial correlation coefficient gives a measure of the phase relationship of the pressures over the surface of the test structure. It is important in determining which types of modes of vibration will be excited by the sound pressures. Fig. 19.3 shows typical narrow-band pressure correlations on a structure attributable to a jet efflux and also to boundary layer fluctuations for a centre frequency of 500 c/s. With a normal incidence facility the siren pressure correlation is unity in both directions over the whole area of the test section. With the grazing incidence configuration, pressure correlation is approximately unity in the lateral direction and takes the form of a cosine wave in the longitudinal direction. In the jet noise case it is seen

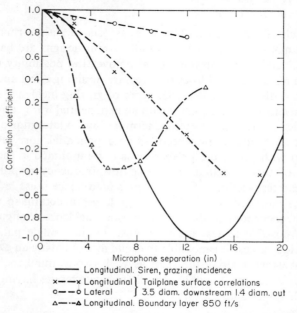

Microphone separation (in)

——— Longitudinal. Siren, grazing incidence

×——× Longitudinal ⎤ Tailplane surface correlations
o——o Lateral   ⎦ 3.5 diam. downstream 1.4 diam. out

△—··—△ Longitudinal. Boundary layer 850 ft/s

Fig. 19.3   Comparison of jet boundary layer and siren filtered space correlograms at 500 c/s.

that the lateral correlation is high over 12 in and is well represented either by normal or grazing incidence layouts. In the longitudinal direction the jet noise pressure correlation varies with position near the jet efflux and in some cases the distance to the first zero crossing point would be less than the 11 in shown and in other cases it may be greater. With the possible exception of positions very close to the boundary it will never be less than the $6\frac{1}{2}$ in of the siren case. By orienting the specimen relative to the sound wave the $6\frac{1}{2}$ in length can be increased up to infinity (representing normal incidence) as proposed in the larger working section of the North American facility. The boundary layer case shown is for an aircraft speed of 850 ft/s.

The spatial scale appears to increase in direct proportion to forward speed and thus will approach the siren case at supersonic speeds. In the lateral direction, however, the length over which the pressures are correlated is likely to be relatively small—of the order of a boundary layer thickness. Thus the siren simulation, where the pressures in the lateral direction are in phase over the whole of the test specimen, is considerably in error.

From these curves it can be seen that neither siren configuration approaches closely the actual pressure correlation patterns in general, although in several cases the jet noise distribution may be reasonably well reproduced by the grazing incidence layout. The degree to which the pressure correlations must be reproduced will depend on the structural wavelength of the particular mode of vibration being investigated. If the actual pressure correlation pattern matches approximately the structural mode pattern then it will be essential to endeavour to reproduce the pressure correlation by grazing incidence. In other cases, where the sound wavelength is long compared with the structural wavelength, both systems should give comparable results.

In many cases, therefore, there may be no overriding reason why one configuration should be used rather than the other. In such cases the fact that the sound pressures can be more accurately controlled and monitored makes the grazing incidence layout the most satisfactory from the practical point of view.

19.4 Structural considerations

In the case of electronic and other equipment mounted internally in the flight vehicle the acoustic considerations will generally be the primary ones. The structural vibration transmitted through the mounting may be important in some cases but usually the anti-vibration mounts will isolate the unit adequately from the surrounding structure. For primary structure, however, the mounting of the specimen and the size of section used can have a considerable effect on the results. These additional effects will now be considered.

19.4.1 *Specimen size*

The problem of representing a flight vehicle structure in a small-scale acoustic test rig is aggravated by the fact that the form of the random vibration taking place in the actual aircraft or missile is not known. From strain gauge measurements, Douglas and others have concluded that, in a typical fuselage, skin panels appear to vibrate in a single mode between the frames which are relatively widely spaced (about 20 in). In other cases when the frame spacing is reduced, as in the Caravelle fuselage and control surfaces, strain gauges show a complex multi-modal response. In the multi-modal case it is clearly insufficient to consider only a single panel vibrating in its fundamental mode, and therefore a much more extensive spatial grouping of panels and support structure must be used. As a result of this the tendency in most of the siren test facilities is to move on from the single panel testing, of which a lot was done in the early days, to tests on panel groups and sections of flaps, etc. This trend reaches its limit in the A.S.D. siren facility, now being constructed, where whole wings or large sections of fuselage could be placed near the sirens. Over large areas of structure like this, however, the siren may not reproduce the actual pressure correlations and indeed unrealistic forms of vibration may be set up. As the form of vibration taking place appears to be relatively local rather than overall it is difficult to see what additional significant information can be gained from this type of test as opposed to a programme using a single siren and smaller structural sections.

19.4.2 *Attachments*

In small-scale tests the type of panel mounting has a considerable effect on fatigue life. In the case of single panels most companies have used fully fixed edge conditions partly to get repeatable test results and also because this was considered to be more representative of the aircraft structure. Fig. 19.4 shows a comparison of the effect of three different types of edge condition. The simple bolted edge gave least scatter but the lowest life, as

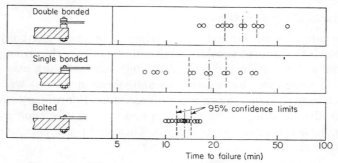

Fig. 19.4   Effect of type of edge fixing on panel life.[7]

failure started at the bolt hole. In the bonded cases the stress concentrations are reduced, mean life increased, but scatter is also increased. Thus the simple bolted configuration is used in many cases. Where multi-panel groups or flat sections are being tested the edge conditions are not quite so critical. In these cases the structure can be mounted from the ribs or frames, care being taken to attach the edges of the skin to prevent flapping. Where skin is attached to rigid supports it is generally necessary to bond a stiffening strip to the skin along the line of the attachment bolts.

19.4.3 *Fatigue aspects*

Because of the difficulty of obtaining high intensity random noise sources, discrete frequency excitation is still used extensively. This immediately raises the serious problem of how the random noise pressures can be represented by a discrete frequency. Strictly a discrete frequency forcing of the structure at its fundamental mode of vibration is only reasonable if the structure is responding in a single mode to the noise environment. If this condition is true, as it may be for some large panels, then the appropriate discrete frequency amplitude can be selected by methods developed from Miles' single-degree-of-freedom analysis by Belcher and others.[12,13, 14,15] Even here the representation cannot be exact, for the uni-modal strain response to noise takes the form of a sinusoidal variation with randomly varying amplitudes. Fatigue under these conditions is related to the constant amplitude sinusoidal siren test by Miner's cumulative damage law, which is known to be only approximately true. Thus, even in the true uni-modal response case, the extrapolation to aircraft lives cannot be exact. In cases where the panel on the aircraft vibrates in a multi-modal form then the extrapolation from discrete to random cannot be carried out other than by empirical methods such as those put forward by Belcher. The N.A.S.A. results (Figure 19.5) illustrate this difficulty when comparison

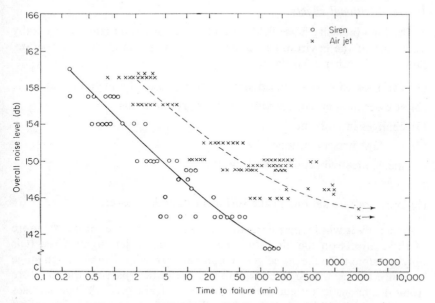

Fig. 19.5 Fatigue life of 0.032 in thick panels under random and discrete frequency loading.[7]

is made between random (by air jet) and discrete frequency testing. Fig. 19.5 shows that, for a given overall noise level, life under siren conditions (discrete frequency) is shorter than that for jet excitation. This is because only a fraction of the energy in the air jet is available at the resonant frequencies of the panel, whereas the siren is tuned to excite one resonance.

One other factor which must be borne in mind is that as a fatigue crack develops the frequencies of the predominant modes change. Thus in the discrete frequency siren tests provision must be made for monitoring the strain response and correcting the siren frequency to keep the excitation on the resonant frequency.

The predominant form of skin vibration under noise excitation occurs in the frequency range from 100 to 1000 c/s. Thus from a fatigue point of view the number of stress reversals which occur during the aircraft life may be as high as 10^9. It is therefore desirable in the development testing to accelerate the test by using increased noise levels. This raises the additional problem of nonlinearities which can occur during the accelerated high level test which may not be present to the same extent under the true environmental conditions. There is also the difficulty of extrapolating down the *S–N* curve for the material to estimate the increased life under the lower stress conditions.

19.4.4 *Additional effects*

The high intensity noise field around the jet or rocket efflux is generally only part of the environment under which the structure has to operate. Some of the other effects are:

(1) variation in noise-induced stresses during the ground run at take-off;

(2) stresses arising from aerodynamic or ground loads;

(3) changes in ambient temperature;

(4) fuselage pressurization;

(5) mechanical vibration environment;

(6) buffet;

(7) pressure fluctuations in the turbulent boundary layer.

Generally it will be necessary to make some allowance for one or more of these effects on the life estimated from siren or jet engine tests. It is only possible to make some very empirical correction in the design stage but once the prototype aircraft is flying, the critical parts of the structure from the acoustic fatigue point should be strain gauged. The stresses occurring during the remainder of the flight regime can then be measured and a more realistic allowance made for their effect.

19.5 Prediction of fatigue life in service

Even when a complete flight vehicle and its correct engines are used for an acoustic fatigue proof test it is not possible to estimate *accurately* the fatigue life of the structure. Apart from the scatter associated with all fatigue testing, the items mentioned under section 19.4.4 will affect the life and the proof test results must be multiplied by a factor less than unity to give an estimate of life under service conditions.

As the test conditions deviate more and more from this complete aircraft proof test the accuracy of any prediction of service life can be expected to deteriorate. Thus some companies use siren tests merely in a comparative way to determine which skin design or attachment is the most suitable for the application. Other companies, notably Douglas and Convair, have devised a method for computing aircraft lives in terms of the life under accelerated discrete frequency testing. The method attempts to allow for the effect of discrete rather than random excitation and the effect of non-linearities and multi-modal response. These allowances, however, are quite empirical and could lead to serious discrepancies. Thus the present trend is towards reducing the number of empirical factors which have to be used. Excitation is changed from discrete to random and larger sections of structure such as a control surface or several frame and stringer bays in a skin specimen are used. In this way a closer prediction can be obtained.

An alternative technique now being considered is to use acoustic test facilities to study and measure the response of the structure. In this case the test levels may be appreciably lower than the anticipated operational levels but the operational levels can be estimated on the assumption of linear response. This procedure will give an estimate of the service stresses in critical regions. Small-scale fatigue tests using electromagnetic vibrators can then be designed to check the lives of the critical parts in a similar way to that described by Townsend and Corke.[16]

In the last resort, however, a check should be made on the stresses in the critical region when the prototype is flying to make sure that an adequate life is likely to be achieved.

Acknowledgment

This chapter is reproduced by permission of the Council of the Institution of Mechanical Engineers from the Proceedings of the Joint Convention on Machines for Materials and Environmental Testing.

References

1. D. M. Forney. Acoustical fatigue test procedures used in aircraft industry and their limitations, *W.A.D.C. T.R.* 59–676

2. R. A. Bianchi and R. T. Bradshaw, and others. Survey and evaluation of sonic fatigue testing facilities, *A.S.D. T.R.* 61–185.
3. B. L. Clarkson. Structural aspects of acoustic loads, *AGARDograph* 65 (1960).
4. G. E. Fitch and others. Establishment of the approach to, and development of, interim design criteria for sonic fatigue, *A.S.D. TDR* 62–26.
5. H. C. Schjelderup. Structural acoustic proof testing, *Aircraft Eng.*, Oct. (1959).
6. Anon. Acoustic problems of flight vehicle design, *Proc. Instn. Environ. Eng.*, Dec. (1959).
7. R. W. Hess, R. W. Herr and W. H. Mayes. A study of the acoustic fatigue characteristics of some flat and curved panels exposed to random and discrete noise, *N.A.S.A. T.N.* D-1.
8. R. Clark Jones. A fifty horsepower siren, *J. Amer. acoust. Soc.*, **18**, No. 2 (1946).
9. I. Rudnick. On the attenuation of a repeated saw-tooth shock wave, *J. Amer. acoust. Soc.*, **25**, No. 5 (1953).
10. J. N. Cole, R. G. Powell, H. L. Oestreicher and H. E. von Gierke. Acoustic siren for generating wide-band noise, *J. Amer. acoust. Soc.*, **35**, 173 (1963).
11. Bolt, Beranek and Newman, Inc. Wide band sound for W.A.D.D. sonic facility, *B.B.N. Report* 685.
12. J. W. Miles. Structural fatigue under random loading, *J. Roy. Aeron. Soc.*, **21**, 753 (1954).
13. G. L. Getline. Correlation of structural fatigue relative to discrete frequency constant amplitude and random excitation, *Convair Report* DG-G-170.
14. A. L. Eshleman, J. D. Van Dyke and P. M. Belcher. A procedure for designing and testing aircraft structure loaded by jet noise, *Douglas Eng. Paper* 693.
15. P. M. Belcher, J. D. Van Dyke and A. L. Eshleman. Development of aircraft structure to withstand acoustic loads, *Aero/Space Eng.*, **18**, No. 624 (1959).
16. N. A. Townsend and D. M. Corke. The fatigue strength between 10^6 and 10^{10} cycles to failure of light alloy skin-rib joints under fully reversed bending loads, *De Havilland Aircraft Co. Note*, Aug. (1960).

CHAPTER 20

Fundamental Aspects of Fatigue

20.1 Introduction

Fatigue is the fracture of crystalline solids under repeated stresses. Since its diagnosis over 100 years ago (the term was first used by Braithwaite in 1854) a vast amount of work has been carried out on fatigue, the main impetus for research coming from the aircraft industry. The two reasons for the highlighting of the problem in aeronautical engineering are:

(1) the high stress levels designed in aircraft structures to increase payload;
(2) the development of high strength age hardening aluminium alloys whose fatigue strength did not keep in step with their improved tensile strength.

20.2 Characteristics of fatigue fractures

20.2.1 *Visual*

(1) Fracture occurs with negligible ductility.
(2) The fracture surface usually reveals a duplex structure consisting of a smooth part, exhibiting tide markings, caused by slow crack propagation in fatigue, and a fibrous or crystalline portion, indicating a rapid tensile fracture.
(3) Fatigue crack propagation occurs in a direction normal to the maximum resolved tensile stress in the system.
(4) The conchoidal marking on the smooth portion of the fracture emanate from the origin of fracture, which is frequently associated with either a geometric stress raiser or fretting.

20.2.2 *Fractographic*

The two features characteristic of fatigue fractures are:

(1) A 'platy' structure of bands parallel to the direction in which the fatigue crack is propagating (best observed at a magnification of ×500).

417

(2) A lamellar or striated pattern in some instances, which lies normal to direction of propagation, and is only just resolvable under the light microscope. These lamellae frequently lie on the tops of the long plateaux.

Work at the R.A.E. Farnborough by Ryder[1] and Forsyth[2] on both service failures and laboratory tests has led to the following conclusions concerning fatigue fractures:

(1) The platy structure is not generally useful in diagnosing fatigue.
(2) The presence of fine striations on a fracture is evidence that failure has occurred by fatigue.
(3) Striations appear to be alternate bands of dark plastic fracture and brittle silvery fracture.
(4) The brittle component increases in width with alloys of decreasing ductility.
(5) Each cycle of stress produces a pair of ductile and brittle striations.
(6) The width of the striations vary with both stress level and frequency of loading.

20.2.3 *Metallographic*

The fatigue fracture path is usually transcrystalline. In corrosion fatigue the most striking feature, apart from the presence of corrosion product, is the multiplicity of cracks only one or two of which propagate to a large extent.

20.3 Mechanism of fatigue

Fatigue in metals is caused by repeated plastic deformation and the basic problem in combating fatigue is the prevention of cyclic plastic deformation. Comparisons of the metallographic features of metals deformed in static and fatigue loading reveal slip lines dispersed regularly through the grain in tensile loading, whereas in fatigue, localized slip bands develop and cracks eventually form within the persistent slip bands. Forsyth[3] has observed 'extrusions' from such persistent slip bands.

Figure 20.1 shows the idealized model of slip during fatigue submitted by Cottrell and Hull.[4] Persistent slip bands form from a pair of complementary sources thus developing the 'extrusions' and 'intrusions' observed metallographically, and which provide the surface roughening of the geometric theory of crack nucleation. Limited amount of grain rotation during primary slip would facilitate this type of fatigue mechanism, and it is significant that the persistent slip bands are characteristic of low amplitude fatigue. The Bauschinger effect will also promote reverse slip on the active slip planes because plastic flow on reversed stressing begins at a load much lower than that for flow under continued loading in the same direction. It would be expected that the backward and forward movement of numerous dislocations in very localized regions of the grain would produce large numbers of vacancies. Thin foil electron microscopy reveals dense clusters of dislocation loops in material stressed at both high and low levels of

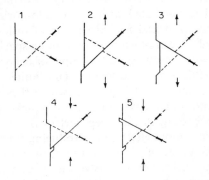

Fig. 20.1   The generation of a crack and an extrusion by slip from Frank–Read sources.

fatigue, thus confirming that a large number of vacancies are nucleated in fatigue. These vacancies then condense to form arrays which collapse, yielding the characteristic dislocation loops of the electron microscope photomicrograph. A high concentration of vacancies can induce structural instability and the various possibilities have been reviewed by Broom[5]:

(a) The creation of vacant lattice sites by moving dislocations with the consequent increase in diffusion rate, causes localized overaging comparable with that occurring isothermally at higher aging temperatures.

(b) An increase in the diffusion rate is brought about in multiplication of dislocations with the consequent increased chances of 'short circuit' diffusion paths (pipe diffusion).

(c) The interaction of mobile dislocations with other dislocations which form the boundaries of coherent precipitates may cause these precipitates to become incoherent, thus reducing elastic strains.

(d) The nucleation of stable precipitates may occur which is a process of overaging.

(e) Sufficiently high temperatures may be generated in slip bands by moving dislocations for normal aging to take place leading to softening and cracking.

(f) Depletion may occur by the re-solution of precipitates. It is suggested that there may be partition of solute atoms between dislocations ('atmospheres') and precipitates. The multiplication of dislocations may necessitate the transfer of solute atoms from precipitates.

20.3.1 *Change in fatigue mechanism with changing initial stress level*

Wood and Bender[6] describe two distinct types of fatigue mechanisms in copper. Region I is associated with the pseudo-horizontal part of the curve where the backward and forward range of dislocation movement during stress reversals does not exceed the average spacing of the dislocation obstacles, so that no progressive 'pile-ups' of dislocations occur, and there is no appreciable strain hardening. The limited strain hardening allows continued slip movement in regions where they are initiated, and therefore slip concentrations develop leading to cracking.

Region II is associated with the steep part of the *S/N* curve and this mechanism is therefore more effective in producing fracture. Wood states that in Region II the range of dislocation movement exceeds the average spacing of lattice obstacles and extensive dislocation interaction occurs causing progressive strain hardening and intense dislocation pile-ups. The dislocation movement is dispersed, thus avoiding slip band cracking, and experimental evidence shows that fracture starts from micro-cracks directly generated by 'pile-ups' in strain hardened regions. Region II type fracture would correspond to a transition between the low amplitude mechanism of fatigue and that associated with ductile static tensile fractures.

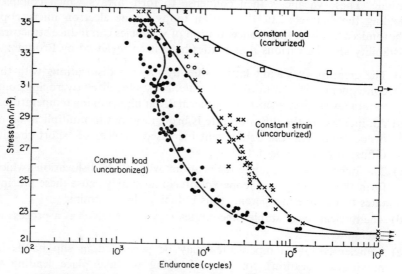

Fig. 20.2 Comparative fatigue testing of 18/8 stainless steel.

A very significant feature of fatigue is that materials which exhibit discontinuous yielding such as mild steel, also show a sharp knee in their *S/N* curve.

Levy[7] has explained the knee of the *S/N* curve in terms of strain aging, which is spontaneous strengthening occurring in mild steel after plastic deformation, due to the diffusion of carbon and nitrogen in the lattice to cause locking of dislocations. At low amplitudes and long life, time is available for the diffusion mechanism to occur, relocking dislocations.

Several investigators have referred to a 'discontinuity' in the *S/N* curve of different materials when tested at high stress levels.[8,9,10,11] Williams and Stevens[12] have considered the significance of the discontinuity in the *S/N* curve obtained with plain specimens (see Fig. 20.2). They suggest that the discontinuity represents a stress level above which no increase in strain hardening can occur due to dynamic recovery effects. The conventional fatigue limit which is approximately 10 ton/in^2 below the stress level of the discontinuity is accounted for by assuming that the stress level for the exhaustion of strain hardening in the weakly supported surface grain is substantially below that required for the core grains.

Williams and Abdilla[13] have subsequently shown that surface treatments aimed at reinforcing the weakly supported surface grains can raise the conventional fatigue limits of both an 18/8 stainless steel and a mild steel to the level of their discontinuities (see Figs. 20.2 and 20.3).

Recent tests at Southampton University[14] utilizing notched fatigue specimens have further emphasized the importance of the discontinuity in the *S/N* curve. Figure 20.4 shows comparative fatigue data in the plain and notched conditions for an 18/8 stainless steel in the cold worked condition. The fatigue limit of the plain specimens has been raised to the level of the discontinuity by the prior strain hardening treatment. Two important features appear in the notched tests. A discontinuity can be seen at a stress level corresponding to the discontinuity in the *S/N* curve of smooth test pieces of the alloy and the fatigue limit in the notched condition is related to the discontinuity by the relationship

$$\sigma \times K_t = D$$

where σ is the general stress, K_t the elastic stress concentration factor and D the stress level of the discontinuity. If therefore the stress level of the discontinuity could be raised a corresponding improvement in fatigue strength could be anticipated and metallurgical effort to improve fatigue resistance in the notched condition should be aimed to this end.

The precise nature of dynamic recovery under cyclic loading conditions has not been established but Feltner[15] suggests that it may involve a detailed balance between pinned dislocations and climbing dislocations.

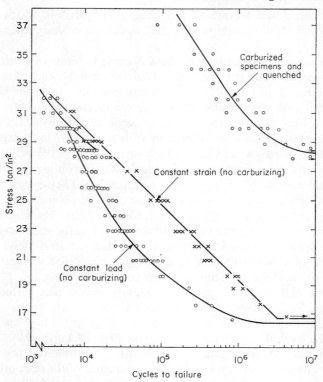

Fig. 20.3   Effect of carburizing on fatigue characteristics of mild steel.

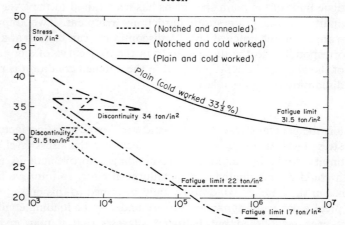

Fig. 20.4   Comparison of notched ($k_t = 2.0$) and plain fatigue testing of 18/8 stainless steel.

The high concentrations of vacancies arising from the movement of vacancy producing jogs in the cyclic movement of jogged dislocations at localized slip planes can afford opportunities for the climb and annihilation of dislocations, thus developing a saturation of strain hardening effects.

20.3.2 *Change in fracture mechanism at constant load*

It is usual to regard fatigue as involving two distinct processes—that of crack initiation followed by propagation. Forsyth[2] has disregarded this division claiming that the crevice deepening due to 'unslipping' in reverse glide may continue to an appreciable depth, and can become a process of crack propagation as shown in Table 20.1 and Fig. 20.5.

Table 20.1

Stage I	Stage II
Fracture occurs along a slip plane	Fracture occurs in a plane perpendicular to the principal tensile stress
No striations on fracture surface	Striations are present on the fracture surface
Stage I is favoured by	Stage II is favoured by
(i) Single slip system	(i) Multiple slip
(ii) High shear/tensile stress system	(ii) Low shear to tensile ratio and therefore by superimposed tensile stress
(iii) Low stresses associated with corrosive environment	

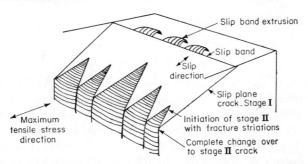

Fig. 20.5   Schematic diagram showing slip plane crack, stage I, the onset of stage II cracking, and the new fracture plane.

Transition from stage I to stage II often occurs at grain boundaries and if a stage I crack is formed which does not receive sufficient tensile stress to propagate it as a stage II crack beyond the grain boundary, a non-propagating crack is formed. The low yield stress of weakly supported surface grains will allow stage I cracking to occur at small amplitude fatigue excitation, and from this model of fatigue, the development of materials resistant to stage II type crack propagation is essential for good fatigue strength.

20.4 Role of active environments in fatigue damage

The role of surface reactions during fatigue is controversial. Work in America[16] has shown that when particular liquids are applied to the surface of a fatiguing test piece, considerable improvement in fatigue life can be obtained (see Fig. 20.6). Nevertheless fatigue damage can occur even in vacuo, and although surface chemical reactions can accelerate fatigue, they do not appear to be essential to the damage process.

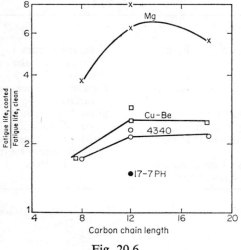

Fig. 20.6

The nature of the striations observed on the fatigue fracture surfaces is also controversial. The component of plastic deformation increases in thickness with increasing stress and with increasing ductility of the material. The brittle shiny component has been identified[17] as cleavage fracture on a cube face in the Al–Zn–Mg alloys. In research at the National Bureau of Standards[18] bubbles of hydrogen gas formed under transparent tape applied to the surface of aluminium alloys during fatigue testing and it may therefore be possible that hydrogen plays an important role in fatigue. The increase in width of the cleavage striations reported by Forsyth[2] with decreasing frequency of fatigue testing suggests that some gaseous diffusion process is occurring in the elastoplastic enclave at the fatigue crack tip. Nevertheless it has not been possible to induce cleavage fracture in the static tensile testing of aluminium despite intense hydrogen charging. The problem of explaining why a crack propagates under a repeated load which it could endure for infinite time under single application of the same load remains to be solved.

20.5 Methods of improving fatigue strength

To offset the types of fatigue damage described by Wood[6] it is necessary to disperse slip movement throughout the crystal so as to avoid high vacancy concentrations in persistent slip bands. It is necessary at the same time to minimize strain hardening and so avoid intense stress concentration at dislocation pile-ups. These requirements appear to be conflicting. Williams and Lowcock[19] have shown that intense strain aging during the blueing of steel is a very effective method of improving fatigue strength and this effect should be further exploited.

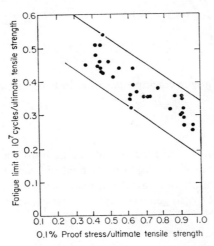

Fig. 20.7   Relation between fatigue ratio and proof stress/ultimate tensile strength ratio for various aluminium alloys.

In smooth test pieces where the stage I crack initiation process is involved, reinforcement of the weakly supported surface grains offers a method of raising the fatigue limit of ferritic and austenitic steels by a substantial amount. For the age hardening aluminium alloys the development of alloys with a more stable precipitate–matrix relationship is desirable and it is significant that aluminium alloys of a low PS/UTS ratio offer improved resistance to crack initiation[20] (see Fig. 20.7). Hardrath[21] has shown, however, that alloys with a high PS/UTS ratio are more resistant to crack propagation at relatively high stress levels (see Fig. 20.8).

The fabrication of duplex structures containing high strength fibres embedded in a crack resisting matrix is a line of development which offers improved methods of design to prevent crack propagation.

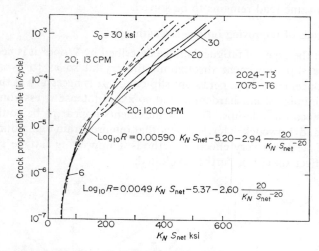

Fig. 20.8   Fatigue crack propagation in two aluminium alloys,
$R = -1$.

20.6 Conclusions

More information is required on fatigue at both the crack nucleation and propagation stages. Microplasticity analysis, and the formulation of the laws of deformation and fracture of restrained enclaves within crystals, is necessary together with general relationships between macro and micro stresses in solids.

References

1. D. A. Ryder. *Metal Ind. (London)*, **89**, 66 (1961).
2. P. J. E. Forsyth. Crack Propagation Conference, Cranfield, September, 1961.
3. P. J. E. Forsyth. *Proc. Roy. Soc.* A **242**, 198 (1957).
4. A. H. Cottrell and D. Hull. *Proc. Roy. Soc.* A **242**, 211 (1957).
5. T. Broom. *J. Inst. Metals*, **86**, 17 (1957–58).
6. W. A. Wood and H. M. Bender. *Trans. Met. Soc. AIME*, **180**, 224 (1962).
7. J. C. Levy. International Conference on Fatigue, London, 1956, p. 887.
8. J. C. Levy and J. Porter. *J. Inst. Metals*, **89**, 86 (1960).
9. J. C. Levy and J. Porter. *Metal Ind. (London)*, **94**, 11 (1959).
10. J. C. Levy and J. Porter. *Metal Ind. (London)*, **99**, 10 (1961).
11. M. T. Lowcock and T. R. G. Williams. *Univ. Southampton A.A.S.U. Rep.* No. 225.

12. T. R. G. Williams and D. T. Stevens. *Nature*, **201**, 1181.
13. T. R. G. Williams and J. A. Abdilla. *The Engineer*, **218**, 325 (1964).
14. G. O'Hanlon. Effect of prior strain hardening on fatigue characteristics of 18/8 stainless steel. Honours project 1966, Southampton University.
15. C. E. Feltner. *Acta. Met.*, **11**, 817 (1963).
16. Organic coatings lessen fatigue damage. *Engineering*, **190**, 7 (1960).
17. P. J. E. Forsyth and C. A. Stubbington. *J. Inst. Metals*, **90**, 238 (1962).
18. Surface reactions during metal fatigue. *Metal Treatment*, **30**, 110 (1963).
19. T. R. G. Williams and M. T. Lowcock. *Iron & Steel*, **35**, 554.
20. P. J. E. Forsyth. *J. Australian Inst. Metals*, **8**, 57 (1963).
21. H. F. Hardrath and A. J. McEvily, Jr. Crack Propagation Conference, Cranfield, September 1961.

CHAPTER 21

The Problem of Cumulative Fatigue

21.1 Introduction

Designing to prevent fatigue failure presents the engineer with problems which can be divided into three groups:

(1) Analysis of the stress patterns in structures into the features important in fatigue damage.

(2) The formulation of the laws of cumulative fatigue damage in a wide range of different materials. It is desirable that the laws utilize the vast amount of S/N fatigue data extant, but if these are inadequate, to specify tests which require the minimum of laboratory testing time.

(3) Compilation of a programme of testing for full-scale assembled structures. The aim in programme testing is to simplify the spectrum experienced in operation into its essential features with regard to fatigue, so as to expedite testing.

21.2 Review of the literature

The cumulative damage theory now widely used is Miner's Rule[1] which is derived as follows:

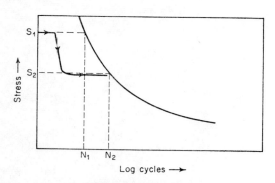

Fig. 21.1   Typical S/N curve of a metal or alloy.

If n_1 cycles of amplitudes S_1 are first applied, then assuming a linear damage theory the proportion of fatigue damage will be n_1/N_1 where N_1 = total number of cycles to failure at amplitude S_1. Let n_2 cycles be the residual fatigue life at a stress level S_2. Then assuming that the fatigue damage at stress levels S_1 and S_2 are both linear and additive, we have:

$$\frac{n_1}{N_1} = \frac{N_2 - n_2}{N_2}, \quad \text{i.e.} \ \frac{n_1}{N_1} + \frac{n_2}{N_2} = 1.$$

In the general case

$$\sum \frac{n}{N} = 1.$$

To facilitate the use of statistical techniques in stress analysis Miner's Rule is often expressed in a form utilizing probability data.

Let N_T be the total number of cycles of varying amplitude to cause fatigue failure.

$$\frac{n_1}{N_1} + \frac{n_2}{N_2} + \cdots + \frac{n_m}{N_m} = 1 \quad \text{(Miner's Rule)}.$$

Then

$$\frac{n_1}{N_1 N_T} + \frac{n_2}{N_2 N_T} + \cdots + \frac{n_m}{N_m N_T} = \frac{1}{N_T}.$$

But $n_m/N_T = p_m$, the probability of occurrence stress S_m in a total of N_T cycles.

Hence

$$\frac{p_1}{N_1} + \frac{p_2}{N_2} + \frac{p_3}{N_3} + \cdots + \frac{p_m}{N_m} = \frac{1}{N_T},$$

i.e.

$$N_T = \frac{1}{\sum \dfrac{p}{N}}.$$

The following are some of the limitations of Miner's Rule:

(i) The chronological sequence in which batches of repeated loads of various magnitudes are applied is discounted.
(ii) Rest periods and stresses below the fatigue limit are ignored.
(iii) The effect of the application of static loads in tension or compression cannot be assessed.

Experimental work has shown all these factors to be important.

Sequence effects

A considerable amount of work has been undertaken on cumulative damage in fatigue at two stress amplitudes both of which lie above the endurance limit. There are several ways of presenting the results and the graphical method is frequently adopted.

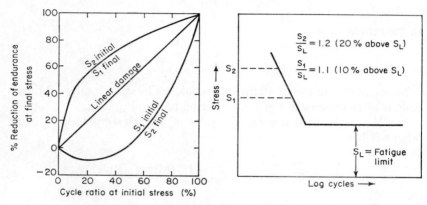

Fig. 21.2   Effect of sequence loading in steel.

Fig. 21.2 shows results from two-level block testing data on mild steel carried out by Kommers.[2] These show that improved fatigue strength can be obtained by applying the low loads first in two level testing.

Similar results were obtained by Pope, Foster and Bloomer[3] in tests on plain mild steel test pieces fatigued in rotating cantilever machines. When the high stresses are applied first, values of $\sum n/N$ less than one were obtained and vice versa. The application of small blocks of stresses alternating between the two stress levels produced $\sum n/N$ approximately equal to one.

Lowcock and Williams[4] have reported the opposite effect in sequence loading age hardening aluminium alloy L73 in a push–pull fatigue machine utilizing unnotched sheet test pieces—see Fig. 21.3.

Tests carried out by Wood and Reimann[5] on copper and 70/30 brass also show that the Lo-Hi sequence of testing presents the most severe fatigue damage conditions, agreeing with the aluminium results and contrary to those reported for ferritic steels. Wood and Reimann applied torsion fatigue to plain test pieces and state that the Lo-Hi sequence involves the application of large strains to a structure containing slip zone microcracks which have divided the grain into loosely attached blocks. The high strain loadings open up cracks produced by the prior low strain loadings and a rapid accumulation of fatigue damage occurs. In the Hi-Lo sequence, the high stresses cause a dispersal of slip and a delay in the crack initiation processes.

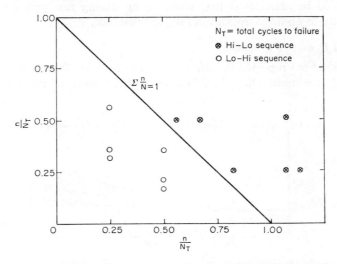

Fig. 21.3   Effect of sequence loading on L73 aluminium alloy.

Levy[6] has suggested that strain aging in the ferritic steels, involving the anchoring of dislocations by solute atoms, could explain the fatigue limit in steels. The same phenomenon could explain the reversal of the sequence effects in ferritic steels compared to the face-centred cubic materials. The development of micro-cracks in the slip bands of the surface grains would be offset by the anchoring of the dislocations in strain aging, and the crack initiation process during low stress cycling would be retarded if not inhibited.

It is possible that the application of a few high stress cycles initially, especially in the notched condition, could reverse the trend in ferritic steels making the Hi-Lo sequence less injurious by the combined action of strain hardening, blunting of notch profile and superimposed strain aging.

Effect of rest periods

Deaves, Gerald and Schultz[7] state that for steels in which free ferrite is present, there is an increase in life when they are rested. Their results show that for an 0.7% C steel loaded in rotating bending to a stress of ± 35 tons/in^2, the mean life of ten specimens was 65,000 cycles. If the specimens were rested for 12 hours after every 6,500 cycles, i.e. 10% of the mean life, the mean of ten such periodically rested specimens was raised to

96,000 cycles, an increase of 48%. One conclusion was that the smaller the number of cycles between successive rest periods the more marked was the increase in life.

It would be anticipated that strain aging during rest periods would contribute to an improvement in fatigue resistance.

Effect of frequency of loading

Serensen[8] has reported the effect of frequency on the fatigue strength of steel and aluminium alloy. Fig. 21.4 shows the results for mild steel and

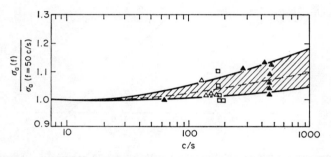

Fig. 21.4   Influence of frequency on the fatigue limit of mild steel.

it can be seen that an increase in frequency of loading in the range 50–1,000 c/s can raise the fatigue limit of steel by as much as 20%.

From Fig. 21.5 it can be seen that increasing the frequency of testing

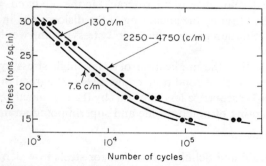

Fig. 21.5   *S/N* curves for light alloys produced at varying rates of loading.

from 7.6 to 2,250 c/min can increase the fatigue endurance of an aluminium alloy type (24S-T) sixfold at a stress of 15 tons/in². No significant increase in fatigue endurance was observed by increasing the frequency from 2,250 to 4,750 c/min.

Finney[9] has reported inflections in the fatigue curves of an aluminium alloy 2024-T4 when tested in the frequency range 100–12,000 c/min (see Fig. 21.6).

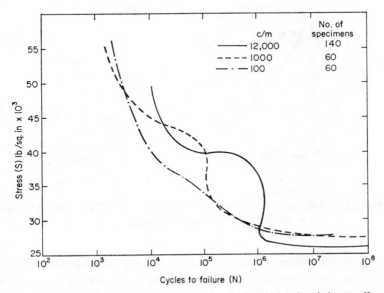

Fig. 21.6 Rotating cantilever tests on extruded aluminium alloy 2024-T4 (notched, $K_t = 1.5$) (after Finney[9]).

Effect of residual stresses

The effects of pre-loading notched test pieces in tension or compression are shown in Fig. 21.7.[10] Residual compressive stresses are beneficial whereas residual tensile stresses are detrimental.

Frost[11] has shown that strain hardening prior to fatigue testing can raise the fatigue strength. He applied compressive stresses to mild steel and then prepared smooth test pieces from the compressed block. Figure 21.8 shows the results obtained. A progressive improvement in fatigue endurance limit is attained with increasing amounts of prior plastic deformation up to 50–60%, but above this range, collapse of the strain hardening effect occurs.

The occasional application of an unusually large static load during the life of a component can therefore have a very significant effect on its fatigue behaviour. Local plastic yielding at notches resulting in residual

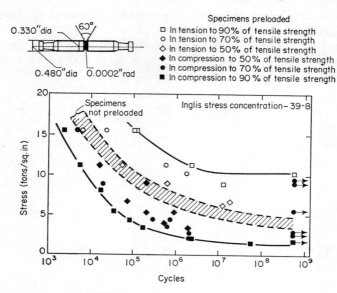

Fig. 21.7   Effects of internal strain on the rotating-beam. Fatigue
strength of 755-T6 (after Templin[10]).

compressive forces can be very beneficial, but the extent of such effects is
difficult to gauge and makes fatigue prediction very hazardous. The wide
discrepancy in the cumulative damage tests reported in the literature is

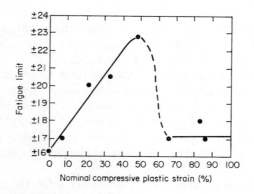

Fig. 21.8   Fatigue limit vs. nominal compressive plastic strain (mild
steel).

undoubtedly due to the variation in the type of stressing employed in the experimental work and the variations in the associated stress fields.

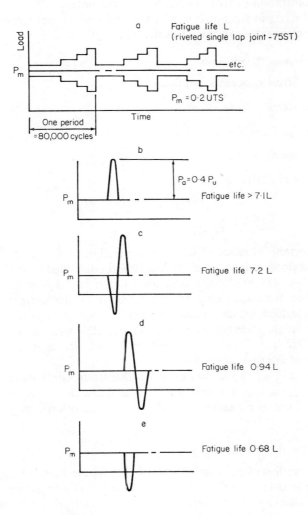

Fig. 21.9   Schematic picture of five test series.

Schijve[12] has investigated cumulative damage in riveted lap joints and has revealed the extreme importance to fatigue life of the residual stress pattern resulting from the loading adopted (Fig. 21.9a and b).

Effect of notch

The presence of a notch introduces three effects:

1. Concentration of stress at the base of the notch.
2. Yield stress at the notch is raised by induced triaxial stresses.
3. The strain rate is increased at the base of the notch.

Now $\sigma_{\text{maximum}} = K\sigma_{\text{nominal}}$

where K = stress concentration factor

and σ = stress.

Hence $\epsilon_{\text{maximum}} = K\epsilon_{\text{nominal}}$

where ϵ = strain

and $\dot{\epsilon}$ = strain rate.

Therefore

$$\dot{\epsilon}_{\text{maximum}} = K\dot{\epsilon}_{\text{nominal}}.$$

The important influence of stress concentration on fatigue resistance is therefore obvious. It is an intensely local effect controlled by design detail. Like buckling strength, fatigue strength is determined by the geometry of design rather than the basic fatigue strength of the material. Joints, riveted and welded, are discontinuities of necessity, and they will inevitably have lower fatigue strength compared to monolithic structures of uniform shape and free from joints. A rivet hole will intensify stress in its vicinity by a factor of approximately 3.0.

Formulae are available for assessing the stress concentration factor $K_t = \sigma_{\text{maximum}}/\sigma_{\text{nominal}}$. For round bar containing a circumferential groove of large root radius r Neuber[13] has derived the equation:

$$K_t = 1 + 2\left(\frac{t}{r}\right)^{1/2}.$$

When the radius of the notch approaches zero, the K_t value approaches infinity and to overcome this difficulty Neuber has developed a concept of elementary blocks so that fatigue resistance is governed not by the maximum stress but the average stress over an elementary block, and the technical stress concentration factor K incorporates this concept:

$$K = 1 + \frac{K_t - 1}{1 + \dfrac{\pi}{\pi - \omega}\left(\dfrac{A}{r}\right)^{1/2}}$$

where ω = one-half flank angle of notch and A = material constant (see Fig. 21.10).

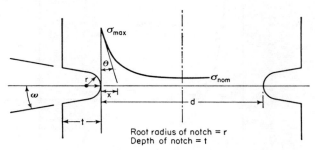

Fig. 21.10   Stress distribution at a notch.

The stress gradient $G = d\sigma/dx = \tan \theta$ and for a round bar with a hyperbolic notch

$$G = \frac{2\sigma_{max}}{r} + \frac{\sigma_{max}}{d}.$$

An experimental or effective stress concentration factor K_f can also be derived, thus:

$$K_f = \frac{\sigma_{R=1(\text{plain})}}{\sigma_{R=1(\text{notched})}} \quad \text{where } R = \frac{\text{compressive peak stress}}{\text{tensile peak stress}}.$$

K_f is usually less than K_t and a qualitative difference can be assessed by the notch sensitivity index $q = (K_f - 1)/(K_t - 1)$ so that q has a range of values 0 to 1.0, i.e. when $K_f = 1, q = 0$, and when $K_f = K_t, q = 1$.

Experimental work at N.E.L. by Phillips, Frost and co-workers[14] has shown that a fatigue crack can be a less severe stress raiser than the original notch and their conclusion is in agreement with Neuber's theory. They differentiate between the stresses necessary to initiate cracks at the root of notches and the stresses required to propagate them. Fig. 21.11 shows some of their results. To a close approximation cracks were found to be initiated if $K_t \times$ nominal attending stress was equal to or greater than the conventional fatigue limit for plain specimens. For crack propagation however a relationship of the form $\sigma^3/l = c$ was deduced where σ is the alternating stress and l is the combined length of the original notch and the crack and c is a constant depending on the material, being 5.5 for mild steel, 0.2 for aluminium alloy BSS L65 and 0.6 for copper. Where however a mean stress is applied sufficient just to keep the crack open through the cycling the value of c becomes $c/8$. For lower mean stress the alternating stress σ_a could be taken as the alternating stress component of the cycle

15

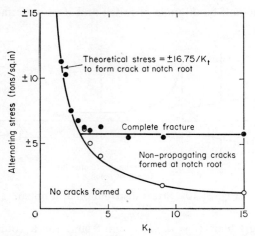

Fig. 21.11   Alternating stress against K_t for mild-steel rotating bending notched specimens.

for which the crack remained open and for higher mean stresses the crack propagation stress could be reduced.

For cases where $K_t\sigma > \sigma_L$ but $\sigma^3/l < c$, non-propagating cracks are formed.

The critical stress for crack propagation is given by

$$\sigma^3_{\text{crit}}l = c$$

$$\text{i.e. } \sigma_{\text{crit}} = \left(\frac{c}{l}\right)^{1/3}.$$

Hence

$$K_{t(\text{crit})} = \frac{\sigma_L}{\left(\dfrac{c}{l}\right)^{1/3}}.$$

Effect of mean stress

Fig. 21.12 shows a collection of a large number of test results made by J. O. Smith[15] for notched test pieces of ductile metals subjected to ranges of repeated axial stress superimposed on a steady stress.

Random loading fatigue testing

Since the London Conference[16] when the work of Hooke and Head revealed the inadequacy of Miner's Rule in predicting fatigue life under random loading conditions, a great effort has been made to derive a

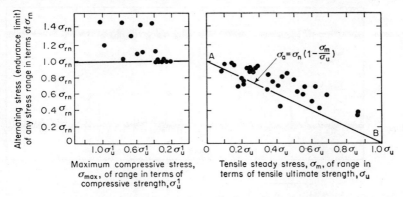

Fig. 21.12   Range of stress diagrams for notched specimens of ductile metals subjected to ranges of repeated axial stress superimposed on a steady stress.

practicable law for assessing cumulative fatigue damage under variable loading conditions.

Most investigators have used a resonance system to amplify the loadings. When excited by a random force input the specimen under these conditions acts as a narrow-band filter at its own natural frequency so that the stress pattern is a modulated amplitude sine wave—see Figs. 21.13a, b and c.

The spectrum of peak stresses under these conditions conforms to a Rayleigh distribution, viz.:

$$P(S) = S\sigma^{-2} \exp\left\{-\frac{1}{2}\left(\frac{S}{\sigma}\right)^2\right\}$$

where P = probability density of peaks

σ = root mean square stress of wave form ($\sqrt{2}\,\sigma$ = r.m.s. of peaks)

S = stress level.

A Rayleigh distribution of peaks may be expressed in terms of $E(S)$, the fractional number of peaks exceeding a stress S, by integrating $P(S)$ between S and ∞, viz.:

$$E(S) = \exp\left\{-\frac{1}{2}\left(\frac{S}{\sigma}\right)^2\right\} \quad \text{i.e. a Gaussian distribution.}$$

Theoretically the values of stress S approaching infinity are included but analysis of the experimental stress distribution shows it to be truncated at values of S/σ in the range 3–3.5 according to the σ level.

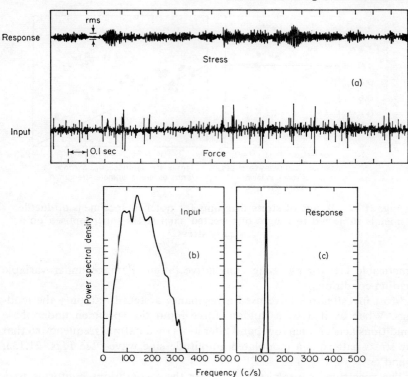

Fig. 21.13   Comparison of input and response for resonance excitation.

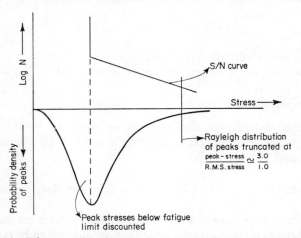

Fig. 21.14   Diagram showing comparative positions of *S/N* curve and distribution of peak stresses in random loading testing.

For the Miner's Rule assessment of fatigue life the fatigue damage contributed by peak stresses from the fatigue limit up to the stress level of truncation is computed assuming a linear damage rule so that the total number of cycles N_T is given by the equation

$$N_T = \frac{1}{\sum \dfrac{p(S)}{N(S)}}.$$

In the work reported by Hooke and Head[16] a notched test piece was subjected to reverse bending and excited at its range of natural frequency. Their results are summarized in Table 21.1 below.

Table 21.1

Average peak stress	11,820 lb/in²	12,900 lb/in²
Average fatigue life cycles:		
Experimental	2.0×10^6	1.15×10^6
Predicted (Miner)	6.9×10^6	3.4×10^6
Ratio: Predicted / Experimental	3.45	2.69

Miner's Rule evidently predicted lives up to 3.5 times the experimental results obtained. The divergence between predicted and experimental life was less at the higher stress levels.

Later work undertaken by Fralich[17] involved the use of notched test pieces of aluminium alloy 7075 excited to reversed bending within their range of natural frequency. Fig. 21.15 shows the results obtained.

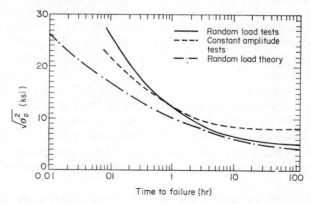

Fig. 21.15 Comparative fatigue data for sinusoidal and random loading.

Random loading studies at Southampton University[18] involve a resonance system, the test piece being subjected to push–pull loadings at a frequency around 90 c/s. Figs. 21.16 and 21.17 show some of the results obtained on aluminium alloy L73 and mild steel.

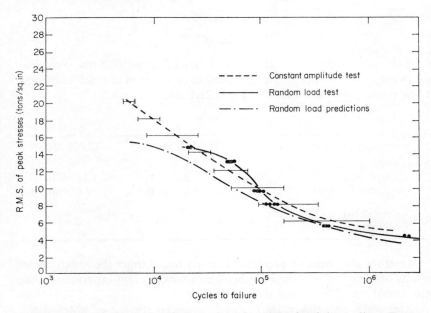

Fig. 21.16   Comparative fatigue data for L73 aluminium alloy. Zero mean load. Push–pull testing. Plain test pieces.

For both the mild steel and the L73 aluminium alloy the application of Miner's Cumulative Damage Rule has provided a conservative estimate of fatigue life. A surprising feature of both sets of results is that the scatter in fatigue life under resonance conditions is very much smaller than that for sinusoidal loading.

There is a discontinuity in the S/N curve of the L73 alloy at a stress level of 16.0 tons/in² and the shape of the variable loading graph is modified when peak stresses in the Rayleigh distribution exceed this stress level. Further evidence of importance of the discontinuity to fatigue life under random load conditions is seen in Fig. 21.18.[19] A discontinuity appears in the random loading curve when peak stresses attain the level of the discontinuity.

The results also show that whereas the discontinuity in the sinusoidal loading curve involves a substantial shift to shorter fatigue life at higher stresses, the discontinuity in the random loading curve reveals the opposite

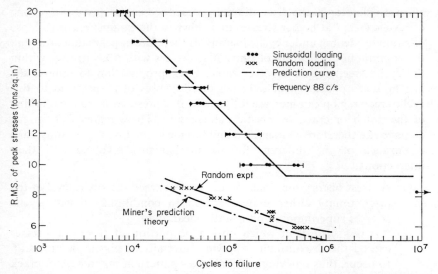

Fig. 21.17 Sinusoidal and random *S/N* curves in mild steel. Notched test-pieces $K_t = 1.5$. Push–pull testing with zero mean load.

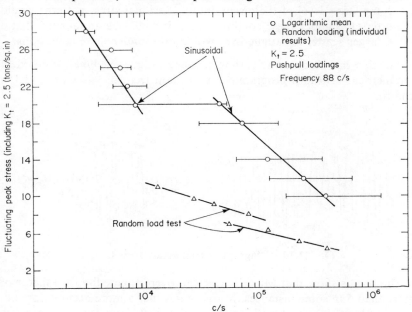

Fig. 21.18 Results on L73 aluminium alloy with a static mean load = 15 tons/in².

effect. In chapter 20 the discontinuity is described as the onset of a regime of 'cyclic creep' at higher stresses due either to the exhaustion or collapse of strain hardening under cyclic loading. It therefore appears that although the continuous application of sinusoidal stresses with peak stresses within the cyclic creep region involves rapid crack propagation to failure, the intermittent application of only the peak stresses of a spectrum to this cyclic creep region can increase fatigue life, due presumably to the blunting of the notch or crack by plastic deformation. Three regimes of fatigue damage can therefore be identified and fatigue prediction theory must take cognizance of the different fatigue mechanisms involved and their interactions:

(1) A stress history for which all the peaks exceed the stress level of the discontinuity either under zero mean conditions or with cyclic stresses superimposed on a static mean load.
(2) A stress history in which only a proportion of the peak stresses exceed the discontinuity allowing intermittent plastic deformation to occur, thus relieving residual stress pattern at notch base or crack tip.
(3) A stress history of low r.m.s. value in which no peak stresses exceed the stress level of the discontinuity. Residual stress patterns would develop under these conditions and it would be anticipated that a larger scatter in fatigue lives would arise under these conditions.

Marsh[20] has modified a rotating–bending fatigue machine to provide a symmetrical sawtooth programme at constant frequency (see Figs. 21.19a and b).

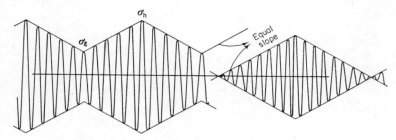

Fig. 21.19   Diagram of triangular stress block.

The results show that Miner's Rule can overestimate life by a factor of more than 5 in some instances. Marsh states that a more accurate prediction of fatigue life can be obtained for mild steel if it is assumed that low amplitude cycles down to 80% of the fatigue limit contribute to fatigue damage in random loading.

Kirkby and Edwards[21] have developed a programme of cyclic stresses to simulate a gust spectrum issued by the Royal Aeronautical Society. A 20-ton Avery Schenck resonant fatigue machine was modified for performing random loading at a testing frequency of approximately 112 c/s. Rayleigh distributions of peaks of different r.m.s. levels were mixed to simulate the first spectrum and applied to simple loaded-hole lug specimens, the lug being machined from BSS L65 bar and the pin being a steel to BSS 94. Fig. 21.20 shows the results obtained and it can be seen that

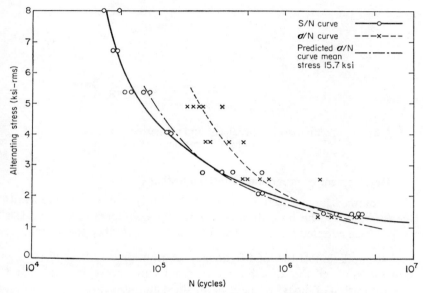

Fig. 21.20   *S/N* and *σ/N* curves for L65 material.

Miner's hypothesis provides an accurate estimate of life at low r.m.s. levels and is a conservative estimate at high r.m.s. levels.

Christensen[22] has applied a linear damage rule to predict the growth of fatigue cracks under random load conditions at resonance. Test panels of aluminium alloy 2024-T3 containing a small starter notch centrally located in the width of the panel were excited in reversed bending at a resonance around 47.0 c/s. Curves of the growth of fatigue cracks in sine wave constant amplitude tests were established and the data was applied in a linear damage rule to predict crack growth for random loadings at resonance. Some of the results are shown in Fig. 21.21 and it can be seen that good agreement is obtained between the predicted and measured crack growth.

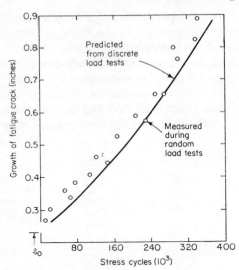

Fig. 21.21   Comparison of predicted and measured crack growth under random vibration.

21.3 Development of programmed fatigue testing

Programmed fatigue testing arises when prototype structures are available for testing. The procedure laid down by Gassner[23] before the last war has been adopted for this test and Haas[24] has described the technique:

(a) Compilation of the load frequency distribution.
(b) Division of the distribution diagram into stress intervals, and allocation of blocks of loads of identical amplitude.
(c) Pairing of positive and negative peaks of equal amplitude.
(d) Arrangement of stress blocks in ascending and descending order.
(e) Commencement of test midway through ascending peak.
(f) Assessment of life in terms of $\sum(n/N)$, the Miner's Rule of cumulative fatigue damage.

Hardrath and Naumann[25] have studied the effect of programme loading on aluminium alloys and critically evaluated the assumptions made by Gassner and Haas. The following are some of their conclusions (see Fig. 21.22):

(1) The sequence of applying loads in a step test has a highly significant influence on fatigue life—see Figs. 21.22a and b.
(2) The mean stress applied in the test has a pronounced effect on the

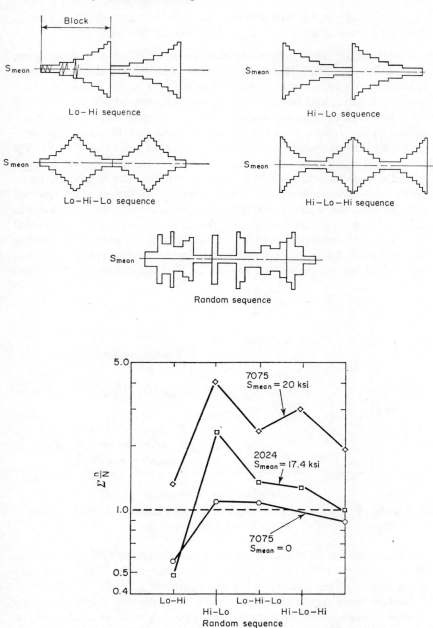

Fig. 21.22   Effect of sequence block loadings on fatigue life.

value of $\sum(n/N)$, tests in which $S_{mean} = 0$ tend to give $\sum(n/N) < 1$, and those in which $S_{mean} > 0$ tend to give $\sum(n/N) > 1$.

Kowalewski[26] has investigated the relationship between programme and random loading. He claims that as every peak value of random loading is substituted in programme loading with half a sine wave of the same peak value, and since a positive peak is combined with a negative of the same distance from the overall mean, the average of all amplitudes is greater in programme than random loading and therefore life under programmed loading based on distributions of peaks will be shorter than life under random loading conditions. Kowalewski carried out his tests on extruded aluminium alloy 2024 test pieces prepared with a circumferential notch of Neuber factor 1.77. The test piece was subjected to reversed bending and the natural frequency of the specimen and vibrator was arranged to be appreciably above the frequency range of excitation so that irregular stress patterns could be applied to the test pieces. By filtering the excitation he developed three wave traces of the same r.m.s. but differing in general shape—see Fig. 21.23. He then distinguished between

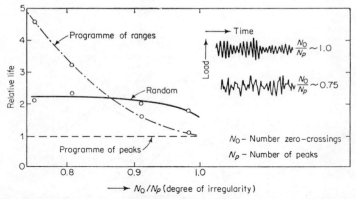

Fig. 21.23   Comparison of fatigue life under random-noise loading and programme fatigue loading.

the wave patterns by calculating the ratio of the number of zero crossings N_0 to the number of peaks N_1; decreasing value of the N_0/N_1 ratio indicated increasing irregularity in the wave pattern.

Kowalewski's results are shown in Fig. 21.23. Surprisingly there is only a small change in the programme life factor despite a variation of N_0/N_1 from 0.75 to 0.98. Some variation in strain range will occur with changing shape of the wave traces but it should not be a large variation and its consequent effect on fatigue strength small. Explanation of the factor of 2 between fatigue lives under random and programme loading conditions must therefore depend on the other significant differences between the two types of stresses, possibly the sequence effect in loading.

21.4 Conclusions

For random loading conditions at resonance the majority of the results show that a linear damage rule can provide a reasonably accurate estimate of fatigue life based on the S/N curve. For conditions where sequence effects apply then Miner's Rule cannot provide a basis for prediction theory.

References

1. M. A. Miner. *J. appl. Mech.*, **12**, 3, A159–A164 (1945).
2. J. B. Kommers. *Proc. Am. Soc. Testing Mater.*, **45**, 532 (1945).
3. J. A. Pope, B. K. Foster and N. T. Bloomer. *Engineering*, **184**, 236–241 (1957).
4. M. T. Lowcock and T. R. G. Williams. *Univ. Southampton, Dept. Aero and Astro. Rep. A.A.S.U.* 225.
5. W. A. Wood and W. H. Reimann. *J. Inst. Metals*, **94** (2), 66–70 (1966).
6. J. C. Levy. International Conference on Fatigue, Inst. of Mech. Eng., London, 887 (1956).
7. H. Deaves, D. Gerald and E. H. Schultz. *Stahl u. Eisen*, **60** (1940).
8. S. V. Serensen. Neschuscaja sposobnost i rascot detalej mashin na procnost, Maschgiz, Moscow (1963): p. 166.
9. J. M. Finney. *J. Inst. Metals*, July, 380–382 (1964).
10. R. L. Templin. *Proc. Am. Soc. Testing Mater.*, **54**, 641 (1954).
11. N. E. Frost. *Metallurgia*, **62**, 85–90 (1960).
12. J. Schijve. International Conference on Fatigue, Inst. of Mech. Eng., London, 754 (1956).
13. H. Neuber. *Theory of Notch Stresses*, J. W. Edwards (Ed.), Ann Arbor, Michigan, 1946.
14. N. E. Frost. *Proc. Inst. of Mech. Eng.*, **173**, 811–827 (1959).
15. J. O. Smith. *Univ. Illinois, Eng. Exp. Sta. Bull. No.* 334 (1942).
16. A. K. Hooke and F. H. Head. International Conference on Fatigue, Inst. of Mech. Eng., London, 301–303 (1956).
17. R. W. Fralich. *N.A.S.A. Memorandum* 4-12-59L (1959).
18. T. R. G. Williams and M. T. Lowcock. *Engineering*, **200**, 571 (1965).
19. P. Mitchell. M.Sc. Thesis, Institute of Sound Vibration Research, University of Southampton (1966).
20. K. J. Marsh. *N.E.L. Rep.*, 204 (1965).
21. W. T. Kirkby and P. R. Edwards. *R.A.E. Tech. Rep.* 66023 (1966).
22. R. H. Christensen. Growth of fracture in metal under random cyclic loading, *Douglas Paper No.* 3279 (1965).
23. E. Gassner. *Konstruktion*, **6**, 97–104 (1954).
24. T. Haas. Simulated service life testing, *The Engineer*, Nov. 14th and 21st, 1958.

25. H. F. Hardrath and E. C. Naumann. N.A.S.A. (paper presented at the Third Pacific Area National Meeting of the American Society for Testing and Materials, October 11–16, 1959).
26. J. Kowalewski. *Full Scale Testing of Aircraft Structures*, Pergamon Press, London, 1961: pp. 60–73.

CHAPTER 22

The Problems of Calculating Internal Noise Levels

22.1 Introduction

The contents of this chapter are divided into two parts. The first part describes briefly the rather crude methods often adopted for the estimation of aircraft internal noise levels and the second part discusses several theoretical research investigations into the radiation of noise from idealized structures excited by boundary layer pressure fluctuations.

22.2 Current practical methods

In low-speed aircraft the internal cabin noise is determined mainly by the noise radiated by external sources. The average internal noise levels can be calculated if the acoustic wave transmission characteristics are known for the aircraft skin and soundproofing. In practice these characteristics are not known in detail and usually a crude estimate is made using a simplified theory based on the mass law. In the absence of a more suitable alternative the above method is often used for higher-speed aircraft where the boundary layer pressure fluctuations are the predominant sources of noise, the boundary layer pressures being expressed in terms of an equivalent noise level.

The transmission loss for acoustic wave propagation through single and double partitions is discussed in chapters 3 and 23. To calculate the transmission loss at all frequencies, including panel resonances, it is necessary to know the stiffness and damping present in the structure. As these coefficients are often difficult to obtain, the theory is usually applied in a restricted frequency range whose lower limit is approximately $4f_0$ (where f_0 is the panel fundamental frequency) and whose upper limit is the critical frequency for coincidence. In this range the transmission loss for a given angular frequency ω is determined solely by the mass/unit area, m, of the panel and is given by

$$r_s = 10 \log_{10}\left[1 + \left(\frac{\omega m}{2\rho a}\cos\theta\right)^2\right] \text{ dB}, \qquad (22.1)$$

where ρ is the density of air and a is the speed of sound in air. In the derivation of equation 22.1 it is assumed that the incident sound is in the form of plane waves at an angle of incidence θ, whose value depends on the relative positions of the external noise source and the region of the cabin under consideration. Comparisons between theoretical and experimental results for incident sound waves show that equation 22.1 is a good approximation to the mean line through the measurements, but it does not predict the detailed variation with frequency which occurs in all practical structures.

When the fuselage construction is in the form of a double partition the transmission loss can be estimated from equation 23.41, where the resonances of the individual panels are again neglected but the mass–air–mass resonances of the system are included. The transmission loss is given by

$$r_d = 10 \log_{10}\left\{1 + 4m_1 m_2 \left(\frac{\omega}{2\rho a} \cos \theta\right)^2 \left[\cos\left(\frac{\omega d}{a} \cos \theta\right) - \frac{\omega m_1}{2\rho a} \cos \theta \sin\left(\frac{\omega d}{a} \cos \theta\right)\right]\right.$$

$$\times \left[\cos\left(\frac{\omega d}{a} \cos \theta\right) - \frac{\omega m_2}{2\rho a} \cos \theta \sin\left(\frac{\omega d}{a} \cos \theta\right)\right]$$

$$\left. + \left(\frac{\omega}{2\rho a} \cos \theta\right)^2 (m_1 - m_2)^2\right\} \text{dB}, \qquad (22.2)$$

where m_1, m_2 are the surface densities of the two panels, which are separated by a distance d. In equations 22.1 and 22.2, the surface densities are those appropriate to the panel alone and exclude the mass of the frames, stringers, rivets, etc., of the structure.

For certain frequency ranges, equation 22.2 can be expressed in a simpler form. At frequencies below the first resonance of the system, i.e. when

$$\frac{\omega m_1 \cos \theta}{2\rho a} \ll \cot\left(\frac{\omega d}{a} \cos \theta\right),$$

equation 22.2 reduces to

$$r_d \approx 10 \log_{10}\left[1 + \left(\frac{\omega(m_1 + m_2) \cos \theta}{2\rho a}\right)^2\right] \text{dB}, \qquad (22.3)$$

which is equal to the transmission loss predicted by equation 22.1 for a single panel of the same total mass.

When

$$\frac{\omega m_1 \cos \theta}{2\rho a} \gg \cot\left(\frac{\omega d}{a} \cos \theta\right), \qquad (i = 1, 2) \qquad (22.4)$$

i.e. at frequencies above the first resonance and not in the neighbourhood of any subsequent resonance, equation 22.2 reduces to

$$r_d \approx 20 \log_{10}\left[\tfrac{1}{2}\left(\frac{\omega}{\rho a}\right)^2 (m_1 m_2) \cos^2 \theta \sin\left(\frac{\omega d}{a}\cos\theta\right)\right] dB$$

$$= 20 \log_{10}\left[\frac{\omega m_1}{2\rho a}\cos\theta\right] + 20 \log_{10}\left[\frac{\omega m_2}{2\rho a}\cos\theta\right]$$

$$+ 20 \log_{10}\left[2 \sin\left(\frac{\omega d}{a}\cos\theta\right)\right] dB$$

$$= r_s(m_1) + r_s(m_2) + 20 \log_{10}\left[2 \sin\left(\frac{\omega d}{a}\cos\theta\right)\right] dB. \qquad (22.5)$$

As an upper limit, take $\sin[(\omega d/a)\cos\theta] = 1$ and equation 22.5 reduces to

$$r_d \approx r_s(m_1) + r_s(m_2) + 6 \text{ dB}, \qquad (22.6)$$

where $r_s(m_1)$ is the transmission loss given by equation 22.1 for a single panel of mass m_1. Thus, under the conditions given by equation 22.4, the transmission loss of the double partition is essentially that due to the two component panels acting independently. It is seen from equation 22.6 that r_d is a function of ω^4 and therefore increases at a rate of 12 dB/octave, whereas r_s increases at 6 dB/octave.

In typical aircraft constructions the sound insulation consists of a combination of impervious panels and porous blankets. Calculations of the transmission loss through combinations of this type require values for the acoustic impedances of the components. In general the parameters are not known in detail and the transmission losses of complex structures are often determined experimentally.

For boundary layer noise the above equations are frequently used in association with an external noise level which is calculated directly from the boundary layer pressure fluctuations. Ideally the pressures should be measured on the aircraft concerned but general values can be taken from existing experimental results. If, for example, the overall r.m.s. boundary layer pressure fluctuation p' is assumed to be given by

$$\frac{p'}{q_0} = 0.006 \qquad (22.7)$$

(where q_0 is the free stream dynamic pressure), then the equivalent sound pressure level is

$$\text{S.P.L.} = 20 \log q_0 + 83 \text{ dB (re 0.0002 dyn/cm}^2) \qquad (22.8)$$

with q_0 expressed in lb/ft^2.

The associated frequency spectrum for the incident sound is obtained in a similar way from the boundary layer pressure spectra (see chapter 8).

When equations 22.1 and 22.2 are adapted for boundary layer noise, it is necessary to assume a value for θ. Transmission losses, measured in the siren tunnel at the University of Southampton, for a single panel excited by grazing incidence sound waves were of similar magnitude to those for a panel excited by normal incidence sound waves (i.e. $\theta = 0$). Although it is possible that these measurements overestimate the transmission loss for boundary layer pressure fluctuations, equation 22.1 with $\theta = 0$ (i.e. transmission loss for normal incidence) has been found in some cases to give reasonable agreement with aircraft measurements. The errors involved can, however, be large and arise from uncertainties regarding the boundary layer pressure spectrum (there is a scatter in the results of Reference 1 of approximately ± 20 dB at the higher frequencies) and the transmission loss of the structure. The above method can be considered as giving little more than a rough estimate of the internal noise levels.

22.3 Review of work on predicting noise levels due to boundary layer induced vibrations

To attempt to solve the problem exactly involves trying to consider the coupled vibrations of the fuselage skin and air inside and outside the fuselage. In practice, to deal with any one of these problems involves a great deal of work and so it is usual to consider the vibrations of the structure and the vibrations of air independently. The fuselage of an airliner is usually so full of sound-absorbent objects that the effects of the fuselage as a closed space can almost certainly be neglected. As far as boundary layer induced structural vibration is concerned, the wavelengths excited will be small in comparison to the fuselage diameter for most cases of interest and so the effects of curvature are also normally neglected.

Papers by Ribner,[2] Corcos and Liepmann,[3] and Kraichnan[4] have been written in attempts to shed light on the particular problem of noise due to boundary layer induced vibrations. Dyer[5] has written a paper which shows, for a particular type of enclosure, how the effects of the air inside the closed space can be coupled to the vibrations of the structures.

In all these papers, the structure is taken to be a simple flat plate in bending. The techniques used in these papers are discussed below together with their conclusions.

22.3.1 *General assumptions*

(i) All papers assume that any interaction between the plate vibrations and the turbulent boundary layer can be neglected.

(ii) All except Dyer assume that they can use the known properties of the boundary layer to calculate the response of the panel and then use this response to calculate the sound due to panel vibrations (without allowing any interaction to occur between the last two steps).

Dyer shows how, for a very particular system, the effects of the sound field can be included in the plate vibration equations without making the problem of plate vibrations much more complicated.

(iii) Apart from Dyer, who considers an enclosed space, it is assumed that the air within the fuselage can be represented as being an infinite space on one side of an infinite plane wall. In view of comments made earlier about airliners this is probably quite justifiable for pressure cabins and the like.

Although all except one of the papers being considered present the structural side of their work first, it seems logical to look at the acoustic side of the problem in order to discover what properties of the plate vibration are needed to calculate the sound field produced by this vibration.

22.3.2 *Method of analysing transmitted sound*

It is assumed that normal acoustic theory for small pressure fluctuations can be used to analyse the sound field. The remaining problem is to find a solution of the wave equation which satisfies the boundary conditions appropriate to the problem in hand. These boundary conditions are normally chosen to ensure that the velocity of air particles adjacent to the walls of the structure are the same as the velocity of the wall. Dyer's model has additional conditions which will be discussed later.

Kraichnan, and Corcos and Liepmann have a similar model to consider. They both wish to find the sound field produced on one side of an infinite plane wall of which part (or the whole) may be vibrating.

Corcos and Liepmann use a solution in the form of an integral relating the sound pressure field to the normal acceleration of the plate normal displacement. This result is well known and can be found in many books on acoustics and mathematics.[6,7] By using this result it is shown that the mean square pressure fluctuations are given as a function of the space–time correlation of the plate normal accelerations. This result is not used in this form. The concept of correlation length is introduced and the result used in an approximate form involving correlation area of plate displacement and mean square plate normal acceleration.

Kraichnan attacks the same basic problem in a different manner. Using three-dimensional Fourier transforms (involving time and the two space coordinates in the plane of the plate) to solve the wave equation for an infinite half-space, Kraichnan relates the Fourier transform of the sound pressure fluctuations to the Fourier transform of the plate normal velocity.

It can be seen from this analysis that the efficiency of the vibrating surface as a radiator of sound varies with its wavelength. The radiation efficiency at some given frequency is greatest when the panel wavelength is the same as the acoustic wavelength corresponding to that frequency, and it falls away as the panel wavelength increases. If the panel wavelength is less than the corresponding acoustic wavelength, however, according to this analysis the sound field is a reactive one where the mean square pressure fluctuation falls rapidly away with increasing distance from the plate and with no radiation of energy.

Kraichnan then goes on to say that a measure of the sound transmitted is given by the 'radiated sound power'. In this context, the 'radiated sound power' is the sound pressure at the plate multiplied by the plate velocity (which is the same as the air velocity at the plate) and integrated over the whole of the plate. This result is expressed in wave number and frequency form omitting from the integration those wave number components which lead to a reactive sound field. The radiated sound power is thus obtained as an integral involving the wave number–frequency spectrum* of the plate normal velocity.

Ribner's work on the sound field, though based on a similar model to that discussed above, is less satisfactory. He says that the sound is generated by waves running along the fuselage wall. If these waves are supersonic, Mach waves are produced and the pressure levels may be calculated according to theory for supersonic flow past a wavy wall. If the waves running along the plate are subsonic, Ribner says that sound will not be radiated unless there are discontinuities or boundaries in the structure to produce reflections and hence standing waves. Since these boundaries always exist in practice, sound is radiated and Ribner produces a result similar to the approximate one used by Corcos and Liepmann except that he has included a 'universal correlation length' whereas Corcos and Liepmann have retained a general correlation length.

The model used by Dyer is considerably different from the infinite half-space model used by the other authors. Dyer considers the sound field within a rectangular space bounded on one side by an elastic wall (the vibrating plate) and on the remaining five sides by 'pressure release surfaces', i.e. by surfaces on which the acoustic pressure fluctuations are identically zero. Such conditions are appropriate to an air-filled rubber bag according to Dyer.

The sound field within the enclosure is expressed as a sum of the normal modes of the room. The coupling of the plate vibrations to the acoustic vibrations within the room is again obtained by equating the velocity of the air particles adjacent to the velocity of the wall.

---

* The wave number–frequency spectrum is the Fourier transform, with respect to the space coordinates and time, of the space–time correlation.

22.3.3 *Analysis of plate vibration*

From the previous section, it is seen that, in order to calculate the noise levels transmitted through an elastic structure, it is necessary to find the space–time correlation of the plate normal acceleration or the wave number–frequency spectrum of the plate normal velocity.

Corcos and Liepmann require for their analysis the space–time correlation of the plate normal acceleration. This is used to find the mean square acceleration integrated over the surface of the plate and can be expressed in terms of the wave number–frequency spectrum integrated over all wave numbers and frequencies.

They argue that the mean square acceleration when integrated over the surface of the plate will not be too dependent on the plate boundary conditions (for typical aircraft structures and boundary layers). Consequently they assume that an infinite plate will be a satisfactory model of the structure.

By means of Fourier transforms, a relation is established between the wave number–frequency spectrum of the plate response and the wave number–frequency spectrum of the excitation—in this case the boundary layer. This result is a generalization of the result obtained for the power spectrum of the response of a single-degree-of-freedom system (chapter 14) and is valid only for an infinite plate.

This result exhibits the coincidence effect. At frequencies and wave numbers corresponding to the speed and wavelength of propagation of free waves in the plate the response would be infinite in the absence of damping.

Ribner's approach to the structural problem is similar to that of Corcos and Liepmann. The only difference is that he treats his structure as an infinite beam rather than as a plate. Ribner's equations are thus analogous to those of Corcos and Liepmann except that they contain only one space dimension instead of two.

Dyer and Kraichnan both consider a finite panel. Thus in Kraichnan's case, only part of his infinite plane wall is vibrating. Owing to the difficulties of dealing with realistic boundary conditions, they both consider a plate which is simply-supported at all its edges.

At first sight, the methods used by these two authors appear to be different but in fact both methods can be classified as 'normal mode methods'.

The method used by Kraichnan is the direct normal mode method as described in chapter 14. Kraichnan makes use of a slight variation in the formal application of this method. It is usual to assume that the boundary layer pressure fluctuations form a homogeneous, stationary process (i.e. that the space–time correlation is a function only of spatial separation and time delay). By expressing the joint acceptance expression in terms of the Fourier transforms of the mode shapes and space–time correlation, and making use of the homogeneity hypothesis, two of the four integrals can immediately be evaluated. This means that the joint acceptance is expressed as a double integral involving the wave number–frequency spectrum rather than as a quadruple integral involving the space–time correlation.

Dyer's analysis is based on finding the response of the panel to a unit impulsive loading at a general point in space and time and then integrating the effect of a large number of impulses to give the response to the given loading. This result is generalized to give the space–time correlation of the excitation. The response of the panel to a unit impulse is expanded as a sum of the plate normal modes, so that Dyer's result for the space–time correlation of the plate response is the Fourier transform (wave number and frequency) of the result obtained by Kraichnan for the wave number–frequency spectrum of the plate response.

Dyer's work shows how the response of his 'enclosure' can be coupled to the vibrations on the plate. Using the result described in section 4 of reference 5, he finds an expression for the acoustic pressure at the plate due to plate vibrations. It can be seen that this is expanded in terms of the plate normal modes and that the coefficient of the nth normal mode is a function of the amplitude of nth normal mode of the plate motion. The reason for this is that the modal pattern on the vibrating plate of the normal modes of the room corresponds exactly to the modal pattern of the normal modes of the plate. This means that the effect of the room is the same as the effect of adding an extra generalized mass to the equation for the response of the plate in each normal mode.

If the plate under consideration was clamped instead of simply-supported this would not be possible. In general, the coupling of acoustic vibration and structural vibrations leads to coupling of the normal modes of the structure—thus leading to an infinite set of *simultaneous* linear equations for the generalized coordinates corresponding to the structural modes.

22.3.4 *Application of methods to real conditions*

The equations obtained in general terms for the noise radiated by boundary layer induced vibrations are extremely involved and the authors under review have all used many approximations and simplifications in attempts to provide simple formulae as a final result.

The most obvious approximations used are those concerned with the representation of the space–time correlation, or cross power spectrum, of the boundary layer pressure fluctuations.

Ribner and Kraichnan both assume that the boundary layer pressure fluctuations can be represented by the convection of a rigid pattern of turbulence which does not decay with time.

Ribner considers a one-dimensional pattern of turbulence with a wavelength spectrum concentrated about a wavelength which is equal to the boundary layer thickness.

Kraichnan assumes a turbulence pattern which is two-dimensional and in which wavelengths in the flow direction have a greater contribution to the mean square pressure fluctuation than the same wavelengths at right-angles to the flow direction. He also assumes that there is a cut-off in the wave-number spectrum. This cut-off is assumed to be such that the wavelengths of the turbulent pressure fluctuations are all greater than the boundary layer thickness. This latter assumption is very difficult to understand; Ribner's wavelength distribution seems much more reasonable on both intuitive and experimental grounds.

Dyer describes two representations of the space–time correlation function. The first one uses exponential functions and allows for both convection and decay of the turbulent pressure fluctuations. This representation will not allow negative correlations, however. A much better representation can be obtained by using a sum of two such correlation functions. These can be chosen to give negative correlations and generally give a good representation.

For use in his analysis, Dyer makes use of another space–time correlation. This is a product of two Dirac δ-functions and an exponential function representing convection, spatial separation and time decay respectively. This corresponds to a turbulence pattern consisting of pressure fluctuations with very small wavelengths.

For aircraft applications the most important properties of the assumed pressure correlations are (i) the effects of convection, (ii) the spatial (i.e. wavelength) distribution, and (iii) the effects of decay. Of these (i) and (ii) are the most important. We see, therefore, that none of these representations are really satisfactory, but a generalization of Ribner's to include another space dimension should be best.

The only assumption made by Corcos and Liepmann about the wave number–frequency spectrum of the pressure fluctuations is that it is symmetrical in its two-wave number components and that it is fairly smooth. By using an extension of the first mean value theorem for integrals to obtain approximate values for their integrals, they leave their results in terms of a function which is the wave number–frequency spectrum of the boundary layer pressure fluctuations evaluated at those wave numbers and frequencies which satisfy the coincidence condition for the plate.

The structural models used by Ribner, and by Corcos and Liepmann, because they have no boundaries, have no discrete resonant frequencies. They do, however, have a continuous range of frequencies at which coincidence can occur. It follows, therefore, that the spectrum of noise produced by such a structural model will be smooth and will follow the general shape of the excitation spectrum.

Approximations introduced by Kraichnan to enable him to sum the structural response over all normal modes are such that, in the special case of low damping, the discrete set of resonant frequencies of the plate are replaced by a continuous distribution. (Kraichnan also considers the case of critical damping but this does not seem to have any practical interest.) The spectrum produced by Kraichnan would, therefore, also be expected to be smooth and not contain the peaks that are found in practice.

Dyer calculates the mean square response in each mode and the sound pressures produced by this response adjacent to the plate. He evaluates this response for two conditions (i) when the convection speed of the boundary layer is less than the coincidence speed for the modal wavelength, and (ii) when the convection speed is equal to the coincidence speed. Dyer is interested in applying this work to the special case of a boundary layer in water flowing at 20 ft/s past a steel plate which is five feet square and $\frac{1}{2}$ in thick. For such a panel, the conditions below coincidence will apply. For aircraft applications the conditions at coincidence will be more realistic.

22.3.5 *Theoretical results*

For thin boundary layers, Ribner, Corcos and Liepmann, and Kraichnan agree that the transmitted noise levels vary with the fifth power of the free stream Mach number. Ribner suggests that this falls to a third power law for thick boundary layers.

The variation of transmitted noise with boundary layer thickness is given as a linear relation by Corcos and Liepmann. Ribner gives a third power law for thin boundary layers and a linear one for thick boundary layers. Kraichnan gives a fourth power law for thin boundary layers.

Ribner also gives a result relating boundary layer thickness and free stream Mach number to transmitted noise level which, it is claimed, is valid for all boundary layer thicknesses. It contains the above relations as special cases.

According to the results given in these papers, the transmitted noise can be reduced by increasing the plate damping coefficient—doubling the damping coefficient can give a noise reduction of 3 dB for a lightly damped system. Increasing the thickness of the plate will also reduce the noise level. Doubling the plate thickness will give a noise reduction which varies from 9 dB to 3 dB in the different papers.

Dyer is in agreement with the trends of these results as regards the effects of damping and plate thickness. He suggests, however, that, if the convection velocity of the boundary layer is below the coincidence velocity of a particular normal mode, the response in this mode above some critical frequency is governed more by the mean eddy lifetime in the boundary layer than by the damping coefficient. He gives a criterion for this frequency. This latter result is more applicable to underwater applications than to aircraft.

Kraichnan's final results also indicate that the noise transmitted is inversely proportional to the length of the plate (in this case, a square plate). This result is difficult to understand but may be true if the length of the plate is greater than the acoustic wavelength corresponding to the lowest frequency of interest.

The differences in the results given by these papers show that the noise transmitted can be sensitive to the shape of the assumed correlation functions and spectra. The results also show how different interpretations of the approximations involved in simplifying integrals and summations can lead to different final results.

22.4 Experimental results

There is a limited amount of experimental data available for comparison with the theoretical results. The noise radiated by single panels excited by the turbulent boundary layer in subsonic wind tunnels has been measured by Ludwig[8] and Maestrello[9] but the results do not show full agreement. For an airflow velocity range of 60 ft/s to 200 ft/s, Ludwig observed that the radiated acoustic power was proportional to the product of the fifth power of the velocity and the square root of the boundary layer thickness, and inversely proportional to the panel thickness, i.e.

$$\text{acoustic power} \propto U^5 \delta^{1/2} h^{-1}.$$

Maestrello measured the radiated power in a higher speed range (300 ft/s to 700 ft/s) and his results can be divided into two categories. When the panel vibration is dominated by a standing wave pattern, the radiated acoustic power relationship is

$$\text{acoustic power} \propto U^5 h^{-1.6},$$

and when the vibration is dominated by running waves

$$\text{acoustic power} \propto U^{2.3} h^{-1.0}.$$

Maestrello did not study the effect of boundary layer thickness. The two investigations appear to give similar U^5 relationships, but measurements by el Baroudi[10] of the vibration of the panels used by Ludwig show the presence of a running wave system. One might expect, therefore, that the results of Ludwig should be compared with the $U^{2.3}$ law of Maestrello.

A second disagreement arises when the radiated power spectra are compared. The measured spectra show a rapid fall-off of spectral density at low and high frequencies and in both investigations the upper cut-off frequency is a function of the boundary layer characteristics and the panel thickness. This dependence does not appear to be true at low frequencies where Maestrello claims that the low frequency cut-off occurs at the panel fundamental natural frequency whereas the results of Ludwig show that the cut-off occurs well above the panel fundamental.

22.5 Conclusions

The existing theoretical and experimental work gives only a limited understanding of the radiation of sound from structures excited by boundary layer turbulence. Further work is required, particularly with reference to complex panel–stringer arrays, and the theoretical investigations should include more realistic representations of the excitation field. Full-scale measurements may be necessary before adequate solutions of the problem, illustrated so clearly by Hay,[11] are found.

References

1. E. J. Richards. Boundary layer noise research in the U.S.A. and Canada; a critical review, *Univ. Southampton Rep.* AASU, 131 (1960).
2. H. S. Ribner. Boundary layer induced noise in the interior of aircraft, *Univ. Toronto Rep.* UTIA 37 (1956).
3. G. M. Corcos and H. W. Liepmann. On the contribution of turbulent boundary layers to the noise inside a fuselage, NACA TM 1420 (1956).
4. R. H. Kraichnan. Noise transmission from boundary layer pressure fluctuations, *J. acoust. Soc. Amer.*, **29**, 65 (1957).
5. I. Dyer. Sound radiation into a closed space from boundary layer turbulence, *Bolt, Beranek and Newman Inc. Rep.* BBN 602 (1958).
6. Lord Rayleigh. *Theory of Sound*, Dover Publications, New York, 1945.
7. I. N. Sneddon. *Elements of Partial Differential Equations*, McGraw-Hill, New York, 1957.
8. G. R. Ludwig. An experimental investigation of the sound generated by thin steel panels excited by turbulent flow (boundary layer noise), *Univ. Toronto Rep.* UTIA 87 (1962).
9. L. Maestrello. Measurement of noise radiated by boundary layer excited panels, *J. Sound Vib.*, **2**, 100 (1965).
10. M. Y. el Baroudi. Turbulence-induced panel vibration, *Univ. Toronto Rep.* UTIAS 98 (1964).
11. J. A. Hay. Problems of cabin noise estimation for supersonic transport, *J. Sound Vib.*, **1**, 113 (1964).

CHAPTER 23

Soundproofing Methods

23.1 Introduction

At an early stage in the design of an aircraft the designer wishes to determine the weight of soundproofing material required in the passenger cabin and crew compartment. The problem of soundproofing aircraft interiors is fundamentally similar to that of soundproofing a room in a building and similar methods of calculating expected noise levels can be used. Owing to the particularly complicated acoustic behaviour of aircraft structures and the character and location of aircraft noise sources these methods, which are at best approximate ones, are not likely to yield results of a high degree of accuracy unless they can be supplemented by full-scale tests and flight measurements using the actual type of aircraft concerned. Estimates of interior noise levels made at the design stage, therefore, may be in error by several decibels which means that the estimate of the weight of soundproofing material required may be in error by as much as a factor of two. Pending further research to develop better prediction methods the accuracy of the estimate can be improved only by the use of previous practical experience with similar aircraft, when available.

This chapter is restricted to a discussion of an approximate theory of the transmission and absorption of sound in an aircraft interior which is useful in practice, and to discussion of other practical aspects of the problem.

23.2 Noise sources

Possible sources of aircraft cabin noise are:
(a) Propeller noise
(b) Piston engine exhaust noise
(c) Jet engine noise (exhaust and compressor)
(d) Boundary layer noise
(e) Aircraft equipment noise, e.g. hydraulic pumps, the air conditioning system, etc.

The generation of noise by the majority of these sources has been discussed in preceding chapters. It cannot be over-emphasized that suppression at the source is one of the most important methods of noise control. The growing development of rear-engined aircraft will help to reduce the effect in the cabin of item (c) but the ever-increasing aircraft cruising speeds will only aggravate the problem of boundary layer noise, the intensity increasing at approximately the fourth power of the equivalent air speed. Until some form of boundary layer control is adopted for the fuselage, the problem of aircraft cabin soundproofing is likely to become increasingly severe.

Noise interference can occur between the different engines of an aeroplane, propeller or jet, creating unpleasant beats inside the cabin. These can usually be eliminated by engine synchronization or by accurate balancing of the engines.

The noise levels are extremely high in the plane of rotation of a propeller and whenever possible passenger compartments should be positioned away from this area. The region could be used for luggage compartments or toilets.

The noise generated by item (e) is often omitted in soundproofing calculations. However, once this equipment is installed it is often difficult to make any great reduction to the noise produced. These noises can be very annoying to passengers and wherever possible the equipment should have noise levels compatible with the design levels for the cabin. Equipment should not be audible in an ideal cabin during the normal cruise of the aircraft. This implies that equipment noise must be at least 10 dB below cabin ambient levels.

23.3 Approximate theory

In considering the acoustic characteristics of an aircraft interior it may be required to estimate the average noise level inside a compartment when the outside surface is exposed to acoustic or other excitation. Alternatively it may be required to estimate the noise level in a cabin when there is an assumed distribution of noise sources on the internal surface of the cabin walls. In the first case the transmission and absorption of the external excitation by the aircraft structure must be taken into account. In the second case it is assumed that these are known.

As cabins are usually fairly absorptive, strong resonances of the air inside the cabin will be exceptional occurrences and in their absence it can be assumed that the motion of the aircraft structure due to the noise sources can be determined independently of the noise level they may cause inside the cabin. Thus the first case above may be treated in two parts. The motion of the structure due to the external sources can be estimated first. The remainder of the problem is identical with the second case as knowledge of the motion of the structure is equivalent to knowledge of the effective distribution of noise sources on the internal surface of the cabin walls.

In general the interior noise level depends on the nature of the external sources, on the nature of the fuselage structure, and on the shape of the cabin and the amount and distribution of absorbent material in it. The fuselage structure has very many modes of oscillation with characteristic frequencies in the audio range, the modes having varying degrees of damping. Similarly the air in the cabin has very many modes of oscillation with characteristic frequencies in the audio range and varying degrees of damping. Fundamentally the noise level in the cabin is a superposition of the contributions from all the various modes of oscillation of the air within it, these modes being coupled to the actual noise sources via the modes of the fuselage structure.

When the modes in the cabin are randomly excited—as by noise sources randomly distributed over the wall surfaces—the average mean square pressure will be fairly uniform throughout the cabin. Points close to the walls which are in the near field of the source distribution are exceptions to this general rule; at those points noise levels higher than the average are to be expected. Also at low frequencies, where only a few modes are available in a given frequency band, there will be a tendency for standing wave patterns to be prominent, having a regular variation of pressure level with position. Apart from these exceptions the acoustic field in the cabin can be thought of as a superposition of plane waves travelling with equal probability in all directions (see e.g. reference 25, chapter 8, pp. 381 ff.). The energy density in the field is then proportional to the mean square pressure and is equal to the sum of the energy densities in the component plane waves.

On the assumption that the sound field in the cabin has statistical uniformity of this kind it is possible to estimate the noise levels if the similarly averaged transmitting and absorbing properties of the fuselage structure and the soundproofing materials are known. With such an estimate as a basis, modifications can be made to take account, approximately, of isolated departures from average conditions—as when, for example, a significant amount of pure tone coherent excitation occurs, or the excitation has a well-defined spatial pattern.

23.3.1 *The average coefficients of transmission, absorption and reflection*

In architectural acoustics sound transmission from one room to an adjoining one through a partition can be characterized by three average coefficients, providing the sound fields in both rooms are statistically uniform. These coefficients are:

$$\text{Sound transmission coefficient} = \tau = \frac{\text{sound energy transmitted}}{\text{sound energy incident}}$$

Sound absorption coefficient $= \alpha = \dfrac{\text{sound energy absorbed}}{\text{sound energy incident}}$

Sound reflection coefficient $= \beta = \dfrac{\text{sound energy reflected}}{\text{sound energy incident}}$.

For a given partition $\tau + \alpha + \beta = 1$, from energy considerations. The *transmission loss* of a partition is given by

$$r = 10 \log_{10}(1/\tau) \text{ dB.} \tag{23.1}$$

For a single plane wave incident on a partition the transmission, absorption and reflection of acoustic energy depend upon the angle of incidence. In keeping with the assumption of statistical uniformity the average coefficients above are obtained for the purposes of architectural acoustics by averaging the corresponding angle-dependent coefficients over all angles of incidence.

In the case of an aircraft it is not necessarily valid to assume that the external excitation is a statistically uniform acoustic field. The assumption is fairly reasonable for the radiated noise from jet engine exhausts and the like, especially if some account is taken of the directionality of the radiation. For example, the coefficients can be averaged over a range of angles corresponding more closely to those at which the actual radiation strikes the fuselage, rather than overall angles with equal weight.

The assumption of a statistically uniform acoustic excitation field is more unrealistic for near field excitation and boundary layer excitation of the fuselage. Unfortunately little evidence is as yet available to indicate the values of the coefficients in such cases. The supposition that the excitation field can be represented by a superposition of plane acoustic waves remains valid in principle, but these waves certainly would not have a uniform distribution in angle. In the case of boundary layer excitation especially, where the pressure fluctuations on the fuselage are associated with convected turbulence, the spatial scale of a given frequency component of the excitation field is much smaller than an acoustic wavelength. Thus it is to be expected that, if the 'sound energy incident' in the above definitions of the coefficients is interpreted as being proportional to the pressure fluctuations on the surface, then the effective values of the coefficients would be appreciably different from the values they have when the excitation has uniform distribution in angle of plane acoustic waves.

It should be borne in mind therefore that values of the average coefficients, appropriate for the particular type of excitation concerned, should be used when available. Calculations based on coefficients obtained from reverberant room tests of the type used in architectural acoustics will not necessarily be valid. When data are not available experimental determinations of the values should be made if possible. In what follows the coefficients are to be interpreted as those appropriate to the problem in hand, unless otherwise specified. The noise levels due to different sources may be added linearly to give the total noise level, unless of course the partition response is predominantly non-linear.

23.3.1.1 *Simple enclosure.* Assume the enclosure has homogeneous walls and is exposed to a statistically uniform acoustic field of plane waves. If

E_e = total sound energy external to the enclosure
I_e = intensity (i.e. energy/unit area) on exterior
E_i = total sound energy inside the enclosure
I_i = intensity inside the enclosure
S = total surface area of the enclosure walls,

then equilibrium between energy entering and leaving the enclosure is achieved when

$$\tau I_e S = \alpha I_i S + \tau I_i S.$$

By definition the sound reduction factor is

$$R = 10 \log_{10}\left(\frac{E_e}{E_i}\right) \text{ dB.}$$

Thus

$$R = 10 \log_{10}\left(\frac{\alpha S + \tau S}{\tau S}\right)$$

$$= 10 \log_{10}\left(1 + \frac{\alpha}{\tau}\right) \text{ dB.} \tag{23.2}$$

This variation of R with α and τ is shown in Fig. 23.1. It should be noted that if $\alpha = 0$ then $R = 0$ regardless of the value of τ, i.e. when the absorption is zero the noise levels inside and outside the enclosure are the same and the transmission loss through the wall has no effect. Such a condition is, however, difficult to achieve in practice; one of the least acoustically absorptive materials is marble slate with an absorption coefficient $\alpha \approx 0.01$.

In a typical aircraft cabin the walls will not be homogeneous as there will be variations in skin thickness or sound insulation and the fuselage panels will contain windows. Equation 23.2 has thus to be slightly modified. Suppose that the walls consist of areas $S_1, S_2, \ldots, S_n$ having coefficients of transmission, $\tau_1, \tau_2, \ldots, \tau_n$, and of absorption $\alpha_1, \alpha_2, \ldots, \alpha_n$ respectively. Then the total number of 'transmission units' for the cabin wall is

$$T = \sum_{j=1}^{n} \tau_j S_j$$

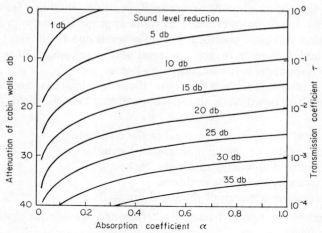

Fig. 23.1   Theoretical sound reduction curves (equation 23.2).

and the total number of 'absorption units' for the internal surface is

$$A_w = \sum_{j=1}^{n} \alpha_j S_j.$$

Further, there will be extra absorption which is not distributed on the cabin walls, e.g. seats and passengers. Let this additional absorption be denoted by A_i. Then

$$A = A_w + A_i$$

and the sound reduction factor becomes

$$R = 10 \log_{10}(1 + A/T)\,\text{dB}. \tag{23.3}$$

Equation (23.3) will provide an average value for the cabin noise level. For statistically uniform fields the energy reduction factor R is equal to the reduction in pressure level.

23.3.1.2 *Effect of additional insulation.* Consider the effect of changes in the acoustical properties of the fuselage skin or insulation. To simplify the notation take mean values of $\bar{\alpha}_w$, $\bar{\tau}$ defined by:

$$\bar{\alpha}_w = \frac{A_w}{S} = \frac{\sum_{j=1}^{n} \alpha_j S_j}{\sum_{j=1}^{n} S_j} \qquad \left(\text{where total area, } S = \sum_{j=1}^{n} S_j\right)$$

and

$$\bar{\tau} = \frac{T}{S} = \frac{\sum_{j=1}^{n} \tau_j S_j}{\sum_{j=1}^{n} S_j}.$$

Let unprimed and primed symbols denote conditions before and after the changes in treatment and assume that the absorption A_i provided by surfaces other than the walls (e.g. by seats) does not change. Then

$$R = 10 \log_{10}\left(1 + \frac{\bar{\alpha}_w S + A_i}{\bar{\tau} S}\right) = r + 10 \log_{10}\left(\bar{\tau} + \bar{\alpha}_w + A_i/S\right) \text{ dB}$$

and

$$(R' - R) = (r' - r) + 10 \log_{10}\left[\frac{\overline{\tau'} + \overline{\alpha'_w} + A_i/S}{\bar{\tau} + \bar{\alpha}_w + A_i/S}\right] \text{ dB.}$$

Except at low frequencies it can be assumed that $\bar{\tau}$, $\overline{\tau'} \ll A_i/S$ and

$$(R' - R) \approx (r' - r) + 10 \log_{10}\left[\frac{\overline{\alpha'_w} + A_i/S}{\bar{\alpha}_w + A_i/S}\right] \text{ dB.} \tag{23.4}$$

The first term on the right-hand side of equation 23.4 is the change in transmission loss and the second term is the change in attenuation due to variations in the absorption. The effect of variation in $\bar{\alpha}_w$ is seen to depend greatly on the degree of absorption provided by seats, etc. Changes in trim giving small improvements in $\bar{\alpha}_w$ may thus have a negligible effect on the overall noise reduction.

23.3.1.3 *General accuracy of theory.* The preceding theory is not directly applicable to cases where the external noise cannot be considered as essentially uniform over the fuselage nor where there are large variations of the absorption and transmission coefficients in the cabin. Such problems will give regions in the cabin having noise levels appreciably higher than in the surrounding areas. The estimates have then to be modified to include transmission of sound inside the aircraft itself. An approach to this problem is discussed in the next subsection.

Inaccuracies also arise at low frequencies if standing waves are set up. When these occur the noise level in the fuselage varies regularly in either a longitudinal or lateral direction depending on the orientation of the waves and there are alternate regions of high and low noise. Standing waves are unlikely to exist when the wavelength of sound is small compared to the cabin dimensions, a condition which can be assumed true in the passenger cabins of large aircraft for frequencies higher than about 600 c/s. Standing waves are likely to be more in evidence if the excitation spectrum contains an effectively continuous pure tone component.

Under normal conditions, equation 23.3 can be simplified still further. For noise reductions greater than 15 to 20 dB, the value of A/T becomes large compared to unity and equation 23.3 can be written

$$R \approx 10 \log_{10}(A/T) \text{ dB.} \tag{23.5}$$

16+

In estimates of aircraft cabin soundproofing this approximation can generally be used over the whole audio frequency range although the validity should be checked at low frequencies.

23.3.1.4 *Modified theory.* In practice the sound field around an aircraft is not uniform but has regions of high intensity. Such areas occur, for instance, in the plane of a propeller or near the exhaust of a turbo-jet. This problem is discussed by Kobrynski[1].

If measurements are made in an unfurnished aircraft and it is required to calculate the noise levels in the cabin after a *uniform* installation of soundproofing, the noise reduction given by equation 23.5 would be

$$R' - R = 10 \log_{10}\left[\frac{A'}{A}\frac{T}{T'}\right] \text{ dB} \tag{23.6}$$

where unprimed and primed symbols denote untreated and treated cabins respectively. The increase in absorption in the cabin will also have reduced the noise transmission inside the cabin from regions of high to regions of low noise level. Equation 23.6 will thus give the reduction in the area of highest noise level. In neighbouring regions of the cabin, however, there will be a further reduction in noise given by

$$10y \log_{10}\left(\frac{E'}{E}\right)$$

where y is the distance from the noisiest part of the cabin (in feet) and E, E' are respectively the acoustical energies absorbed per foot run in the cabin during propagation inside the fuselage before and after insulation.

The total reduction will then be

$$R' - R = 10 \log_{10}\left[\left(\frac{A'}{A}\right)\left(\frac{T}{T'}\right)\left(\frac{E'}{E}\right)^y\right] \text{ dB.} \tag{23.7}$$

From measurements, estimated values of $10 \log_{10}(E'/E)$ in reference 1 are 0.075 dB/ft run for the 37.5–75 c/s frequency band and 0.15 dB/ft run for higher frequency bands.

23.3.2 *Noise in enclosures*

It is assumed here that a panel of the fuselage skin radiates noise of a known pressure and intensity, which can be determined either experimentally or theoretically. It is then necessary to find the noise reduction due to the transmission loss through the soundproofing alone and the internal absorption of the structure as a whole.

Consider first the effect of absorption. Assume that the point of observation is at a sufficient distance from the source ($> \lambda/3$ say) for plane wave conditions to be satisfied. Then the energy density Φ can be written[2]

$$\Phi = \frac{|p|^2}{\rho a^2} \qquad (23.8)$$

where the sound energy density is the sound energy in an infinitesimal part of the gas divided by the volume of that part of the gas, p is the r.m.s. sound pressure, ρ is the density of air and a the speed of sound in air.

Then, at a given point in the fuselage:

Total energy density = direct energy density

+ reverberant energy density

i.e. $\qquad\qquad \Phi = \Phi_1 + \Phi_2$

where, by definition, direct sound reaches the observer before being reflected and reverberant sound is that which has been reflected at least once.

It can be shown that

$$\Phi_2 = \frac{4W(1 - \bar{\alpha})}{aS\bar{\alpha}}$$

where W is the power radiated by the source, $\bar{\alpha}$ is the mean absorption coefficient (including all internal fittings) and S is the wall surface area.

Thus, from equation 23.8

$$|p|^2 = |p_1|^2 + 4W\rho a \frac{(1 - \bar{\alpha})}{S\bar{\alpha}}. \qquad (23.9)$$

Assume that the sound sources are uniformly distributed over the internal surface of the fuselage wall. Further assume that the fuselage is an infinitely long circular cylinder and that the sound sources radiate into a solid angle of 2π. Then, if I is the intensity of the surface distribution of sources, the total direct pressure at a point P is

$$|p_1|^2 = \pi\rho I a \qquad (23.10)$$

and $W = IS_n$ where S_n is the noise-radiating area of the cabin wall.

Thus the total sound pressure at a point is

$$|p|^2 = \left[\pi + 4\frac{(1 - \bar{\alpha})}{\bar{\alpha}} \frac{S_n}{S} \right] \rho I a. \qquad (23.11)$$

A change in absorption coefficient will therefore give a corresponding change in sound absorption of

$$\Delta dB_\alpha = 10 \log_{10} \frac{|p|^2}{|p'|^2}$$

$$= 10 \log \left[\frac{\pi + 4 \left(\dfrac{1 - \bar{\alpha}}{\bar{\alpha}} \right) \dfrac{S_n}{S}}{\pi + 4 \left(\dfrac{1 - \bar{\alpha'}}{\bar{\alpha'}} \right) \dfrac{S_n}{S}} \right] \text{dB}. \qquad (23.12)$$

For $S_n/S \approx 0.79$ equation 23.12 becomes

$$\Delta dB_\alpha = 10 \log_{10} \left(\frac{\overline{\alpha'}}{\overline{\alpha}}\right) \text{ dB.} \qquad (23.13)$$

This is then equivalent to the second term on the right-hand side of equation 23.4. It thus seems plausible that the addition to equation 23.13 of the change in transmission loss through the soundproofing alone should give approximately the total change in cabin noise level.

Several points should be noted:

(a) It is difficult to measure the transmission loss through the sound-proofing alone when it is mounted in conjunction with the fuselage skin.

(b) It is assumed in the above that the noise in the cabin is diffuse, i.e. the sound waves are travelling in all directions. For large values of $\overline{\alpha}$, ($\overline{\alpha} > 0.4$), the analysis loses its accuracy since the travelling waves die out quickly and are not reflected very much. However, the equations should be valid for aircraft soundproofing where $\overline{\alpha}$ is rarely > 0.5.

(c) It is assumed that the point under consideration is at a distance of at least $\lambda/3$ from the noise sources, and will thus be valid at the centre of the cabin for frequencies > 70 c/s. The noise levels will, however, increase near the cabin walls.

23.4 Noise transmission through panels

The soundproofing characteristics of an aircraft structure depend on the noise transmission properties of thin metal plates and the noise transmission and absorptive properties of sound insulation when used in association with such plates. In problems of building acoustics the transmission and radiation of noise by lightweight structures has been extensively investigated[3-7] and some of these results can be applied in work on aircraft.

23.4.1 *Single flat panels*

23.4.1.1 *Elastic waves in plates.* A finite panel can transmit several different wave motions:

longitudinal (compressional) waves
flexural (bending) waves
transverse (shear) waves
torsional (twist) waves
Rayleigh (surface) waves.

In fluids, sound waves are essentially compression waves. Such waves in solid materials usually radiate little noise into the surrounding air (or other fluid). Flexural waves, however, are associated with relatively large transverse displacements which can readily generate compressional waves in surrounding fluids. Conversely they can be excited easily by sound waves in the air. From an acoustic point of view, they are the most important of the five types listed above.

If the acoustic characteristics of a panel are to be determined it is necessary to know the relation between the panel dimensions and the wavelength of vibration. Such a relation can be obtained from the velocities of propagation of the various waves.

The equation of motion for bending waves in a thin plate is

$$\nabla^4 w + \frac{12\rho_p(1 - \sigma^2)}{Eh^2} \frac{\partial^2 w}{\partial t^2} = 0 \qquad (23.14)$$

where w is the displacement of the plate perpendicular to its surface,

h is the thickness of the plate, and

E, σ, ρ_p are, respectively, the Young's modulus, Poisson's ratio, and density of the material of the plate.

A solution of equation 23.14 gives the velocity of propagation c_B of flexural waves of frequency f:

$$c_B = \sqrt{1.8hfc_L} \qquad (23.15)$$

where

$$c_L = \sqrt{\frac{E}{\rho_p(1 - \sigma^2)}} = \sqrt{\frac{12D}{mh^2}} \qquad (23.16)$$

is the velocity of propagation of longitudinal waves, D is the flexural stiffness, $Eh^3/12(1 - \sigma^2)$, and $m = h\rho_p$ = mass of panel/unit area. The wavelength of a flexural wave is

$$\lambda_B = \frac{c_B}{f} = \sqrt{\frac{1.8hc_L}{f}}. \qquad (23.17)$$

It is seen from equation 23.16 that the velocity of longitudinal waves is independent of frequency, as are longitudinal waves in a fluid, but flexural waves (equation 23.15) are propagated at speeds which depend on frequency and are therefore called 'dispersive' waves.

The derivation of the above velocities assumes that the panel is thin with respect to the wavelength of the waves, the condition for validity of equation 23.15 being, approximately, $6h < \lambda_B$, a condition satisfied by aircraft panels throughout most of the audio-frequency range. This condition implies that $c_B < 0.3\ c_L$ which is a more stringent requirement than the limit $c_B \leq c_R$, where c_R (=0.57 c_L) is the velocity of propagation of Rayleigh waves.

16*

23.4.1.2 *Plate resonances.* When a plate of finite area is excited by sound waves the resultant vibration can be represented by a combination of two families of waves. The sound waves will impose a 'forced' flexural wave pattern on the plate, as they would on an infinite plate, and this pattern is superposed on a pattern of 'free' flexural waves which may be considered to be released from the boundaries of the plate as a consequence of the disturbance produced in the forced wave by the change in structure. The system of free waves allows the plate boundary conditions to be satisfied.

Under certain conditions the free flexural waves released from opposite edges of a panel can combine to give a standing wave pattern and the panel is then in a resonant condition. The frequencies (called normal frequencies) at which resonance can occur depend on the dimensions and materials of the panel and the constraints at the boundaries. For a rectangular plate, simply supported at all edges, the standing wave pattern for a normal frequency $f_{m,n}$ can be assumed to be (for suitable coordinate axes)

$$w_{m,n} = A_{m,n} \sin \frac{m\pi x}{L_x} \sin \frac{n\pi y}{L_y} e^{i\omega_{m,n}t} \qquad (m, n \geq 1) \qquad (23.18)$$

where L_x, L_y are, respectively, the length and breadth of the panel and m, n are integers denoting the number of half-waves in the standing waves in the directions of L_x and L_y. The normal frequency is then given by

$$f_{m,n} = 0.454 \, hc_L \left[\left(\frac{m}{L_x} \right)^2 + \left(\frac{n}{L_y} \right)^2 \right] \text{c/s} \qquad \text{where } m, n \geq 1. \qquad (23.19)$$

Putting $m = n = 1$ gives the fundamental frequency of the plate.

General formulae for the normal frequencies of plates with free, simply supported or fixed boundary conditions have been determined by Warburton[8]. For a rectangular plate having all sides fixed

$$f_{m,n} = \frac{\pi \lambda_{mn} h}{4 L_x^2} \sqrt{\frac{E}{3\rho_p(1 - \sigma^2)}} \text{c/s} \qquad \text{for } m, n \geq 1 \qquad (23.20)$$

When $m, n \geq 2$,

$$\lambda_{m,n}^2 = (m + \tfrac{1}{2})^4 + (n + \tfrac{1}{2})^4 \left(\frac{L_x}{L_y} \right)^4$$

$$+ 2 \left(\frac{L_x}{L_y} \right)^2 (m + \tfrac{1}{2})^2 \cdot (n + \tfrac{1}{2})^2 \left[1 - \frac{2}{(m + \tfrac{1}{2})\pi} \right] \left[1 - \frac{2}{(n + \tfrac{1}{2})\pi} \right].$$

The panel fundamental frequency ($m = n = 1$) is obtained when

$$\lambda_{1,1}^2 = (1.506)^4 \left[1 + \left(\frac{L_x}{L_y} \right)^4 \right] + 2 \left[1.248 \frac{L_x}{L_y} \right]^2.$$

As a general guide the fundamental frequency of a fully fixed plate is approximately twice that of a simply-supported plate of the same dimensions.

From equations 23.19 and 23.20 it is obvious that there is an infinite number of possible normal frequencies for a given panel. However, there are usually energy losses in the plate and at the boundaries which reduce considerably the effect at higher frequencies, the magnitude of the panel response being determined by the damping present.

23.4.1.3 *Critical frequency.* It has been seen in equation 23.17 that free flexural waves in a plate have wavelengths which are inversely proportional to the square root of the frequency. When a plate is excited at frequency f by a plane acoustic whose trace wavelength on the surface of the plate is equal to λ_B, the flexural wavelength for frequency f, there will be complete matching between the free and forcing waves. Such a matching is called 'coincidence'.

When a plate is excited by sound waves, coincidence can occur for a certain combination of frequency and angle of incidence provided that the frequency lies within specified limits.

Consider an incident plane sound wave travelling with speed a and frequency f, which strikes a plate at an angle of incidence θ (Figure 23.2).

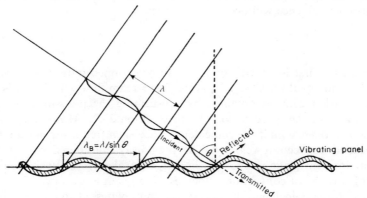

Fig. 23.2 Diagram illustrating wave coincidence.

The trace waveform over the surface of the plate will then have a frequency f, a wavelength $\lambda/\sin \theta$ and a trace velocity $a/\sin \theta$. Coincidence will occur when

i.e.

$$\left. \begin{array}{c} c_B = a/\sin \theta \\ \\ \lambda_B = \dfrac{c_B}{f} = \dfrac{a}{f \sin \theta} \end{array} \right\}$$ (23.21)

Thus the coincidence frequency is given by

$$f_\theta = \frac{a^2}{1.8hc_L \sin^2 \theta} \text{ c/s.} \qquad (23.22)$$

From equation 23.22 it appears that the frequency limits within which coincidence can occur are given by the condition $0 \le \sin \theta \le 1$. The upper limit $f = \infty$ (when $\sin \theta = 0$) is not in fact achieved because the assumptions for the existence of free flexural waves are no longer valid (e.g. λ_B is no longer large compared with h). The true upper limit for f_θ is given by the condition for Rayleigh waves. Taking the maximum velocity of free flexural waves to be c_R, then the range of values for $\sin \theta$ becomes

$$\frac{1.75a}{c_L} < \sin \theta \le 1.$$

The lower limit for f_θ determines the critical frequency f_c

$$f_c = \frac{a^2}{1.8hc_L} \text{ c/s.} \qquad (23.23)$$

This is the coincidence frequency for grazing incidence sound waves.

For boundary layer excitation, coincidence occurs under rather different conditions. If the pressure field is convected over the plate with a velocity U, then coincidence will occur at a frequency

$$f' = \frac{U^2}{1.8hc_L} \text{ c/s,} \qquad (23.24)$$

where, as a first approximation, U can be assumed independent of frequency and equal to $0.8 \, U_0$. U_0 is the free stream flow velocity. It is seen that equation 23.24 is obtained by substituting the pressure convection velocity U for the speed of sound a in equation 23.23. However, f' is not a lower frequency limit for coincidence since each frequency has a unique associated convection velocity. The previously considered case of randomly incident plane waves was, in fact, equivalent to an excitation field having a range of convection velocities for a given frequency.

In addition coincidence can occur on finite plates, at frequencies lower than those given in equations 23.23 and 23.24 for infinite panels, if the free and forced waveforms match at a panel natural frequency. As an example of this for boundary layer excitation consider a simply supported panel. At resonance the panel mode shapes in the x and y directions can be considered as components of a standing wave whose direction makes an angle $\phi_{m,n}$ with the x-axis, where, for a simply supported panel

$$\phi_{m,n} = \tan^{-1}\left(\frac{nL_x}{mL_y}\right) \qquad (23.25)$$

and where (m, n) give the mode order in terms of the number of half-waves in the respective directions. The standing wave is composed of two progressive waves of equal magnitude, propagated in opposite directions with velocity c_B given by

$$c_B = \sqrt{1.8hc_Lf_{m,n}}.$$

The coincidence condition is therefore satisfied, for a wave pattern of frequency $f_{m,n}$ and velocity V along the x-axis, when the component of V in the direction of the standing wave is equal to the free flexural wave velocity, i.e. when

$$V = c_B/\cos \phi_{m,n} = c_B\left[1 + \left(\frac{nL_x}{mL_y}\right)^2\right]^{1/2} \tag{23.26}$$

23.4.1.4 *Noise transmission through a single panel.* From equation 23.1 the transmission loss of a panel is given by

$$r = 10 \log_{10} \frac{1}{\tau} = 10 \log_{10} \left|\frac{p_1}{p_2}\right|^2 \tag{23.27}$$

where p_1, p_2 are the root mean square values of the incident and transmitted sound pressure amplitudes. In general, p_1 is assumed to be the sound pressure at the measuring point before the panel is inserted. If, however, p_1 is taken to be the pressure at the surface of the panel it will include incident and reflected pressure waves and the apparent transmission loss will be greater than for the generally accepted definition.

Consider an infinite panel and assume p_1 is the sound pressure before the insertion of the panel. It can then be shown[5] that the acoustical energy loss through the plate is

$$\frac{1}{\tau} = \left|\frac{p_1}{p_2}\right|^2 = \left[1 + \frac{\eta\pi fh}{a}\frac{\rho_p}{\rho}\left(\frac{c_B}{a}\sin\theta\right)^4 \cos\theta\right]^2$$

$$+ \left[\frac{\pi fh}{a}\frac{\rho_p}{\rho}\left\{\left(\frac{c_B}{a}\sin\theta\right)^4 - 1\right\}\cos\theta\right]^2 \tag{23.28}$$

where η is the panel damping in the form of an internal loss factor, from the complex Young's modulus $E^* = E(1 + i\eta)$.

If damping is neglected, $\eta = 0$ and equation 23.28 becomes

$$\frac{1}{\tau} = 1 + \left[\frac{\pi fh}{a}\frac{\rho_p}{\rho}\left\{\left(\frac{c_B}{a}\sin\theta\right)^4 - 1\right\}\cos\theta\right]^2 \tag{23.29}$$

with total transmission $(1/\tau = 1)$ occurring when $(c_B/a)\sin\theta = 1$. From equation 23.21 this is seen to be the condition for coincidence.

From the general equation $1/\tau$ is again unity when $\theta = 90°$. However, as θ tends to $90°$, the approximate theory of bending waves, on which equation 23.28 is based, breaks down and the limit $\theta = 90°$ is not permissible. Thus, for non-zero η, $1/\tau > 1$ and complete transmission does not occur in practice.

At frequencies close to and greater than the critical frequency, the transmission loss is a function of panel density, thickness, stiffness and damping. At frequencies well below critical, where it can be assumed that $(c_B/a)^2$ is small compared with unity, terms containing $[(c_B/a) \sin \theta]^4$ can be neglected, and equation 23.28 reduces to

$$\frac{1}{\tau} \approx 1 + \left(\frac{\pi f h}{a} \frac{\rho_p}{\rho} \cos \theta\right)^2 \quad \text{or} \quad \frac{1}{\tau} \approx 1 + \left(\frac{\omega m}{2\rho a} \cos \theta\right)^2 \quad (23.30)$$

in terms of angular frequency ω $(= 2\pi f)$ and panel surface density m.

The panel transmission loss below coincidence is then given approximately by

$$r = 10 \log_{10}\left[1 + \left(\frac{\omega m}{2\rho a} \cos \theta\right)^2\right] \text{ dB} \quad (23.31)$$

and, at a given frequency, depends only on the surface density of the panel. Although derived for an infinite panel, equation 23.31 applies equally well for a panel of finite dimensions provided that the frequency lies above the range containing the lower normal frequencies.

For normal incidence equation 23.31 becomes

$$r_{(\theta = 0)} = r_0 = 10 \log_{10}\left[1 + \left(\frac{\omega m}{2\rho a}\right)^2\right] \text{ dB}. \quad (23.32)$$

This reduces to

$$r_0 \approx 20 \log \left(\frac{\omega m}{2\rho a}\right) \text{ dB} \quad (23.33)$$

for $(\omega m/2\rho a)^2$ large compared with unity, a condition usually satisfied for aircraft structures at frequencies for which equation 23.31 is valid. Equation 23.33 gives the well-known 'mass law', a doubling of frequency or surface density producing a 6 dB increase in transmission loss.

In consistent units m is measured in slug/ft^2 and ρ in slug/ft^3. It is usual, however, to express m in lb/ft^2 and the numerical value of ρa is then 84, for a velocity of sound $a = 1100$ ft/sec.

In many problems the incident noise field is diffuse and the waves are incident at all angles. Since τ is a function of θ, the overall transmission loss is then given by a mean value

$$\tau_m = \frac{\int_0^{\pi/2} \tau(\theta) \cos \theta \sin \theta \, d\theta}{\int_0^{\pi/2} \cos \theta \sin \theta \, d\theta}$$

$$= \int_0^1 \tau(\theta) \, d(\cos^2 \theta). \quad (23.34)$$

From equation 23.31,

$$\tau(\theta) = \frac{1}{1 + \left(\dfrac{\omega m}{2\rho a} \cos \theta\right)^2},$$

Hence

$$\tau_m = \left(\frac{2\rho a}{\omega m}\right)^2 \log_e \left[1 + \left(\frac{\omega m}{2\rho a}\right)^2\right] \tag{23.35}$$

and the transmission loss for random incidence is

$$r_m \approx r_0 + 10 \log_{10} (0.23 r_0) \text{ dB} \tag{23.36}$$

where r_0 is given by equation 23.33.

Because of the inaccuracy of the theory at $\theta = 90°$, equation 23.35 will underestimate the transmission loss. It is therefore reasonable to exclude angles in the integration for which the condition $(\omega m/2\rho a)^2 \cos^2 \theta \gg 1$ does not hold.

Then

$$\tau_1 = \left[\tau\right]_0^{\theta_1} = \int_{\cos^2 \theta_1}^1 \left(\frac{2\rho a}{\omega m \cos \theta}\right)^2 d(\cos^2 \theta)$$

$$= \left(\frac{2\rho a}{\omega m}\right)^2 \log_e \left(\frac{1}{\cos^2 \theta_1}\right) \tag{23.37}$$

giving

$$r_1 = r_0 - 10 \log_{10} \left[\log_e \frac{1}{\cos^2 \theta_1}\right]. \tag{23.38}$$

The appropriate value of θ_1 has now to be determined. For $\theta_1 = 82.5°$ (which satisfies the condition where $r > 25$ dB)

$$r_1 \approx r_0 - 6 \text{ dB} \tag{23.39}$$

a value which could also be obtained assuming a mean angle of incidence of 60°. This is sometimes called the '60° mass law'. Equation 23.39 bears a close resemblance to measured transmission losses in buildings, which give an empirical 'field incidence' transmission loss[3], r_f, given by

$$r_f = r_0 - 5 \text{ dB}. \tag{23.40}$$

This relation could be obtained from equation 23.38 assuming an upper limit of integration $\theta_1 = 78°$.

23.4.1.5 *Application to aircraft*. The transmission spectrum for a panel can now be divided into four regions shown diagrammatically in Fig. 23.3, the boundaries of the regions being determined by the natural and critical frequencies of the particular panel considered. At low frequencies (region 1) the noise transmission is controlled by panel stiffness (see chapter 3), mass and damping being unimportant, and the transmission loss decreases at 6 dB per octave as frequency increases to within the neighbourhood of the panel fundamental natural frequency.

In an aircraft structure the boundary conditions of the individual panels will lie between simply supported and fully fixed. Thus the natural frequencies will lie within limits determined by equations 23.19 and 23.20 and will determine region 2 of Fig. 23.3. Within this region the transmitted noise is primarily a function of panel damping (chapter 3). In addition to these single panel resonances an aircraft fuselage will have overall structural resonances of groups of panels and of circular cylinders (chapter 15).

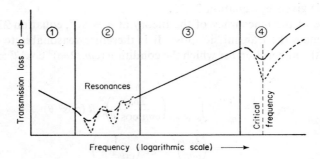

Fig. 23.3   General response of a panel.

The transition from region 2 to region 3 takes place when the resonance effects become negligible. This occurs at frequencies which are approximately three to four times the panel fundamental natural frequency. In region 3 the transmission loss is determined by equations 23.31 to 23.40 and the plate can be considered to move as though it is essentially mass-controlled, i.e. as though it is composed of a large number of masses free to slide relative to each other without inter-mass constraining forces.

At higher frequencies panel stiffness and damping again become important and the transmission loss is determined by equation 23.28. This is shown as region 4 in Fig. 23.3. The critical frequency which determines the approximate boundary between regions 3 and 4 is given by equation 23.23. Beranek[3] has presented general curves for the transmission loss of building constructions at frequencies above the critical value and has shown that the loss increases initially at 10 dB per octave, the slope gradually reducing to 6 dB per octave.

23.4.2 *Curved panels*

The preceding discussion of transmission loss refers directly to flat panels and little information is available to indicate how accurate these results are when applied to curved panels. It has been suggested by von Gierke[9] that the increased stiffness due to panel curvature will increase the low frequency transmission loss whilst at higher frequencies the loss will still be determined by the mass law. Theoretically Cremer[10] has shown that, for cylindrical structures, the transmission loss is essentially flat with frequency, with perhaps lower values at the high frequencies. This rather extreme result does not seem to be realized in practice although measurements on test panels have shown increased losses at low frequencies as predicted by von Gierke.

23.4.3 *Double partitions*

The double wall is the simplest form of multiple partition and consists of two impervious sheets or plates, separated by an air gap, and isolated from each other. For incident plane waves at a frequency below the critical frequency of the panels, and for panels which are mass-controlled so that plate resonances do not occur, the transmission loss provided by a double wall is:

$$\frac{1}{\tau} = \left|\frac{p_1}{p_2}\right|^2 = 1 + 4q_1q_2 (\cos \delta - q_1 \sin \delta)(\cos \delta - q_2 \sin \delta) + (q_1 - q_2)^2 \tag{23.41}$$

where p_1 is defined as for equation 23.28,

$$q_j = \frac{\omega m_j}{2\rho a} \cos \theta \qquad (j = 1, 2)$$

and

$$\delta = \frac{\omega d}{a} \cos \theta.$$

θ is the angle of incidence of the plane waves and d is the distance between the two plates whose surface densities are m_1 and m_2.

The transmission loss of a double partition is, in general, greater than for a single panel of the same total surface density, but equation 23.41 predicts a series of mass–air–mass resonances of the system for which the attenuation is considerably reduced.

The value of $1/\tau$ given by equation 23.41 also depends on the value of the air gap d, the equation having a series of maxima and minima at frequencies corresponding to the (frequency) solution of

$$\cos 2\delta = \pm \left[\frac{q_1q_2 - 1}{\sqrt{(1 + q_1^2)(1 + q_2^2)}}\right]. \tag{23.42}$$

The $-ve$ sign gives the maximum values of $1/\tau$ and the $+ve$ sign gives the minima. The value of d can thus be selected to give the maximum attenuation at the appropriate frequency.

17

In the particular case of two panels of equal mass, equation 23.41 reduces to

$$\frac{1}{\tau} = 1 + 4q^2 (\cos \delta - q \sin \delta)^2 \qquad (23.43)$$

where $q = (\omega m/2\rho a) \cos \theta$ and m is the mass of each panel.

The resonances now occur when

$$\tan \delta = 1/q \qquad (23.44)$$

and when $2\pi d/\lambda$ is small so that

$$\tan \delta \approx \delta,$$

the lowest resonant frequency is given by

$$f_1 = \frac{1}{2\pi \cos \theta} \left(\frac{2\rho a^2}{md}\right)^{1/2}. \qquad (23.45)$$

When the walls are of equal mass the resonance condition (equation 23.44) gives complete transmission, i.e. $1/\tau = 1$ from equation 23.43.

Also the conditions on d for maxima and minima of $1/\tau$ (equation 23.42) reduce to

$$\cos 2\delta = \pm \left(\frac{q^2 - 1}{q^2 + 1}\right), \qquad (23.46)$$

the condition for minima reducing further to equation 23.44 and the condition for maxima giving

$$\tan \delta = -q. \qquad (23.47)$$

A comparison of theoretical and experimental transmission losses for single- and double-pane windows of equal total surface density is shown in Fig. 23.4, the double windows having panes of equal mass. It is assumed

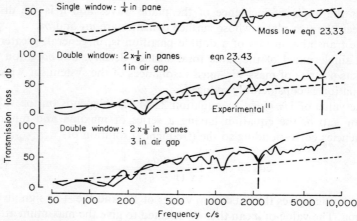

Fig. 23.4   Transmission loss through single and double windows[11].

that the sound waves are normal in all cases. The experimental results[11] although containing individual plate resonances which were excluded from theory by the initial assumptions, show a general trend which is similar to that predicted by the theory.

From equation 23.41 it is seen that τ is a function of the angle of incidence θ. The average value τ_m for randomly incident waves can be calculated, as in the case of a single panel, from equation 23.34. Since the condition for complete transmission through a double structure at a given frequency is satisfied for particular angles only (specified by equation 23.44), the average transmission loss for diffuse noise will never reach zero, even when the panels are of equal mass.

The above results for the theoretical transmission loss of a double wall with panels of equal mass have been extended by London[12] to panels of arbitrary impedance, thereby removing the restriction that the panels are mass-controlled.

A general theory for the transmission loss of multiple structures composed of porous and impervious materials is given by Beranek and Work.[13] The results can be applied to aircraft soundproofing and show that the presence of sound absorbing material provides damping of the resonances in the gap. The absorbent material is particularly successful in increasing the transmission loss if the individual panels are light in weight.[12]

23.4.4 *Boundary layer noise*

The transmission losses determined for single and double panels (sections 23.4.1.4 and 23.4.3) refer specifically to incident sound waves. For excitation by boundary layer pressure fluctuations the outer skin of the fuselage acts as a transducer converting the turbulent pressure fluctuations in the boundary layer into sound waves in the cabin and, as was mentioned in section 23.3.1, the transmission loss will be different from that for a statistically uniform acoustic field. Therefore it is necessary to determine the noise radiated by a panel vibrating under boundary layer excitation and then to establish the noise reduction due to multiple structures and cabin absorption.

As a first approximation it can be assumed that the panel is radiating into free space and the transmission loss for the remaining structure can then be estimated from a modification of the theory of multiple structures. Further research on boundary layer noise transmission loss is needed (see chapter 22).

23.5 Soundproofing treatments

23.5.1 *Requirements*

The sound insulation of an aircraft cabin has to satisfy a large number of requirements, the most important of which are the following:

(a) It must reduce the sound transmitted from the exterior of the aeroplane.

(b) It must effectively absorb sound.

(c) It must be light in weight.

(d) It must have a high resistance to flame.

(e) It must not readily absorb moisture and must dry out quickly when wet.

(f) It must have adequate mechanical strength.

A commonly used material, fibreglass in the form of blankets, does not satisfy conditions (e) and (f) since it disintegrates under repeated vibration and it collects moisture between the fibres. Thus it is enclosed in lightweight bags which must be completely sealed to prevent the entrance of water vapour. The problem of water condensation became severe in several airliners when extremely cold conditions were experienced on polar air routes. In aircraft of one airline[14] the fibreglass was carefully sealed in bags of vinyl-coated nylon cloth, but when two of these aircraft were inspected after one year in service it was found that 80% of the blankets were wet and the total weight of the blankets had increased by 77%.

If natural fibres—kapok, cotton or paper—are used they must be adequately flameproofed and treated to prevent moisture absorption and rot.

Acoustically the material must satisfy two conditions—it must reduce the sound transmitted into the cabin and it must absorb the noise in the cabin. Under the restriction of limited weight these two requirements are, to a certain extent, in opposition and the best treatment is usually a compromise. Furthermore the material is required to be efficient over the frequency range 20 c/s to 10 kc/s, a condition which it is almost impossible to fulfil.

23.5.2 *Typical aircraft acoustic insulation*

In practice aircraft soundproofing usually consists of a combination of fibrous blankets, air gaps and impervious sheeting. Theoretical and experimental studies have been made[3,11,13,15,29] to determine the optimum positioning of the air gaps and impervious and porous materials, but in practical cases the installation is usually determined by other considerations. The fibrous materials have to be placed in bags to prevent the collection of moisture and to keep the fibres in place. The air spaces are often dictated by stringer and frame depth and by cabin air conditioning requirements. The cabin internal trim, the part of the soundproofing which determines its absorptive properties, is usually chosen for the ease with which it can be cleaned and is therefore generally impervious. All these requirements have to be considered when selecting the optimum installation.

23.5.3 *Transmission loss*

Sound transmission through six multiple structures suitable for aircraft soundproofing has been investigated[3,13] for normally incident sound waves. The six types of construction considered are shown in cross section in Fig. 23.5 and consist of a metal panel, an air space, a flexible porous blanket, and an impervious sheet or septum.

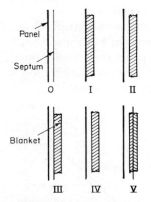

Fig. 23.5 Types of multiple structure.

A measure of the efficiency of a particular arrangement can be obtained from the difference between the theoretical transmission loss and that predicted by the mass law for a single panel of the same total mass. On this basis, systems III, IV and V were found to give similar attenuations at the higher frequencies and were all more efficient than treatments I and II. Systems III and V, however, presented an absorptive surface to the noise in the cabin and hence had a higher absorption coefficient than IV. The selection of a suitable treatment depends therefore on the absorption provided by seats, carpets, etc., in the cabin (equation 23.4). If this is high enough to permit the use of a non-absorbent finish for the cabin walls, then structure IV is the best choice.

At low frequencies the comparison is more difficult because the fundamental resonance of the system has a significant effect, the estimated transmission loss at the fundamental frequency being generally less than that predicted for the equivalent single panel. The system must therefore be chosen so that the lowest resonant frequency does not coincide with the frequency of the predominant noise. At low frequencies away from resonance the highest transmission losses are usually obtained with structures having the greatest depth.

The above comparison assumes that the weight of the septum is light relative to the fuselage skin and the validity of the assumption should be checked when heavier inner sheets are used.

The design of multiple structures is of major importance for the external wall of an aircraft compartment, the design of bulkheads and floors being of secondary importance unless the noise levels in the adjacent compartments are significantly higher than those in the cabin under consideration. If the noise levels in adjoining compartments are similar to those in the cabin, the transmitted noise is negligible when compared with that from sources external to the fuselage and the bulkheads and floor can be assumed to have zero transmission coefficient.

Fig. 23.6 shows typical values for the transmission loss of a bare panel and a panel–soundproofing combination when measured in a reverberant chamber. It is seen that the bare panel transmission loss obeys the field incidence mass law (equation 23.40) but the combination has an increased transmission loss, especially at high frequencies.

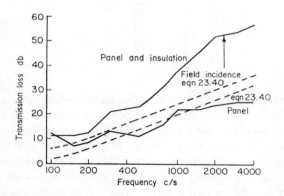

Fig. 23.6   Transmission loss for typical structure: laboratory
measurements.

23.5.4 *Absorption*

Sound can be absorbed in several ways but the two methods of importance in aircraft soundproofing are the use of a diaphragm for low frequencies and of porous material for high frequencies.

The sound insulation of an aircraft compartment is usually covered by a cabin trim which significantly affects the wall absorption. When uncovered, the porous material would have a high absorption coefficient at high frequencies but this absorption is reduced by the trim, the reduction depending on the porosity of the covering. An example of this effect[11] is shown in Fig. 23.7, where the curves are for the following conditions:

(a) Impervious trim.
(b) Trim perforated with $\frac{1}{64}$ in diameter holes at $\frac{1}{4}$ in spacing.
(c) Trim perforated with alternate $\frac{1}{16}$ in and $\frac{1}{64}$ in diameter holes at $\frac{1}{4}$ in spacing.
(d) No trim.

At low frequencies the trim acts as a diaphragm and increases the absorption.

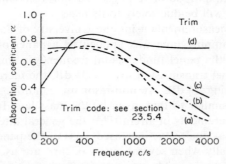

Fig. 23.7   Effect of trim on absorption coefficient of a porous material.[11]

The absorption coefficient of the insulation is also affected by the relative positioning of the panel, porous material and impervious septum.[11] This dependence is illustrated in Fig. 23.8 for several of the structures in Fig. 23.5. It is seen that, as the porous material and septum are moved away from the backing panel the peak absorption shifts to lower frequencies, the insulation then vibrating as a whole.

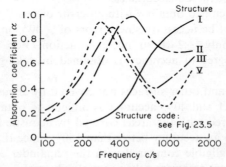

Fig. 23.8   Effect of position of porous and impervious materials on absorption coefficient.[11]

Typical aircraft soundproofing installations have, in general, poor high frequency absorptive characteristics. However, this can often be improved by absorption provided by carpets, seats and passengers.

23.5.5 *Damping tape*

The use of damping tape has been suggested for cabin soundproofing but there is little available information regarding the efficiency of the tape in reducing transmitted noise. The main purpose of damping tape is the control of resonant vibration where damping is the predominant factor in determining the amplitude of response. At non-resonant frequencies the effect of the tape will be due solely to its mass.

In a series of measurements using the University of Southampton siren tunnel, sound damping tape was applied to a panel of typical aircraft construction. At the panel fundamental frequency a single layer of tape increased the panel transmission loss by $3\frac{1}{2}$ dB, the increase predicted by the mass law being $1\frac{1}{2}$ dB. At non-resonant conditions the increase in attenuation was similar to that given by the mass law. In published data taken from measurements in aircraft[16,17] the general attenuation is equal to or less than that due to the increased mass of the panel, the latter case being true especially when several layers of tape are used. Therefore the use of sound damping tape does not appear to be justified unless a severe problem of structural resonance arises, the difficulties of installation and subsequent inspection outweighing the benefits of the treatment.

23.5.6 *Accuracy*

The experimental values obtained for transmission and absorption coefficients of particular treatments will depend to a certain extent on the method used. The absorption coefficient is particularly difficult to measure accurately. However equation 23.2 or Fig. 23.1 will give an indication of the importance of these inaccuracies.

It is often difficult to obtain α with an error of less than 10% but such accuracy may not be necessary since errors of $\pm 40\%$ in total absorption give errors of only ± 2 dB in noise reduction. For materials having $\alpha > 0.44$ this degree of accuracy is obtained by taking α constant and equal to 0.72.

For windows, and other surfaces not covered by acoustic material, α is approximately 0.1 and 40% accuracy is difficult to obtain. If these areas are small compared with the areas covered by acoustic materials, α need not be known to an accuracy better than 150% since the contribution of these areas is negligible compared with the remainder of the cabin. It is thus often sufficient to know only nominal values of α.

Values of α and τ are usually measured at discrete frequencies whereas noise attenuation is often calculated in octave or third-octave frequency bands. In such cases the values of the coefficients should be chosen to give conservative answers. Since variations of α and τ with frequency are often as shown in Fig. 23.9 the following rules should be applied:

(a) Porous trim: Take the value of α and τ for the lowest frequency of the band.

(b) Non-porous trim: Use the frequency giving the most conservative results. Since the effect of τ is usually greater than that of α, the condition is usually satisfied by the lowest frequency in the band.

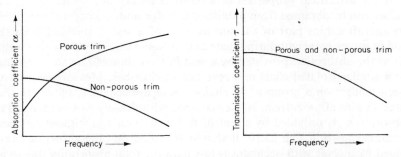

Fig. 23.9   Typical variation of absorption and transmission coefficients with frequency.

23.6 Problems in specific compartments

23.6.1 *Passenger compartments*

23.6.1.1 *Windows.* In common with other multiple partitions, double-pane windows give a higher attenuation than do single-pane windows of the same total surface density. This increased efficiency is predicted by theory and borne out by measurement.[11,16,18,19] An example is given in Fig. 23.4 where theoretical and practical results are compared.

The problem of mass–air–mass resonances is important in the case of windows because absorptive materials cannot be placed in the intervening gap. The dimension of the gap should therefore be designed so that the resonant frequencies do not occur in the neighbourhood of predominant exciting frequencies.

However, the installation of double panes introduces extra complication and weight in the window frames and in certain cases[11,17] where the noise levels transmitted through the windows were not critical, single-pane windows were fitted. The design of windows should therefore depend on the transmission loss required.

23.6.1.2 *Seats.* When seats are placed in an aircraft cabin they will absorb sound and hence increase the overall absorption of the compartment. The average effect of this change can be expressed as an increase in the effective absorption coefficient of the cabin wall (equation 23.3).

If a single chair provides A_c units of absorption, then the total absorption provided by n chairs will be nA_c. The effective increase in the wall absorption coefficient can then be expressed as

$$\Delta\alpha_c = \frac{nA_c}{S},$$

where A_c and $\Delta\alpha_c$ are functions of frequency, and S is the total wall area.

If the acoustical properties of aircraft seats are not available, typical values can be obtained from published data for auditoria seats.[18,20,21,22,23] In aircraft cabins part of the surface area of a seat is shielded from the sound waves by neighbouring seats and, in some cases, by the cabin walls. Thus the absorption provided by a seat in the cabin will be less than that of a similar isolated chair in a reverberant chamber. Measurements have been made[23] on a group of 10 chairs whose upholstery is similar to that used in aircraft, viz. foam rubber covered with densely woven fabric. The absorption A_c provided by a seat of this type, with and without an occupant, is shown in Fig. 23.10. It should be noted that the absorption provided by a chair with occupant is less than the total absorption due to an unoccupied chair and a standing person.

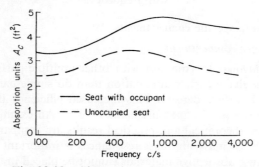

Fig. 23.10   Absorption of sound by seats.[23]

From Fig. 23.10, assuming a wall area per seat of

$$\frac{S}{n} = 26 \text{ ft}^2,$$

typical values of $\Delta\alpha_c$ for an occupied seat are 0.13 at a frequency of 125 c/s and 0.18 at 1,000 c/s, with corresponding values of 0.09 and 0.12 for an unoccupied seat.

23.6.1.3 *Ducts, hatracks, etc.* Noise can be transmitted along the structure and radiated into the cabin by ventilation ducts, hatracks, etc. Adequate sound reduction should be provided in the ducts, and the hatracks should be covered to prevent the radiation of noise by the vibrating panels. Alternatively the racks can be attached to the fuselage structure by means of isolation mounts. This method has been used in a helicopter[17] to provide isolation from gear and engine noises. Mounts having natural frequencies of approximately 200 c/s were found to provide satisfactory attenuation without being too flexible.

23.6.1.4 *Leaks.* The transmission loss of a soundproofing treatment will be reduced considerably if acoustic leaks allow the sound waves to by-pass the insulation. Such leaks can occur as air leaks around poorly fitting components of unpressurized aircraft or as acoustic leaks around sections of the soundproofing. To avoid the latter the insulating material must be fitted around all stringers and frames and must have no gaps between adjacent sections.

In many cases leaks will not be discovered until in-flight measurements are made.[24] Elimination of these sound paths will reduce the local internal noise levels.

23.6.1.5 *Standing waves.* The average absorption coefficients of an aircraft cabin are usually in the range 0.4 to 0.5, with associated reverberation times of approximately 0.2 s. It is therefore unlikely that standing waves will be a problem in passenger compartments. Standing waves may, however, occur in other compartments.

23.6.2 *Helicopters*

The major sources of helicopter cabin noise are engines, rotor blades and transmission systems. The positioning of these sources relative to the passenger cabin depends on the number of rotors but in general the rotor blades pass directly over the cabin. Thus the rotor noise passing through the top of the fuselage is considerably higher than that passing through the remaining areas. The soundproofing along the top of the cabin must therefore provide a higher transmission loss than in other regions—except perhaps near the engine exhaust. A practical example is shown in reference 17 in which the density of soundproofing is varied with position. In this case the high roof transmission loss is achieved by means of a double aluminium skin with the intervening gap filled with fibreglass. On the cabin side-walls where the required transmission loss is lower, the inner aluminium sheet is not fitted.

In the case of twin-rotor helicopters the engines can be mounted at one end of the aircraft and the fireproof bulkhead provides a good basis for soundproofing. However, the propulsive drive has to be carried along the fuselage to the front (or rear) rotor by means of a drive shaft which will radiate noise transmitted from the engine and gear box. If the drive shaft is inside the fuselage it has to be enclosed in a soundproofed duct to reduce the noise radiated into the passenger cabin. A more efficient insulation would be obtained if the shaft was placed outside the fuselage so that full use could be made of the transmission loss provided by the fuselage skin. A design change of this nature was made during the development leading to the Vertol 107 helicopter,[17] the change being an excellent example of the way in which soundproofing principles can be included in basic aircraft design.

23.6.3 *Crew compartments*

Noise control in the flight deck is, in general, more complex than in passenger compartments since the presence of aircraft controls, radio equipment and other instruments makes the estimation of transmission and absorption coefficients very difficult. Furthermore, the equipment itself often generates noise.

In cases of this type, where there is a limited area available for acoustic treatment, it is of interest to consider the effect of partial coverage with absorptive treatment. The average reduction in cabin noise level can be estimated from equation 23.12 but the local effect will depend on the positioning of the absorptive treatments. As an example of the use of equation 23.12, Fig. 23.11 has been plotted assuming $S_n/S = 1$ and an

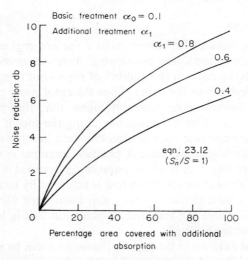

Fig. 23.11   Noise reduction by partial coverage with absorptive material.

initial mean absorption coefficient in the compartment of $\alpha_0 = 0.1$. It is seen that the highest rate of increase in noise reduction occurs when the area of additional treatment is small, e.g. for $\alpha_1 = 0.8$ the noise reduction is improved by 4 dB for 20% coverage of increased absorption, but a further 50% of area has to be covered for a further 4 dB reduction. Limited application of absorptive treatment should therefore be considered for the crew compartments.

Flight deck noise can often be reduced by the correct design and positioning of the many small detectors, aerials, etc., which stick out into the airstream near the nose of an aircraft. It will also be necessary to ensure that the noise generated by the undercarriage, when lowered or retracted, does not create a nuisance. In all these instances reduction of noise at the source will be more efficient than increased cabin insulation.

Because of the multiplicity of sources it is not possible to make detailed estimates of sound reduction for aircraft crew compartments without in-flight measurements, since the theory provides no more than a general guide to the behaviour of a particular compartment.

23.6.4 *Light aircraft*

The enclosed cabins of light aircraft contain large areas of Perspex and equipment which provide little acoustic absorption. Thus the problem is similar to that for the flight deck of a larger aircraft and the limited use of absorptive treatment should again be considered.

The structure of a light aeroplane often provides only a small transmission loss and further attenuation is then necessary. The extra weight of additional soundproofing can be kept to a minimum by treating only the areas where the outside noise levels are the highest. Such areas occur near the engine exhausts and, for multi-engined propeller-driven aircraft, near the plane of rotation of the propellers. In single-engined aircraft the predominant noise in the cabin will probably be generated by the engine exhaust. Hence the end of the exhaust pipe should not be placed near the cabin.

High frequency noise can be generated by ill-fitting doors and poorly designed struts and fairings. Light aircraft are especially prone to air leaks which provide direct acoustic transmission paths. Obviously the best form of sound control in these cases depends on the design of the components concerned.

23.7 Measure of acoustic coefficients

The general properties of a soundproofing treatment can be obtained from theoretical studies, but more detailed values are given only by experiment. In a logical sequence of measurements the transmission and absorption coefficients are measured in the laboratory for a series of treatments and the noise reduction associated with each treatment is then calculated. Several of the more suitable treatments are then installed in full-scale aircraft and the final choice of material is made after noise measurements have been made during flight.[16,17,19,27] The flight measurements do not distinguish between transmission loss and absorption but give the overall noise reduction provided by the insulation. The individual coefficients are given only by the laboratory measurements[26].

Laboratory measurements of transmission loss are normally made in a reverberant chamber or an 'acoustic tube'. The siren tunnel at the University of Southampton has also been used. In all these methods the specimen is rigidly clamped at its edges and, in this respect, is unrepresentative of aircraft conditions.

The differences between the three laboratory methods can be expressed in terms of the pressure correlations over the outer surface of the specimen. In the acoustic tube the direction of propagation of the sound waves is normal to the panel surface and the pressures are thus in-phase (i.e. there is unit correlation) over the panel. The siren tunnel provides grazing incidence sound waves which have unit correlation over the panel in a direction perpendicular to the direction of propagation of the waves and have a cosine correlation function in the direction parallel to the direction of flow. The third method, the reverberant chamber, provides randomly incident noise which is statistically uniform over the panel but the correlation function has the form of a Bessel function.

The siren tunnel method gives an approximate representation of boundary layer excitation, the convected boundary layer pressure field being simulated by the system of progressive plane waves radiated by the siren. The representation is by no means complete but preliminary comparisons show a general agreement between transmission losses measured in aircraft and in the siren tunnel.

Absorption coefficients are measured in either normally or randomly incident sound fields but the results can vary widely from laboratory to laboratory.[28]

Noise reduction measurements in full-scale aircraft can be carried out during ground or flight testing. Ground testing can be carried out using sirens, loudspeakers or aircraft as noise sources, the noise from the sirens or loudspeakers being either discrete or random in frequency, or a recording of the noise of the particular type of aircraft under investigation. When complete aircraft are used the noise source can be the test aircraft itself or a second similar aeroplane,[27] the latter method being of value if it is necessary to distinguish between airborne noise and mechanical vibration transmitted through the structure.

Flight measurements offer the most satisfactory method of determining the overall acoustic properties of a treatment for use in an aircraft cabin. In this case the investigation would consist of a determination of the major areas for treatment and of the optimum materials for the individual regions concerned. When searching for the areas of high noise level the cabin is heavily soundproofed and small areas of insulation are then removed in turn, the associated changes in noise level being measured in each case. The fuselage can then be divided into several small compartments and a particular treatment tested in each section. Details of investigations carried out in this manner can be found in references 16, 17 and 19.

References

1. M. Kobrynski. Calcul approché de l'éfficacité du traitement acoustique des cabines d'avions, *Ann. Télécomm.*, **8**, No. 1 (1953).
2. L. L. Beranek. *Acoustics*, McGraw-Hill, New York (1954).
3. L. L. Beranek (Ed.). *Noise Reduction*, McGraw-Hill, New York, 1960.
4. Bolt, Beranek and Newman Inc. Handbook of acoustic noise control, Vol. 1, Supplement 1, *W.A.D.C. Tech. Rep.* 52–204.
5. A. Schoch and K. Fehér. The mechanism of sound transmission through single-leaf partitions, investigated using small scale models, *Acustica*, **2**, 189 (1952).
6. L. S. Goodfriend. Lightweight partitions, *Noise Control*, **2**, No. 6 (1956).
7. G. Kurtze. Lightweight walls with high transmission loss, *Acustica*, **9**, 441 (1959).
8. G. B. Warburton. The vibration of rectangular plates, *Proc. Inst. Mech. Eng, London*, **168**, 371 (1954).
9. C. M. Harris (Ed.). *Handbook of Noise Control*, McGraw-Hill, New York, 1957.
10. L. Cremer. Theorie der Luftschalldämmung zylindrischer Schalen, *Acustica*, **5**, 254 (1955).
11. R. H. Nichols, H. P. Sleeper, R. L. Wallace and H. L. Ericson. Acoustical materials and acoustical treatments for aircraft, *J. acoust. Soc. Amer.*, **19**, 428 (1947).
12. A. London. Transmission of reverberant sound through double walls, *J. acoust. Soc. Amer.*, **22**, 270 (1950).
13. L. L. Beranek and G. A. Work. Sound transmission through multiple structures containing flexible blankets, *J. acoust. Soc. Amer.*, **21**, 419 (1949).
14. J. Jensen-Gaard. Moisture accumulation in aircraft soundproofing and insulation, *Am. Soc. Mech. Eng.*, Paper No. 60-Av-7 (1960).
15. L. L. Beranek, and others. Principles of sound control in airplanes, *O.S.R.D. Rep.* 1543 (1944).
16. Bolt, Beranek and Newman Inc. Noise reduction programme for the Convair-Liner 340, *Bolt, Beranek and Newman Inc., B.B.N. Rep.* 257 (Nov. 1954).
17. H. Sternfeld. New techniques in helicopter noise reduction, *Noise Control*, **7**, No. 3 (1961).

18. N. Fleming. The silencing of aircraft, *J. Roy. Aeron. Soc.*, **50**, 639 (1946).
19. G. S. Hunter. Sound reduction programme for Convair-Liner 340, *Noise Control*, **2**, No. 1 (1956).
20. R. N. Lane. Absorption characteristics of upholstered theatre chairs and carpet as measured in two auditoria, *J. acoust. Soc. Amer.*, **28**, 101 (1956).
21. R. N. Lane and J. Botsford. Total sound absorption for upholstered theatre seats with audience, *J. acoust. Soc. Amer.*, **24**, 125 (1952).
22. H. W. Rudmose and L. L. Beranek. Noise reduction in aircraft, *J. Roy. Aeron. Soc.*, **14**, 79 (1947).
23. T. D. Northwood and E. J. Stevens. Acoustical design of the Alberta Jubilee Auditoria, *J. acoust. Soc. Amer.*, **30**, 507 (1958).
24. P. Liénard. Les études ONERA relatives aux problèmes acoustiques. Application à l'avion Caravelle, European Congress of Aeronautics, Cologne (Sept. 1960).
25. P. M. Morse. *Vibration and Sound*, 2nd ed., McGraw-Hill, 1948.
26. C. T. Molloy. Electra acoustical programme, *Noise Control*, **5**, No. 1 (1959).
27. L. N. Miller, L. L. Beranek and H. Sternfeld. Acoustical design for transport helicopters, *Noise Control*, **5**, No. 2 (1959).
28. H. J. Sabine. A review of the absorption coefficient problem, *J. acoust. Soc. Amer.*, **22**, 387 (1950).
29. R. A. Mangiarotty. Optimization of the mass distribution and the air spaces in multiple-element soundproofing structures, *J. acoust. Soc. Amer.*, **35**, 1023 (1963).

Additional References

C. Zwikker and C. W. Kosten. *Sound Absorbing Materials*, Elsevier, New York, 1949.
A. London. Principles, practice and progress of noise reduction in airplanes, *N.A.C.A. T.N.* 748 (July 1940).
O. R. Rogers and R. F. Cook. Aerodynamic noise and the estimation of noise in aircraft, *W.A.D.C. Tech. Rep.* 52–341 (Dec. 1952).
M. Kobrynski. Doctrine actuelle d'insonorisation des cabines d'avions, *Acustica*, **6**, 393 (1956).
M. Kobrynski. Méthodes d'insonorisation des avions de transport et spéciale-ment du diréacteur SE120 Caravelle, *Rech. Aeron.* No. 63, 33 (1958).
L. Cremer. Calculation of sound propagation in structures, *Acustica*, **3**, 317 (1953).
C. M. Harris and C. T. Molloy. The theory of sound absorptive materials, *J. acoust. Soc. Amer.*, **24**, 1 (1952).
C. W. Kosten and J. H. Janssen. Acoustic properties of flexible and porous materials, *Acustica* **7**, 372 (1957).
G. T. Gebhardt. Acoustical design features of Boeing Model 727, *J. Aircraft*, **2**, 272 (1965).

APPENDIX

A Note on the Decibel

Throughout this book, the numerical level of a sound has been quoted in terms of 'decibels'. This is a standard practice which is described in most textbooks on acoustics, but for the sake of the completeness of this book, we append this section.

Audible sounds cover an enormous range of pressure magnitudes. For instance, the noise pressure near a large jet engine may be ten million times that of a softly whispering voice. The quantifying of a noise pressure by a decimal number of pressure units is therefore likely to involve many zeros, either before or after the decimal point. Errors will easily slip in as a zero is accidentally inserted or dropped! If the noise pressure is measured on a logarithmic scale (to the base of 10) this enormous range of noise pressures can be condensed to numbers which are much more manageable and less susceptible (on the whole) to errors due to dropped zeros. The decibel system is such a logarithmic scale.

Another reason for the use of the decibel scale in acoustics is that the human ear subjectively assesses a sound pressure magnitude in terms of its ratio to other sounds. In effect, the ear/brain measuring system has a logarithmic response, and supplies the senses with a signal approximately proportional to the logarithm of the incoming sound pressure.

These two reasons show the desirability of using a logarithmic scale to measure noise, quantifying it in terms of the logarithm of its ratio to another standard noise. The decibel scale does just this. It is, in fact, ten times the logarithm (to base 10) of the ratio of two *power* quantities, or quantities proportional to power. It involve essentially a *standard reference quantity* which should, strictly, always be quoted when the scale is used.

Consider a sound source emitting an acoustic power of W_1, and another standard, reference sound source emitting a power of W_{ref}. In decibels (dB) the SOUND POWER LEVEL (PWL) of the first source relative to the reference source is

$$\text{PWL} = 10 \log_{10} \frac{W_1}{W_{ref}} \, \text{dB} \qquad re \ W_{ref}.$$

So long as the power of the reference source is quoted alongside the number of dB's obtained from this, *any* value for W_{ref} could be chosen.

497

However, to establish a generally-accepted decibel PWL scale, W_{ref} is usually taken to be 10^{-13} W—a very small power source indeed. In some countries, however, it has been the practice to make $W_{ref} = 10^{-12}$ W, which emphasizes the importance of quoting the reference level alongside the PWL value. Referred to the 10^{-13} W source, the total sound power emitted by a large rocket engine is found to be about 200 dB, and a loud continuous shouting voice is about 100 dB.

Far from the source, its total power output is of less importance than the actual power flux past a given point. This, of course, is represented by the acoustic intensity (I_1, say) which is the acoustic power flow across a unit area. Being a power quantity, this can quite properly be measured on a decibel scale by comparing it with a reference acoustic intensity, I_{ref}. We then define the INTENSITY LEVEL (IL) at a point in the sound field by

$$IL = 10 \log_{10} \frac{I_1}{I_{ref}} \text{ dB} \qquad re \ I_{ref}.$$

In this context, I_{ref} is commonly taken to be 10^{-12} W/m² ($= 10^{-16}$ W/cm²). Hence, when I_1 is measured in units of W/m², we have

$$IL = 10 \log_{10} I_1 + 120 \text{ dB} \qquad re \ 10^{-12} \text{ W/m}^2.$$

Now the response of the human ear or of a flexible structure depends upon the incident sound *pressure* rather than on the power emitted by the source or the acoustic intensity. When we set up the decibel scale for sound pressure we must remember that the decibel system applies basically to *power* quantities or to quantities proportional to power. Now the intensity is a power quantity, and in a radiating field is proportional to the square of the pressure amplitude. In fact, we have shown on page 7 that

$$\text{Intensity, } I = \frac{p^2}{\rho a}$$

where p is the r.m.s. acoustic pressure. p^2 is therefore proportional to a power quantity, so we can define the SOUND PRESSURE LEVEL (SPL) in decibels for a sound pressure p_1 in the following way:

$$SPL = 10 \log_{10} \frac{p_1^2}{p_{ref}^2}$$

$$= 20 \log_{10} \frac{p_1}{p_{ref}} \text{ dB} \qquad re \ p_{ref}.$$

The value of p_{ref} is usually taken to be 0.0002 microbar ($= 0.0002$ dyn/cm² $= 0.00002$ N/m²). With this value, we find

$$SPL = 20 \log_{10} p_1 + 74 \text{ dB} \qquad re \ 0.0002 \ \mu\text{bar}$$

where p_1 must be measured in microbars. Fortunately, all commercial noise measuring apparatus effectively performs the conversion into microbars and also adds the 74 dB to give us the actual value of SPL in dB's.

The value of p_{ref} was originally chosen as being the minimum sound pressure perceptible by a good human ear. The value of I_{ref} in IL ($= 10^{-12}$ W/m^2) was chosen such that the numerical value of IL should be almost the same as that of SPL for plane or spherical waves under NTP atmospheric conditions. The exact relationship under these conditions is

$$\text{SPL} = \text{IL} + 0.2\,\text{dB} \qquad re\ 0.0002\,\mu\text{bar.}$$

0.2 dB is usually a negligible quantity in acoustic measurements for they can seldom be made to anything approaching this degree of accuracy. The values of SPL and IL are therefore virtually identical at NTP.

If we wish to work back from the measured SPL to find the pressure in Newton/m^2 ($p_{1,\ N}$, say) we may use the relationship

$$\text{SPL} = 10\log_{10} p_{1,\ N} + 94\,\text{dB} \qquad re\ 0.0002\,\mu\text{bar.}$$

Alternatively if p_1 is required in Lb/ft^2 ($p_{1,\ Lb}$, say), then use

$$\text{SPL} = 10\log_{10} p_{1,\ Lb} + 127.6\,\text{dB} \qquad re\ 0.0002\,\mu\text{bar.}$$

When the sound pressures from two independent sound sources add together, it is important to remember that we must *not* add the two SPL values to get the total SPL value. If the two independent r.m.s. sound pressures adding at a point are p_1 and p_2 then the total mean square sound pressure is

$$p_t^2 = p_1^2 + p_2^2$$

so that

$$\text{SPL}_{total} = 10\log_{10}\frac{p_1^2 + p_2^2}{p_{ref}^2}\,\text{dB} \qquad re\ p_{ref}$$

$$\neq \text{SPL}_1 + \text{SPL}_2.$$

If, in fact, the two sound pressures are equal, so that $p_1^2 = p_2^2$, then

$$\text{SPL}_{total} = 10\log_{10}\frac{2p_1^2}{p_{ref}^2}$$

$$= \text{SPL}_1 + 3\,\text{dB} \qquad re\ p_{ref}$$

or

$$= \text{SPL}_2 + 3\,\text{dB} \qquad re\ p_{ref}.$$

The above definitions of the dB scales have been applied to the *overall* sound power and pressure being measured. They may, however, easily be extended to enable spectrum levels to be quoted in dB's relative to a given reference power or pressure. Suppose the analysis of a random pressure shows that in the frequency band of bandwidth Δf, centred on f_c, the pressure has a mean square component of $p_{\Delta f}^2$. Then the SPL contributed by that band is

$$\text{SPL}_{\Delta f} = 10 \log_{10} \frac{p_{\Delta f}^2}{p_{\text{ref}}^2} \text{ dB} \qquad re\ p_{\text{ref}}.$$

In practice, Δf is likely to be $\frac{1}{3}$ octave, or a constant fraction of the varying f_c (1%, 5%, say), or a constant frequency bandwidth (5 Hz, 10 Hz, etc.). The spectral density of the pressure has the average value of $p_{\Delta f}^2/\Delta f$ in the band Δf, and this again may be converted to the dB scale by writing

$$\text{Spectrum level, } S(f) = 10 \log_{10} \frac{p_{\Delta f}^2/\Delta f}{p_{\text{ref}}^2}$$

$$= \text{SPL}_{\Delta f} - 10 \log_{10} \Delta f \text{ dB per unit frequency,}$$

$$re\ p_{\text{ref}}.$$

As a final point, it will be noticed that the symbol "dB" has been used for the decibel, and not "db", i.e. a capital 'B' has been used, to represent the basic named unit 'Bel', and a lower case 'd' to represent the fraction 'deci'. This system is now standard practice with all named units and their multiples and sub-multiples. The move from db to dB has accelerated during the years of preparation of this book. We began by using db's, but in the proof-reading stage it was decided to change completely to dB in the main text. However, to avoid the expense of changing all the figure blocks in like manner, the db's were left in the figures. We apologize for this inconsistency but encourage our readers to use dB's forthwith!

Author Index

501

Subject Index